D1259527

Polymer Association Structures

ACS SYMPOSIUM SERIES 384

Polymer Association Structures

Microemulsions and Liquid Crystals

Magda A. El-Nokaly, EDITOR
The Procter & Gamble Company

Developed from a symposium sponsored
by the Macromolecular Secretariat
at the 194th Meeting
of the American Chemical Society,
New Orleans, Louisiana,
August 30-September 4, 1987

American Chemical Society, Washington, DC 1989

Library of Congress Cataloging-in-Publication Data

American Chemical Society. Meeting (194th: 1987: New Orleans, La.)
Polymer association structures: microemulsions and liquid crystals
Magda A. El-Nokaly, editor

p. cm.—(ACS Symposium Series, ISSN 0097-6156; 384).

"Developed from a symposium sponsored by the Macromolecular Secretariat at the 194th Meeting of the American Chemical Society, New Orleans, Louisiana, August 30–September 4, 1987."

Includes bibliographies and indexes.

ISBN 0-8412-1561-8

1. Emulsions—Congresses. 2. Liquid crystals—Congresses.

I. El-Nokaly, Magda A. 1. II. American Chemical Society. Macromolecular Secretariat. III. Title IV. Series.

TP156.E6A52 1987
660.2'94514—dc19

88-39192
CIP

PRINTED IN THE UNITED STATES OF AMERICA

Contents

Preface

ADVANCES IN THE FIELDS OF POLYMERS AND AMPHIPHILIC colloidal association structures have progressed in a most pronounced manner in the past decade. However, the interaction between these two disciplines and the benefits of cross-fertilization have been limited.

This book on polymeric microemulsions is an attempt at a rapprochement of the methods and structures encountered in the two disciplines. The purpose of this book is to investigate polymer–polymer or polymer–surfactant interactions in solution leading to association structures with properties such as solubilization and anisotropy. These properties are useful in a wide variety of industries such as pharmaceutics, cosmetics, textiles, detergents, and paints.

Each author treated his or her chapter as self-explanatory entities, emphasizing potential and existing applications. The book reviews the association behavior of the different polymer types in isotropic solutions and in the liquid crystalline phase.

The first part of the book discusses formation and characterization of the microemulsions aspect of polymer association structures in water-in-oil, middle-phase, and oil-in-water systems. Polymerization in microemulsions is covered by a review chapter and a chapter on preparation of polymers. The second part of the book discusses the liquid crystalline phase of polymer association structures. Discussed are mesophase formation of a polypeptide, cellulose, and its derivatives in various solvents, emphasizing theory, novel systems, characterization, and properties. Applications such as fibers and polymer formation are described. The third part of the book treats polymer association structures other than microemulsions and liquid crystals such as polymer–polymer and polymer–surfactant, microemulsion, or rigid sphere interactions.

This book will be useful for academic researchers and, even more so, for technology-focused investigators in industry.

Acknowledgments

I thank Howard Needles, who was the ACS Cellulose, Paper, and Textile Division 1987 Program Chairman, for inviting me to organize the symposium; Raymond MacKay, the co-chairman of the Macromolecular Secretariat in 1987, for his assistance during the session preparation; and Jan Bock and Frank Blum for preparing excellent sessions for the symposium, from which some papers were taken for this book.

Working with all the contributing authors was an invaluable experience. I thank each of them for their cooperation.

I thank the editorial staff of the American Chemical Society's Books Department, especially Cheryl Shanks, for their patience and professionalism. I also acknowledge with thanks the support of my colleagues and management at The Procter & Gamble Company.

Finally, I dedicate this book, with the contributors' permission, to the memory of a great woman and achiever, Malak Hussein Fahmy and tell her "De nouveau à toi ma mère."

MAGDA A. EL-NOKALY
The Procter & Gamble Company
P.O. Box 398707
Cincinnati, OH 45239–8707

September 16, 1988

Chapter 1

Association and Liquid Crystalline Phases of Polymers in Solution

H. Finkelmann and E. Jahns

Institut für Makromolekulare Chemie, Universität Freiburg, Stefan–Meier–Straße 31, D–7800, Freiburg, Federal Republic of Germany

The solution behavior of polymers has been intensively investigated in the past. Dilute solutions, where polymer-polymer interactions may be excluded, have become the basis for the characterization of the primary structure of macromolecules and their dimensions in solution. Besides this "classical" aspect of macromolecular science, interest has focussed on systems, where - due to strong polymer/polymer interactions - association of polymers causes supermolecular structures in homogeneous thermodynamically-stable isotropic and anisotropic solutions or in phase-separated multi-component systems. The association of polymers in solutions gives rise to unconventional properties, yielding new aspects for applications and multiple theoretical aspects.

The association and the generation of supermolecular liquid crystalline organizations of polymers in solution strongly depend on the molecular architecture of the macromolecules. From low molar mass liquid crystals it is well known that only particular molecular architectures cause the liquid crystalline state: the molecules have to exhibit either a rigid rod- or disc-like molecular geometry or an amphiphilic chemical constitution. Introducing these particular molecular structures into the primary structure of macromolecules, liquid crystalline phases may exist in defined temperature regimes of the polymers in bulk or in defined temperature and concentration regions of the polymers in solution.

This paper gives a brief survey regarding the association of polymers in solution with respect to the ability to form liquid crystalline structures. Three different types of polymers are considered, which differ in their molecular architecture:

0097–6156/89/0384–0001$06.00/0

i) Non-amphiphilic polymers having a rigid rod-like backbone
ii) Amphiphilic polymers with amphiphilic monomer units and
iii) Blockcopolymers having an amphiphilic backbone.

Non-Amphiphilic Polymers

For non-amphiphilic molecules, the nematic liquid crystalline phase was theoretically explained by Maier and Saupe (1-3) in 1956 on the basis of the calculation of the anisotropic dispersion interaction. In the nematic state, the longer molecular axes are orientationally ordered with respect to the director n , where n denotes the symmetry axis of the orientation distribution function of the long molecular axis (Figure 1). Increasing molecular anisotropy of polarizability, which is nearly always directly correlated with increasing anisometric shape of the molecules, stabilizes the "thermotropic" liquid crystalline state. On the other hand, as early as 1949 Onsager (4), Isihara (5) and, later, Flory in 1956 (6), theoretically predicted, purely on the basis of packing arguments (repulsive interactions) that anisometric rigid rod-like molecules form a "lyotropic" nematic phase in solution above a critical concentration. The critical concentration, where the anisotropic phase begins to separate from the isotropic solution (coexistence line of the biphasic isotropic-l.c. gap) is directly related to the anisometric shape of the rod-like molecules. With increasing axial ratio of the long axis and the diameter of the molecules, the critical concentration decreases. Increasing molecular flexibility, which is reflected in a drop of the statistical chain segment length of a macromolecule, causes an increase of the critical concentration.

Both theoretical approaches qualitatively describe the "thermotropic" and "lyotropic" liquid crystalline state of rod-like molecules (see also D.B. DuPre, R. Parthasarathy, this book). Combination of both theories (Flory, Ronca)(7) slightly improves the predictions compared to the experimental findings. Anisotropic dispersion interactions and/or anisometric molecular shape can thus be the basis for explaining theoretically the appearance of "lyotropic" and "thermotropic" liquid crystalline phases.

The differentiation between "lyotropic" and "thermotropic" liquid crystalline phases of rod-like molecules is rather artificial, because in bulk as well as in solution the same physical origin causes the anisotropic phase. Originally, the term "lyotropic l.c" was clearly related to micellar liquid crystalline

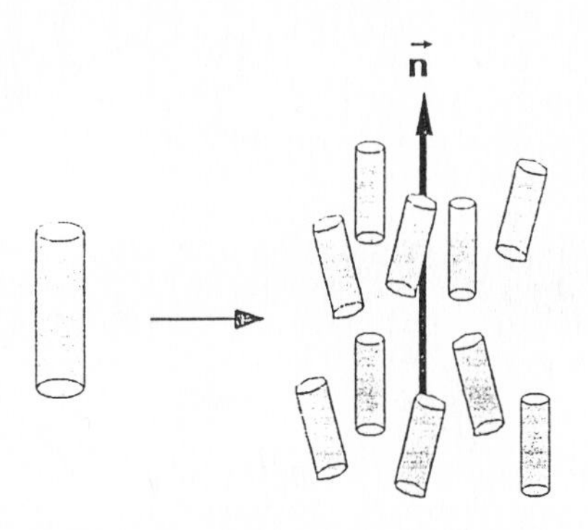

Fig. 1: Nematic ordering of rod-like molecules (n=director)

phases, which does not hold for rod-like molecules in solution.

The existence of liquid crystalline phases of rod-like molecules in solution becomes obvious if we simply consider the binary phase diagram of rigid rod-like low molar mass molecules or macromolecules A in a suitable solvent S (Figure 2). Because of its chemical constitution, A exhibits the nematic phase below the isotropic to liquid crystalline phase transformation temperature $T_{lc,I}$. $T_{lc,I}$ depends - as has been theoretically and experimentally proved - on the chemical constitution of A, as does the axial ratio (which is directly proportional to the degree of polymerization) and the molecular flexibility (which is related to the statistical segment length of the macromolecule). Adding the non-liquid crystalline solvent S to A, $T_{lc,I}$ is lowered, the slope of the isotropic-l.c coexistence line being strongly dependent on the chemical constitution of S and the A/S interactions. The isotropic-l.c. coexistence lines end at c_1 and c_2 and contact the biphasic crystalline phase regime. Consequently c_1 represents the lowest critical concentration (as function of temperature), where the nematic phase begins to separate from the isotropic solution with increasing concentration of A. In the concentration region c_2 to pure A, a homogeneous liquid crystalline phase can be observed. This holds for all thermotropic liquid crystalline monomers, oligomers or polymers, assuming pure A exhibits a stable l.c. phase.

Without solvent many of the rigid rod-like macromolecules do not exhibit a stable l.c. phase above their melting temperature T_m, because chemical decomposition takes place before melting at T_d (Figure 2). If, however, a suitable solvent is added, the instable melting temperature is lowered. Below T_d and in the concentration region c_3 to c_2, a homogeneous ("lyotropic" ?) liquid crystalline phase is stable.

The phase diagram in Figure 2 is related to systems or macromolecules, which do not change their conformation or the statistical segment length by adding the solvent essentially. On the other hand, if a flexible macromolecule is converted into a rigid rod-like conformation by adding the solvent, the pure melt of the macromolecule may not exhibit the liquid crystalline state but solutions with S.

The formation of liquid crystalline solutions of rigid rod-like polymers comprises a broad variety of polymers (see this book e.g. DuPre and Parthasarathy, Fellers and Lewis, Gilbert et.al., Ambrosio and Sixou) such as cellulose derivatives, poly-γ-benzylglutamate and aromatic polyester and polyamides (8-11).

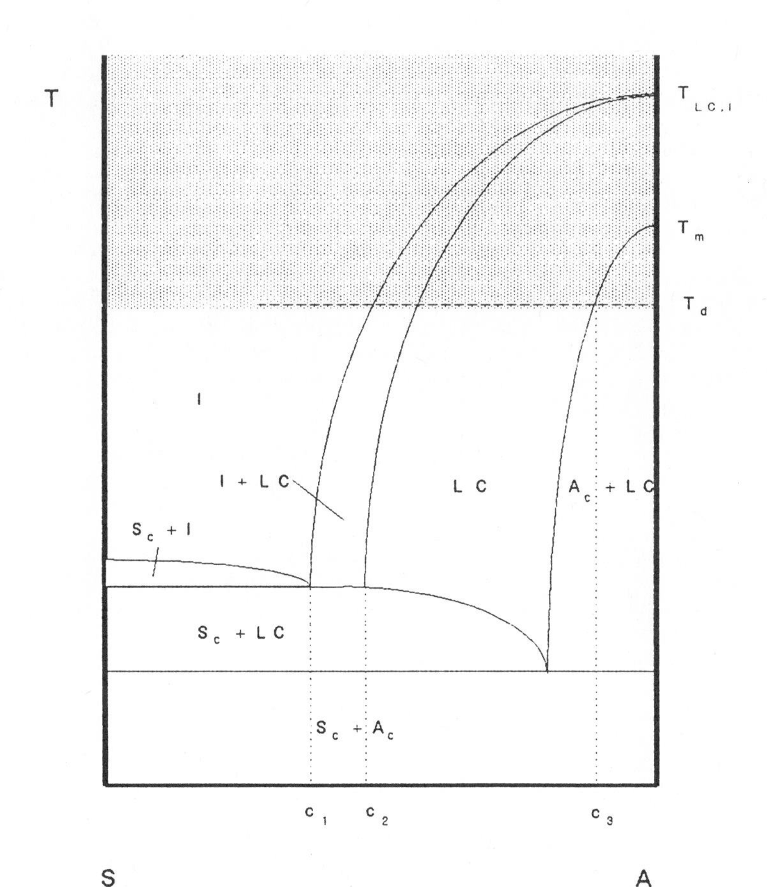

Fig. 2: Schematic binary phase diagram of the component A having a rigid rod-like molecular geometry and a solvent S. (T_d = temperature of chemical decomposition of A; T_m = melting temperature; $T_{LC,I}$ = liquid crystalline to isotropic phase transformation temperature; I = isotropic solution; LC = liquid crystalline; A_c , S_c = crystalline state of A and S; c_1 , c_2 , c_3 refer to text)

Compared to conventional flexible polymers, two aspects are of commercial interest. In the nematic state, a shear-induced macroscopic orientation of the director can be achieved which may cause a decrease of the (anisotropic) viscosity under defined flow processes (11). Polymer solutions can be processed containing a higher polymer concentration in the nematic solution than in the isotropic solution of the same viscosity. Secondly, the macroscopic orientation of the polymer chains, e.g. in the direction of a fiber, strongly improves the (anisotropic) mechanical properties. This is illustrated in Figure 3, in which the mechanical properties of some relevant materials are compared. The theoretical properties of polyethylene are calculated for fully-extended chains in fiber direction, where the tensile strength is given by the C-C bond strength. The poor mechanical properties of conventional polyethylene are due to the disorder of the backbones. For the aramid fiber which possesses the ordered liquid crystalline state during the spinning process, the backbones are highly ordered and the mechanical properties in fiber direction nearly approach the theoretical values. It has to be noted that, due to the macroscopic anisotropic organization of the macromolecules, the mechanical properties also become anisotropic. While in the direction of the orientation of the backbones excellent properties are observed, they are poor perpendicular to this direction.

Amphiphilic Polymers

In this chapter lyotropic liquid crystalline polymers are considered, where micelle-like organizations of the macromolecules cause the formation of mesophases in defined concentration and temperature regions. For these polymers it has to be assumed that the polymer backbone or the monomer units must contain an amphiphilic character.

The solution behavior of low molecular weight amphiphilic molecules has been intensively investigated in the past (12-16) with respect to the formation of liquid crystalline phases. In very dilute aqueous solutions, the amphiphiles are molecularly dispersed dissolved. Above the critical micelle concentration (CMC), the amphiphiles associate and form micelles (Figure 4) of spherical, cylindrical or disc-like shape. The shape and dimension of the micelles, as a function of concentration and temperature, are determined by the "hydrophilic-hydrophobic" balance of the amphiphilic molecules. The formation of spherical aggregates is preferred with increasing volume fraction of the hydrophilic head group of the amphiphile, because the curvature of the micellar surface rises (17,18). Depending on the chemical constitution of the amphiphiles, in defined concentration and temperature regions, the micelles can aggregate to mesophases (Figure 4). For spherical micelles, it is assumed that they form cubic mesophases (I); while rod-like and disc-like micelles aggregate to hexagonal (H) and lamellar (L_α) liquid crystalline phases.

To obtain amphiphilic polymers, different concepts are conceivable to introduce amphiphilic moieties into the polymer backbone. They are schematically summarized in Figure 5. Polymers of type A and B can be realized, if a polymerizable group is introduced into the hydrophobic group (type A) or hydrophilic group (type B) of a conventional surfactant, which exhibit the liquid crystalline state in solution. Copolymerization of a hydrophilic and a hydrophobic comonomer yields amphiphilic copolymers of type C. According to the convention, these polymers may be called "amphiphilic side-chain polymers"

A completely different polymer structure is obtained, if the hydrophilic and hydrophobic moieties are within the polymer backbone, type D ("amphiphilic main-chain polymer").

While only one paper describes the appearance of an anisotropic aqueous solution for a polymer of type D (19), polymers of type A to C have been investigated in more detail recently with respect to formation of liquid crystalline phases (20-23). In the following two sections, the association behavior in dilute isotropic solution and the liquid crystalline phase region is discussed on the basis of some experimental results. The dilute isotropic solutions are of interest with respect to the question whether polymers of type A and B form micelles similar to the corresponding monomeric amphiphiles. The liquid crystalline phase regime gives information whether the linkage of the amphiphiles via a polymer backbone influences the stability of the anisotropic phases and whether the same polymorphism occurs as is known for monomeric amphiphiles.

Dilute Isotropic Solutions

For low molecular mass amphiphiles, hydrophobic interactions and surface effects determine the critical concentration at which micellar aggregates are favored over the molecularly dispersed amphiphilic solutes. For polysurfactants, however, the amphiphiles are linked together and the dynamic exchange of associated and non-associated amphiphilic monomer units is prevented. Consequently the micelle formation does not only depends on the hydrophilic/ hydrophobic balance of the monomer

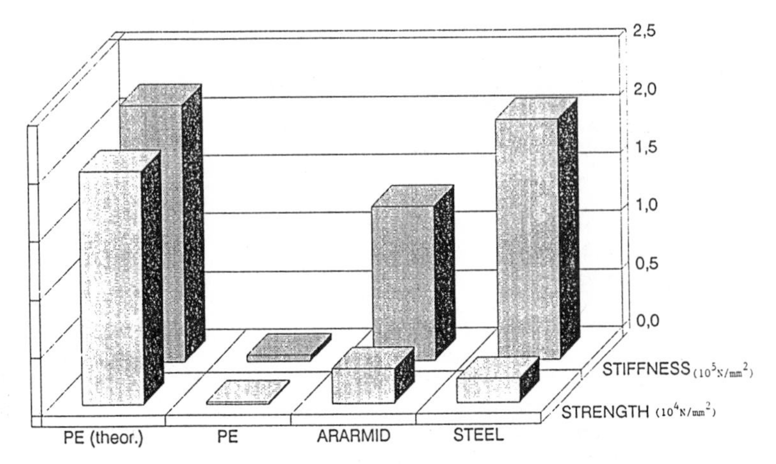

Fig. 3: Stiffness and strength of some materials

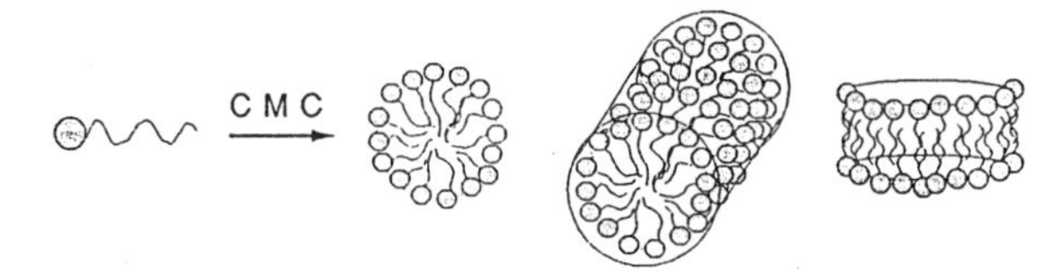

Fig. 4: Micellar association of amphiphiles

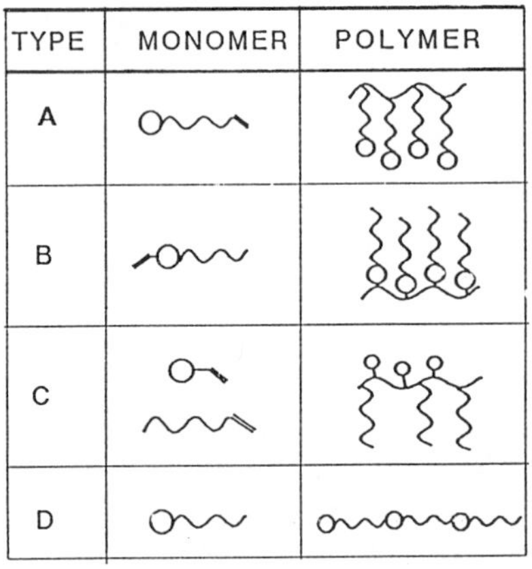

TYPE	MONOMER	POLYMER
A		
B		
C		
D		

Fig. 5: Amphiphilic polymers

units but also on the primary structure of the polymer backbone.

Let us consider the following simplified example: Assuming a monomeric surfactant forms a spherical micelle with the association number N above the CMC. At higher concentrations - due to the hydrophilic/hydrophobic balance of the amphiphile - rod or disc-like micelles appear. If the same surfactant forms the monomer unit of the polysurfactant having the degree of polymerization r (Figure 6), then one polymer molecule gives a micelle-like aggregate in solution. For $r \lesssim N$ a spherical aggregate should be observed, and for $r > N$ an anisometric rod- or disc-like aggregate has to be assumed. Consequently, no "CMC" will be observed for the polysurfactants, at least for the situation $r > N$.

The experiments prove this assumption in practice. In Figure 7a, the solubilization of Eosin in the monomeric aqueous surfactant solution indicates the CMC at $1.55 \cdot 10^{-1}$ mol/l while the corresponding polysiloxane ($r=95$) (Figure 7b) solubilizes Eosin without an indication of a CMC (24). For this polymer r approximately equals N of the monomeric surfactant. Furthermore it must be noted that the ability for solubilization of the monomer and the polymer are equal, which can be easily calculated from the slope of the (c)-curves.

An important aspect to be considered concerns the chemical constitution of the polysurfactants. While monomeric surfactants associate to form micelles in aqueous solutions regardless of whether the hydrophilic group is an anionic, cationic, zwitterionic or non-ionic group, it is more complex for the ionic compared to the non-ionic polysurfactants. This is clearly reflected in the rheological properties of ionic polysurfactants(26).

In Figure 8, the reduced viscosity of aqueous solutions of the monomeric cationic surfactant is plotted against the concentration(26a). The linear relationship can be analyzed on the basis of the simple model of hard spheres(25) and yields reasonable values for the hydrated volume of the micelles. The rheological properties, however, completely change when this monomeric surfactant is polymerized (Figure 9)(26a).The reduced viscosity of the quasi-binary polymer/water solutions rapidly increases with falling polymer concentration. Due to the dissociation of the ionic groups at the side chains of the polymer, the charge density along the polymer rises with falling concentration and causes a change of the backbone conformation towards an extended chain. The extended chain is no longer compatible with a micelle-like aggregate. Only the addition of a strong electrolyte (in this example 0.1 KBr) which increases the counter-ion

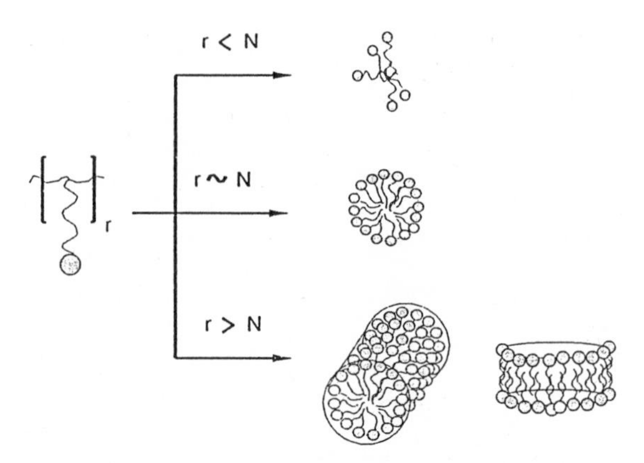

Fig. 6: Polymer aggregates as function of the degree of polymerization r (N = aggregation number of monomeric amphiphiles)

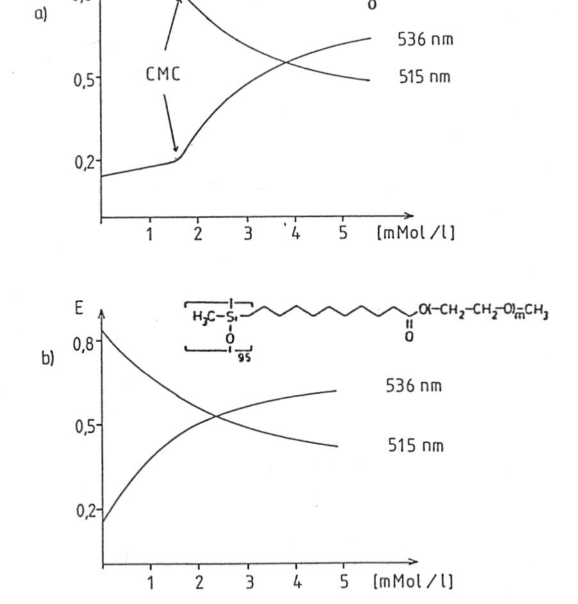

Fig. 7: Solubilization of Eosin (10^{-5} mol/l) for a a) monomeric surfactant, b) polysurfactant

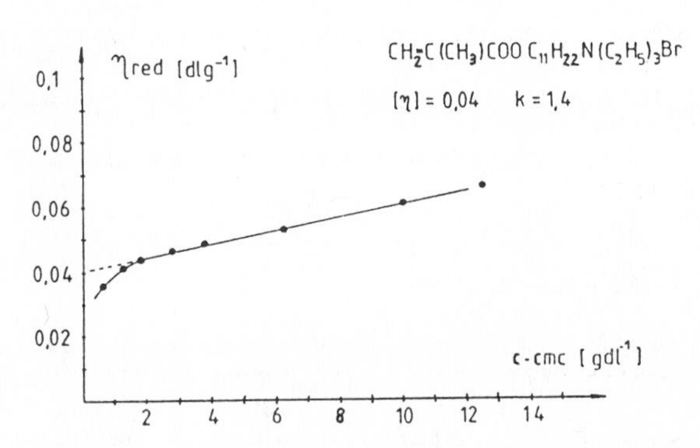

Fig. 8: Reduced viscosity of an ionic surfactant

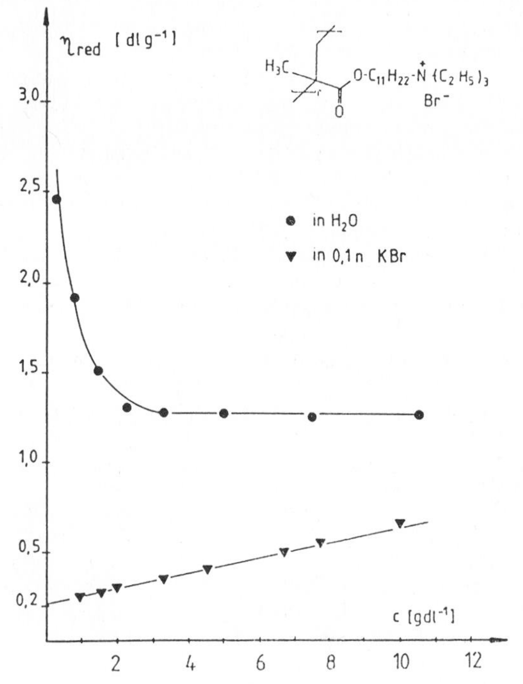

Fig. 9: Reduced viscosity of an ionic polysurfactant

concentration, causes a rheological behavior as expected for uncharged macromolecules.

This example clearly demonstrates, that with respect to a micelle-like association of ionic polysurfactants, the more complex ternary mixtures containing an electrolyte must be used to avoid extended polymer chains. This is in contrast to non-ionic polysurfactants, in which these effects do not occur in binary mixtures.

Liquid Crystalline Phases

The experiments mentioned above indicate that non-ionic polysurfactants associate to micellar-like aggregates in dilute solutions. Therefore, liquid crystalline phases should be observed at higher concentration in analogy to the low molar mass surfactants. Due to the polymerization, the dynamic exchange of amphiphilic moieties between the aggregate and the solution is prevented, which may cause changes in the stability of the liquid crystalline phase. Additionally, for A-C type polymers, the mobility of the hydrophilic and hydrophobic parts of the monomer units (see Figure 5) differ because of their proximity to the polymer backbone. This might affect the packing within the aggregate and cause differences in the liquid crystalline-phases between polymers of A-C type.

Although numerous polysurfactants have been synthesized in the past, systematic investigations comparing the detailed phase diagrams of solutions of a monomer surfactant and the corresponding polysurfactant are still in an early stage. In the following, some basic aspects are mentioned concerning the ability of A-C type polymers to form liquid crystalline phases.

In Figure 10a, the binary phase diagram is shown for a non-ionic amphiphile in aqueous solutions(20). Besides the miscibility gap, the crystalline region and the isotropic solution, this surfactant exhibits a hexagonal liquid crystalline phase. The addition reaction of this amphiphile (via the C-C double bond at the end of the hydrophobic group) to the hydrophobic poly(methylhydrogensiloxane) yields the corresponding polysurfactant of type A. The quasi-binary phase diagram of the polydisperse polymer ($\bar{r}$=95) with water is shown in Figure 10b. Actually, a hexagonal liquid crystalline phase also exists. Furthermore - separated by a bicontinuous cubic phase - at higher concentrations a lamellar liquid crystalline phase also appears.

Polymers of type B have directly linked the hydrophilic group at the polymer backbone. This might cause a difference between polymer type A and B to form liquid crystalline phases due to the difference of the

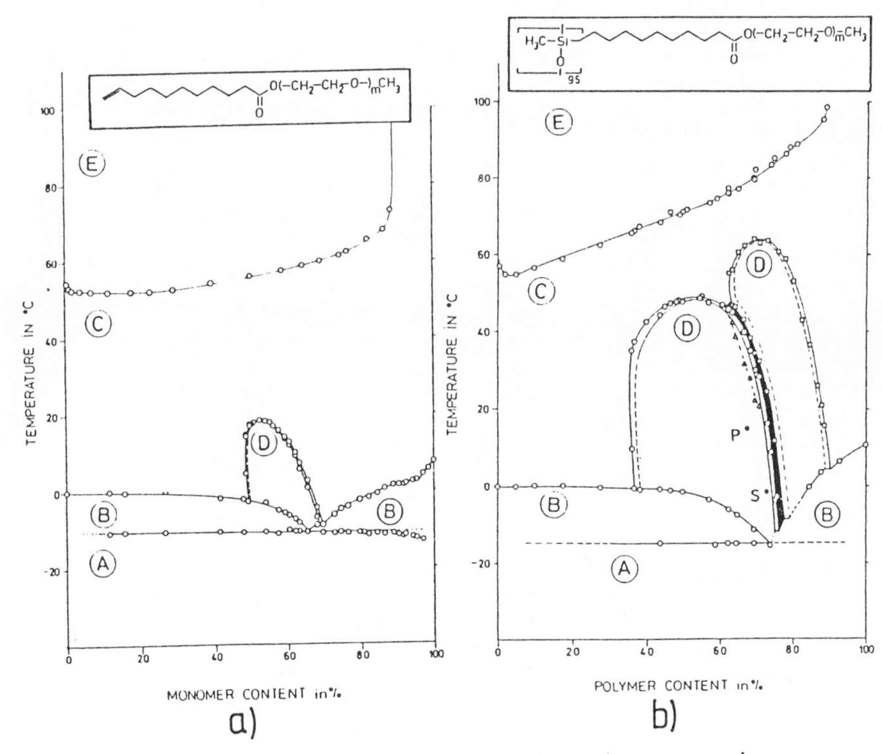

Fig. 10: Binary phase diagram of a a) monomeric non-ionic surfactant, b) non-ionic polysurfactant in aqueous solution
A: heterogeneous mixed crystals
B: heterogeneous melt
C: homogeneous isotropic solution
D: homogeneous mesomorphous phases (mesophase 1: lower polymer content mesophase; 2: higher polymer content mesophase)
E: heterogeneous isotropic liquids, miscibility gap with lower consolute point

Between C and D: heterogeneous region of isotropic liquid and liquid crystal

mobility of the hydrophilic group. The hydrophilic groups might no longer be able to be regularly arranged at the micellar surface. On the other hand, at concentrated solutions, inverted micelles might become stable, because less packing restraints are expected for the hydrophobic ends. Only one example of type B has been described(23). The important result is that polymers of type B also exhibit liquid crystalline-phases. The hexagonal and lamellar phases have been identified, but not the formation of anisotropic phases of inverted micelles. Obviously, the linkage of the amphiphiles via the hydrophilic group to the backbone essentially does not give rise to packing restraints.

Copolymers of type C offer an additional feature: changing the composition of the hydrophilic and the hydrophobic comonomers, the hydrophilic/hydrophobic balance of the copolymers can be easily varied without changing the chemical structure of the starting material(28). Table 1 displays the results of non-ionic copolymers. In the case of a suitable monomer composition, a lamellar liquid crystalline-phase appears.

These examples clearly show that:

i) Polysurfactants form liquid crystalline phases in water

ii) The phase structure is similar to the phase structure of low molecular weight surfactants and

iii) The well known polymorphism occurs.

More detailed measurements(24) additionally proved thatthe liquid crystalline-phase structure depends on the hydrophilic/hydrophobic balance of the amphiphilic monomer unit. One aspect concerning the cubic mesophase has to be mentioned. Recalling Figure 6, a degree of polymerization larger than the association number of N should prevent the formation of spherical aggregates. Consequently, if suitable monomers or oligomers exhibit a cubic mesophase, for polymers the cubic mesophase should vanish. In Figures 11a and b, experimental results are shown which confirm this consideration(24). The polymer (Figure 11a) with r=95 shows no cubic phase while in the oligomer region (Figure 11b) a cubic phase becomes stable. In this way the primary structure of the polymer main chain prevents spherical aggregates which, consequently, is also reflected in the mesophases.

The amphiphilic polymers and copolymers offer broad aspects for applications similar to low molar mass surfactants. Modifications of solutions, e.g. with respect to their viscosity, can be performed by the variation of the degree of polymerization of the polysurfactants easily. The liquid crystalline phase

Table 1: Phase transformation temperatures of the lamellar phase of copolymers as function of the monomer feed

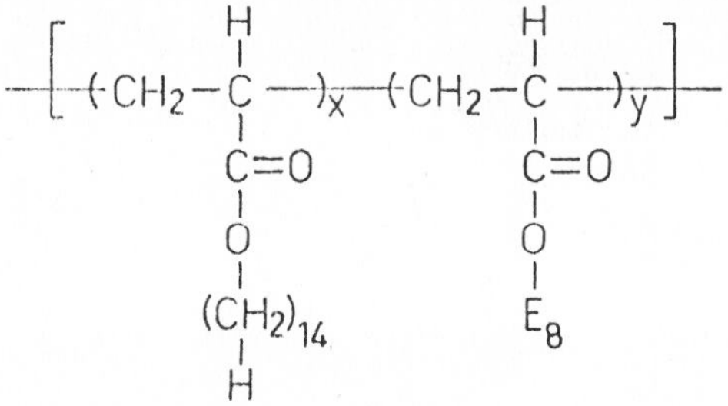

$$\left[(CH_2 - \underset{\underset{\underset{\underset{\underset{\underset{H}{|}}{(CH_2)_{14}}}{|}}{O}}{|}}{\overset{H}{\underset{|}{\overset{|}{C}}}}\!\!\!\!\!\!\!\!\underset{}{})_x\!\!\! - (CH_2 - \overset{H}{C})_y \right]$$

Side groups: x-unit: C(H)–C=O–O–$(CH_2)_{14}$–H; y-unit: C(H)–C=O–O–E_8

COMPOSITION OF MONOMERS y / x	SOLUTION	
60 / 40	T_C > 100 °C	–
50 / 50	T_C = 73 °C	LC
45 / 55	T_C < 0 °C	LC

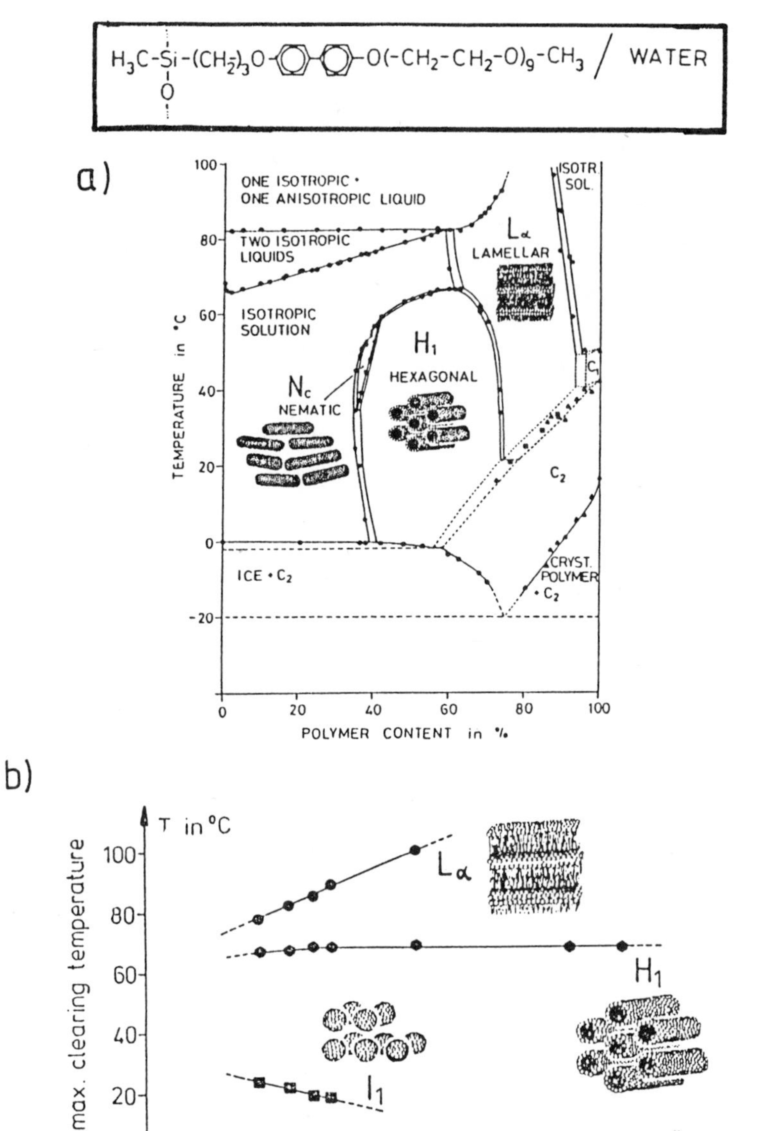

Fig. 11: a) Binary phase diagram of a non-ionic polysurfactant
b) maximum clearing temperature of the liquid crystalline phase as function of the degree of polymerization

behavior of these systems, however, has not received much attention up to now. Both theoretical and experimental research are still in their infancy. On the other hand, the anisotropic structures of the liquid crystalline phases provide interesting aspects e.g. for polymerization processes or rheology.

Block Copolymers

Let us assume that for a conventional surfactant the hydrophobic chain (e.g. an alkyl chain) and the hydrophilic group (e.g. a non-ionic oligo(ethylene glycol) chain) is continuously lengthened. Then we have a continuous transition from a monomeric amphiphile to an AB block copolymer. Similarly, ABA or multiblock copolymers can be realized according to Figure 5, D. These block copolymers show interesting properties in solution as well as in bulk. They are extensively reviewed by Riess(29).

In solution, the macromolecules associate to micellar-like aggregates(30,31). With increasing length of the hydrophilic block, or decreasing length of the hydrophobic block the CMC rises. In this way they behave similarly to conventional low molecular weight surfactants. Interestingly, with these systems the volume fraction of the hydrophobic core can be continuously varied. A continuous change from homogeneous micellar solution to a (at least) microphase-separated solution is conceivable.

In more concentrated solutions, these AB or ABA block copolymers also form anisotropic structures(32). Lamellar as well as hexagonal structures have been identified. The problem may arise, whether the anisotropic solution exists as homogeneous anisotropic phases or as structured phase separated systems. As indicated above, with increasing chain length, e.g. in a lamellar structure, the polymer backbone tends to adopt a statistical coiled chain conformation and only the chain segments of the interlayers are ordered. A continuous departure from the organization of "normal" micelles takes place, although the macroscopic lamellar architecture remains unchanged.

The bulk properties of these block copolymers are also unusual and are not limited only to hydrophilic /hydrophobic systems(33). The only requirement is that the homopolymers of the A-block and B-block are not miscible, which holds for nearly all polymers deviating in chemical constitution. With the variation of the A- or B-block length, amorphous and structured microphase-separated systems occur, as summarized in Figure 12. The cubic, hexagonal, lamellar and the inverse structures are similar to the molecular organization of surfactants

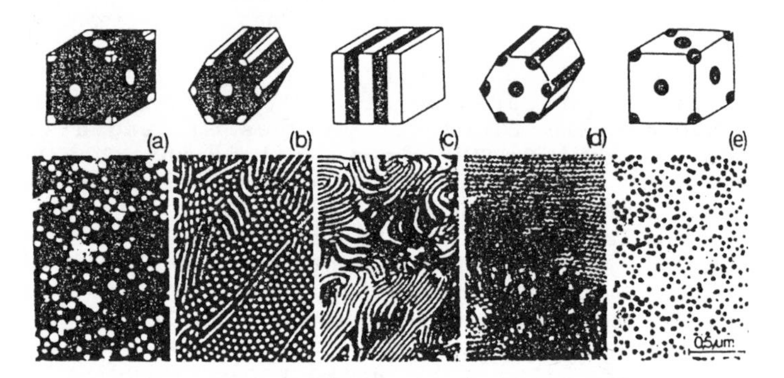

Fig. 12: Morphologies of block copolymers

in aqueous solutions. The driving force for the formation of the structured morphology is due to minimizing the surface to volume ratio of the phase separated, incompatible microdomains.

The possibilities for applications of these block copolymers are extremely broad. Amphiphilic systems are of interest for all applications similar to low molecular weight surfactants. Copolymers without a water soluble block can be used e.g. for the compatibilization of polymer blends. The bulk materials offer broad applications in the field of advanced materials having outstanding mechanical properties.

The intention of this brief survey has been to demonstrate that besides the "classical" aspects of isotropic polymer solutions and the amorphous or partially crystalline state of polymers, a broad variety of anisotropic structures exist, which can be induced by definable primary structures of the macromolecules. Rigid rod-like macromolecules give rise to nematic or smectic organization, while amphiphilic monomer units or amphiphilic and incompatible chain segments cause ordered micellar-like aggregation in solution or bulk. The outstanding features of these systems are determined by their super-molecular structure rather than by the chemistry of the macromolecules. The anisotropic phase structures or ordered incompatible microphases offer new properties and aspects for application.

Literature Cited

1) Maier, W.; Saupe, A. Z. Naturforschg. 1958, 13A, 564
2) " " " 1959, 14A, 882
3) " " " 1960, 15A, 287
4) Onsager, L. New York Acad. Sci. 1949, 51, 627
5) Isihara, A. J. Chem. Phys. 1951, 19, 1142
6) Flory, P.J. Proc. Roy. Soc. 1956, A234, 60
7) Flory, P.J., Ronca, G. Mol. Cryst. Liq. Cryst. 1979, 54, 289
8) Ober, Ch.K.; Jin, J.-I.; Lenz, R.W. In Liquid Crystal Polymers I; Adv. in Polym. Sci. No. 59: Berlin, 1984; p 103
9) Liquid Crystal Polymers I; Adv. in Polym. Sci. No. 59: Berlin, 1984
10) Liquid Crystal Polymers I/III; Adv. in Polym. Sci. No. 60/61: Berlin, 1984
11) Samulski, E.T. In Liquid Crystalline Order in Polymers; Blumstein, A., Ed.; Academic: New York, 1978; Chapter 5

12) Winsor, P.A. Chem. Rev. 1968, 68(1), 1
13) Winsor, P.A. Mol. Cryst. Liq. Cryst. 1971, 12, 141
14) Tiddy, G.J.T. Physics Reports 1980, 57, 1
15) Wennerström, H.; Lindman, B. Physics Reports 1979, 52, 1
16) Lindman, B.; Wennerström, H. Topics in Current Chemistry 1980, 87, 1
17) Israelachvili, J.N.; Mitchell, D.J.; Ninham, B.W. J. Chem. Soc., Farady Trans. II 1976, 72, 1525
18) Israelachvili, J.N. In Surfactants in Solution 4; Mittal, K.L.; Bothorel, P., Eds.; Plenum: New York 1986, Vol. 4, p 3
19) Yu, L.P.; Samulski, E.T. In Liquid Crystals and Ordered Fluids; Johnson, Ed.; Plenum: New York 1984, Vol. 4, p 697
20) Finkelmann, H.; Lühmann, B.; Rehage, G. Coll. Polym. Sci. 1982, 260, 56
21) Lühmann, B.; Finkelmann, H.; Rehage, G. Makromol. Chem. 1985, 186, 1059
22) Finkelmann, H.; Schafheutle, M.A. Coll. Polym. Sci. 1986, 264, 786
23) Jahns, E.; Finkelmann, H. Coll. Polym. Sci. 1987, 265, 304
24) Lühmann, B.; Finkelmann, H. Coll. Polym. Sci. 1987, 265, 506
25) Ionic Polymers; Holliday, L., Ed.; Applied Science: London, 1975
26) a) Wagner, D. Ph.D. Thesis, TU Clausthal, Clausthal-Zellerfeld, FRG, 1985
b) Tanford, C. Physical Chemistry of Macromolecules; Wiley: New York, 1961; p 489 ff
c) Fuoss, R.M.; Strauss, U.P. J. Polymer Sci. 1948, 3, 246, 602
27) Sheraga, H.A.; Mandelkern, L. J. Am. Chem. Soc. 1953, 75, 1979
28) Finkelmann, H. Abstracts of Papers, Makromolekulares Kolloquium, Freiburg, FRG, 1987; p 80
29) Riess, G.; Hurtrez, G. In Encyclopedia of Polymer Science and Engineering; Wiley: New York, 1985; Vol. 2, p 324
30) Tuzar, Z.; Kratochvil, P. Adv. Colloid Interface Sci. 1976, 6, 201
31) Price, C. In Developments in Block Copolymers; Goodman, I., Ed.; Applied Science: London, 1982, pp 39-80
32) Shibayama, M.; Hashimoto, T.; Kawai,H. Macromolecules 1983, 16, p 16, 361, 1093, 1427, 1434
33) Gallot, Y. Colloq. Nat. CNRS 1979, 938, 149

RECEIVED August 16, 1988

WATER-IN-OIL MICROEMULSIONS

Chapter 2

Formation and Characterization of Water-in-Oil Microemulsions Stabilized by A–B–A Block Copolymers

Tharwat F. Tadros[1], P. F. Luckham[2], and C. Yanaranop[2]

[1]ICI Agrochemicals, Jealotts Hill Research Station, Bracknell, Berkshire RG12 6EY, United Kingdom
[2]Department of Chemical Engineering and Chemical Technology, Imperial College of Science and Technology, Prince Consort Road, London SW7 2BY, United Kingdom

The phase diagram of the quaternary system, n-tetradecane, water, A-B-A block copolymer (where A is poly-(12-hydroxystearic acid) and B is poly(ethylene oxide)) and n-butanol was investigated at 7, 23 and 47°C. Two A-B-A polymer concentrations of 10 and 20% were used. The various phases found were studied using phase contrast and polarising microscopy. Particular attention was paid to the L_2 (microemulsion) region which was found to increase with decrease of temperature and increase of polymer concentration. The microemulsion droplet size was determined using time average light scattering. The droplet interaction was taken into account by calculating the structure factor S(Q) using a hard-sphere model. The droplet radius of the microemulsion gradually increased with increase in water volume fraction, at constant polymer concentration.

It is now fairly established that microemulsions are isotropic systems that are thermodynamically stable consisting of oil, water and surfactant(s) (1). The origin of thermodynamic stability of microemulsions arises from the small positive interfacial energy term (arising from the ultralow interfacial tension) which is balanced by the negative entropy of dispersion terms, such that the net free energy formation of the system is zero or negative (2). This low interfacial energy is usually achieved by the use of a surfactant and cosurfactant mixture, which differ in their nature, one being predominently water soluble and the other oil soluble. For example, in the preparation of water in oil microemulsion, one may use a combination of an anionic surfactant such as sodium dodecyl sulphate with an alcohol such as butanol, pentanol or hexanol. The literature on microemulsions is vast and various review articles are available (3-6) that summarise the mechanism of formation of microemulsions and their stability. Due to their thermodynamic stability, microemulsions are very attractive in various industrial applications, eg in the

0097–6156/89/0384–0022$06.00/0

formulation of pesticides and pharmaceuticals. As a result of the small droplet size that is produced and the dynamics of the structural units formed, microemulsions may be applied as delivery systems for optimisation and enhancement of biological activity.

Compared to the research effort that has gone into studying microemulsions stabilised by ionic and nonionic surfactant systems, there is relatively less research using macromolecular surfactants. However, Reiss and coworkers (7-9) used block copolymer consisting of polystyrene (PS) and polyethylene oxide (PEO), whereas Marie and Gallot (10) have reported microemulsions formed with A-B block copolymers polystyrene - poly(vinyl-2-pyridinium chloride). In addition, Candau et al. (11-13) used graft copolymers having a PS backbone and PEO side chains. More recently Barker and Vincent (14) prepared microemulsions using A-B block copolymers of PS/PEO in the presence of propanol as a cosurfactant. They assumed a model in which water is solubilised into the PEO sheath of the inverse PS/PEO micelles that are dispersed in an oil such as toluene.

In this paper, we will present some data on the formation of water in n-tetradecane microemulsions using an A-B-A block copolymer. The A portions are poly(12-hydroxystearic acid) whereas the B is PEO. A cosurfactant, namely n-butanol, was used for preparation of these microemulsions. For a systematic study of microemulsion composition, it is essential to establish the phase diagram of the system under investigation. From these one can identify the extent of the microemulsion region and its relation to other phases (15). The various phases found were identified qualitatively using phase contrast and polarising microscopy. The microemulsions were further characterised using time-average light scattering, as described earlier (16).

EXPERIMENTAL MATERIALS

All water was twice distilled from an all glass apparatus and its surface tension was 72 ± 0.5 mN m^{-1} at 25°C. The n-tetradecane was British Drugs Houses (BDH) reagent grade and was used as received; butan-1-ol an Analar grade (ex BDH) was also used without further purification. The surfactant was an industrial polymer surfactant B246. This material is an A-B-A block copolymer of poly-12-hydroxy stearic acid/polyethylene oxide. The polymer has a number average molar mass of 3543 ± 30 g mol^{-1} (obtained using vapour pressure osmometry) with the ethylene oxide chain having a molecular weight of 1500 g mol^{-1}.

Preparation of Samples for Phase Diagrams. Stock solutions of appropriate concentrations of the A-B-A block copolymer in tetradecane and butan-1-ol were made up. All samples were prepared by weight in glass, screw capped vials and in all cases mixing was carried out using a laboratory "Whirlimixer". The order of mixing the components was as follows, surfactant in tetradecane, tetradecane, butan-1-ol and finally water. All samples were sealed and kept at 23°C for one month to reach equilibrium. The samples were then studied, as described below, mixed resealed and kept at 47°C for a further month. The samples were then studied and then the procedure repeated at 7°C. Samples were prepared at fixed A-B-A block copolymer concentrations of 0%, 10% and 20%.

Assessment of Samples. The different phases were assessed visually, by optical and polarising microscopy. Each phase, unless transparent, was observed under phase contrast and between crossed Nichols using a Leitz microscope. The appearance of each phase under the microscope was noted. To determine whether the phases were oil or water continuous, a dye test was used. A drop of the phase under investigation was placed on a clean microscope slide and a small amount of the two dyes Waxoline red (oil soluble) and Lissamine Scarlet (water soluble) was placed at two separate edges of the droplet. Continuity of the fluid phases was readily determined by observing which dye dissolved in the droplet. For the more viscous phases the mixing had to be aided by a small spatula.

Construction of the Phase Diagrams. The various phase regions were represented on a two dimensional triangular phase diagram. These were constructed by making a large number ($\sim$ 250) of samples to extend over a wide concentration range of each ingredient. Once the phase, or phases, of a sample had been identified, the point on the phase diagram representing that sample, was coded according to the phases found. When this procedure had been completed for all the samples, the extent of each phase region, and the phase boundaries were determined by drawing a line through the points with the same code.

Light Scattering Measurements. Some of the microemulsion samples previously made in the determination of the phase diagram were remade in a class 1000 clean room. Any dust particles were removed from the samples by filtering twice through a 0.2 μm Millipore filter. All glassware used in the preparation and filtration of the samples was thoroughly cleaned using chromic acid and rinsed copiously with water.

A Sofica Photo-Gonio Diffusometer (model 42,000) was used for the light scattering measurements. The polarisation ratio of benzene was checked at a scattering angle of 90° and this was found to be 0.408 which is comparable to the literature values of 0.433 (16) and 0.430 (17). Since the frequent use of benzene was undesirable, the instrument was calibrated only once with benzene (the primary standard) and a secondary standard namely a glass cylinder (supplied with the instrument) was used on a day-to-day basis. The ratio of intensities of the two standards glass:benzene was 1.12. The dissymetry of the instrument was checked using the glass standard, benzene and tetradecane and found to be close to unity.

The refractive index of each sample used in the light scattering experiments was measured using an ABBE 60 refractometer calibrated for use with the mercury green line, 436.1 nm (the wavelength used in the light scattering experiments).

The measured intensity of the scattered light at 90° to the incident beam, i^{sample}, was corrected for; (i) the scattering due to the solvent, $i^{solvent}$; (ii) the refractive index change between the sample and the medium surrounding it, (toluene); (iii) for depolarisation using the Cabannes factor, ρ_c (18); (iv) the dissymetry ratio D_r. Thus the corrected i_{90} value was given by

$$i^{Corr} = (i^{sample} - i^{solvent}) \quad n_{sample}, \quad \rho_C \quad D_r$$

where n_{sample} is the refractive index of the sample.

The Rayleigh ratio, R_{90} is then given by

$$R_{90} = \frac{i_{90}^{corr} \; R_S^2}{I_o} \qquad (1)$$

where R^2 is the distance the scattering volume is from the dector and I_o is the incident beam intensity. R^2 and I_O are machine constants and were measured using a sample of brown turbidity eg. benzene.

RESULTS AND DISCUSSION

Phase Diagrams. The phase diagrams for the systems studied are shown in Figures 1 and 2 at a polymer concentration of 10% or 20% for the three temperatures studied (7°, 23° and 47^oC). These diagrams show the typical phases previously described by many authors (15). A qualitative description of the phases observed and the nomenclature used is given in Table I. Below a brief description of the various regions is given.

TABLE I. Designation and Nomenclature of the Observed Phases

Designation	Structure	Notation
Water Continuous Isotropic Liquid	Normal micelles	L_1
Oil Continuous Isotropic Liquid	Reversed micelles	L_2
Liquid Crystalline phases	Generally two dimensional hexagonal	LC
Emulsions	Droplets	E
Gel	-	Gel

Isotropic Microemulsion Phase, L_2

The microemulsion phase exists in the region near the water lean side of the diagram. The L_2 phase extends over the whole range of tetradecane concentrations but the amount of water solubilised in the inverse micelles changes with both temperature and polymer concentration. Higher amounts of water being solubilised at low temperatures and high polymer concentrations. The maximum amount of water which could be solubilised was 26%. It was noted that as the amount of water solubilised increased then the L_2 region became slightly turbid. The L_2 region became smaller as the temperature was increased, the effect being particularly noticeable at ∿ 30-45% butan-l-ol where a distinctive bulge in the L_2 region was noted.

The extent of the oil continuous transparent region i.e. L_2 for each concentration of the polymer surfactant, can be represented in a

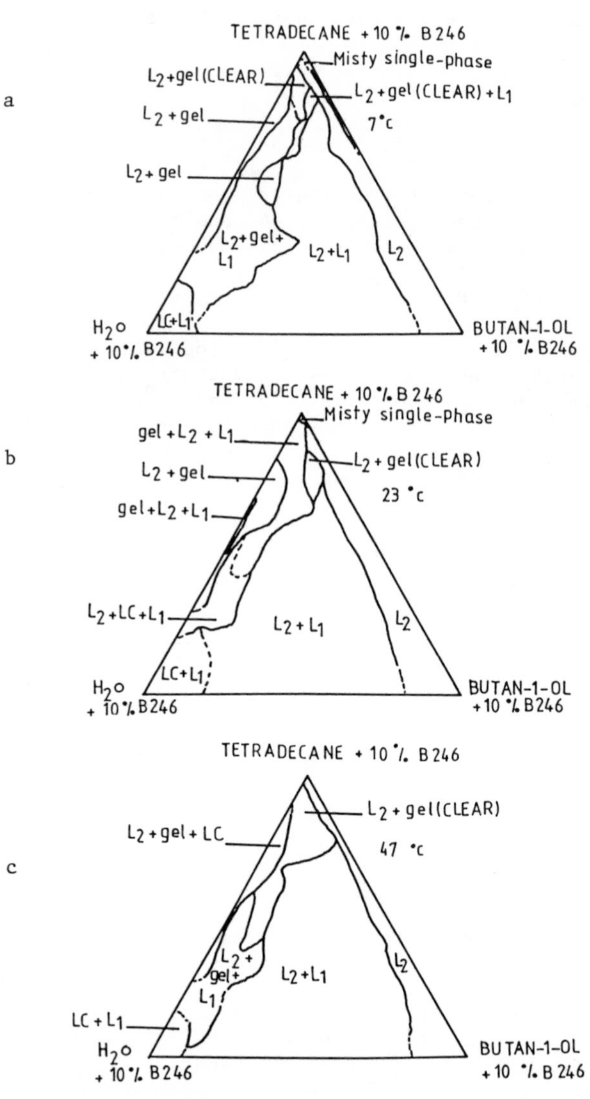

Figure 1 Phase diagrams for the quaternary system tetradecane, butan-1-ol, water, B246 at 7°C (a), 23°C (b) and 40°C (c). Total concentration of B246 is 10%.

a

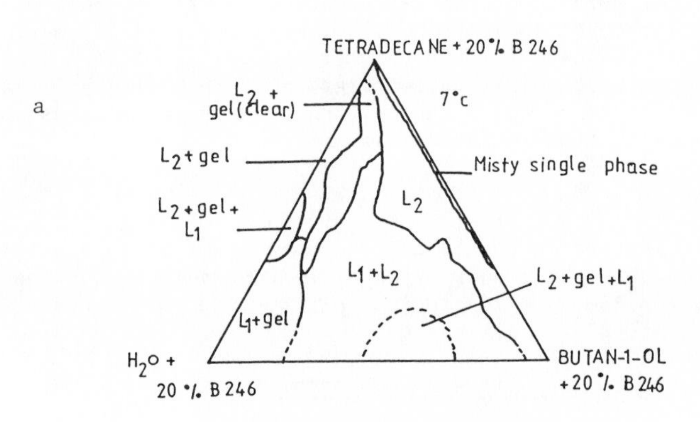

b

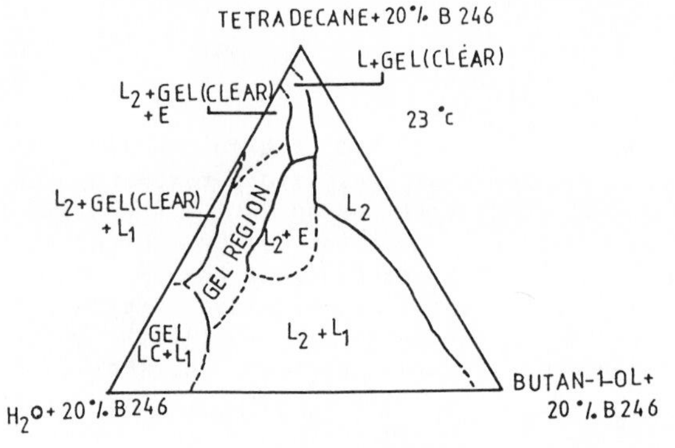

c

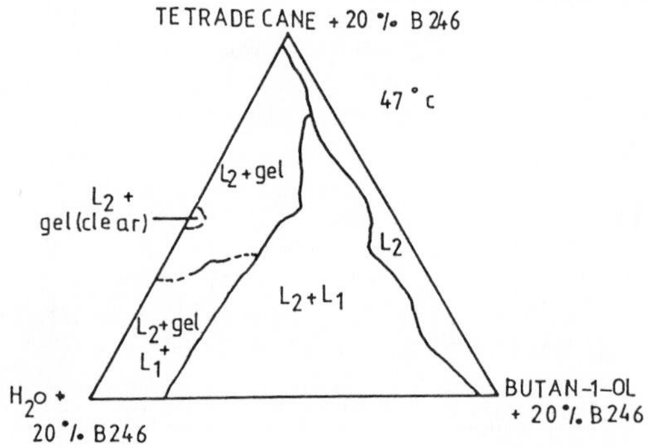

Figure 2 Phase diagram for the quaternary system tetradecane, butan-1-ol, water, B246 at 7°C (a), 23°C (b) and 40°C (c). Total concentration of B246 is 20%.

tetrahedron in Figure 3, for the results obtained at 23°C. Similar representations may be made for the other temperatures. The front of the tetrahedron represents the ternary water-n-butanol-tetradecane system and the corners of the triangle represents the pure components. The fourth corner of the tetrahedron represents pure polymer. An expanded scale is used for the polymer composition to allow a clear representation of the transparent regions. This diagram clearly shows the increase of the extent of the transparent regions as the polymer concentration is increased.

L_1 + L_2 region

This region appears in the 10 and 20% surfactant phase diagrams at all temperatures when the amount of water is increased beyond the solubilisation limit. The region is formed up to a concentration of tetradecane of ∿80% in all cases.

L_2 + gel + L_1, L_2 + E, (inverse emulsion)

L_2 + gel Regions

Bordering the L_2 + L_1 regions were several distinct regions which also contained gel and L_2 phase. The L_2 + gel phases occured at low concentrations of butan-1-ol and at high concentrations of tetradecane. The gel phase was transparent and highly viscous. At higher butan-1-ol concentrations an L_2 + E phase was formed at 23°C for 20% polymer only. The emulsion was determined to be a water in oil emulsion. At high water concentrations and high temperatures an L_2 + gel + L_1 phase was formed; decreasing the temperature moved this phase to higher tetradecane concentrations whilst the water corner of the phase diagram became simply an L_1 + gel phase or L_1 + LC phases.

L_1 + gel phase, L_1 + LC phase

These phase regions were formed in the water rich areas of the phase diagrams. The LC phases were identified using a polarising microscope and determined to be similar to an inverse middle phase observed in ternary systems.

Gel Regions

In addition to the phases already described above, other gel regions were observed. These gels were generally transparent highly viscous materials which became more turbid at higher water concentrations.

The Effect of Temperature on the Phase Behaviour

Increasing the temperature of the microemulsion region leads to a reduction of the size on the phase diagram where they are formed. This is a result in the dehydration of the ethylene oxide part of the copolymer resulting in smaller volumes of water to be solubilised in the microemulsion droplets. The effect of temperature on the stability of the other regions of the phase diagram is complex; to summarise though, it seems that increasing the temperature reduces the variety of phases formed.

Light Scattering

Figure 4 shows plots of R_{90} versus volume fraction of water, ϕ_{H_2O} for the two surfactant concentrations studied, namely 20% B246 and 10% B246. R_{90} increases with increasing water concentration for both surfactant concentrations. This increase in R_{90} indicates the size or number of the microemulsion droplets are increasing as the volume fraction of water is increased. It is also clear from Figure 4 that the R_{90} values are higher for the higher surfactant concentration again indicating the formation of a larger number of droplets. The analysis of the light scattering data was carried out using the same procedure as described by Cebula et al (19) and Baker et al (16). The Rayleigh ratio R_{90} is given by

$$R_{90} = K_o \; MC \; P(Q) \; S(Q) \tag{2}$$

where M is the molecular mass of the scattering units, C their concentration in g. cm^{-3}, Q is the scattering vector given by

$$Q = \left[4 \pi n_o \; \sin (90/2)\right] / \lambda_o \tag{3}$$

λ_o is the wavelength of the light in a vacuum.

K_o is an optical constant given by

$$K_o = \frac{9 \pi^2 n^4}{2 N_A \lambda_o^4 \rho^2} \left[\frac{n_c^2 - n^2}{n_c^2 + 2n^2}\right] \tag{4}$$

where n is the refraction index of the medium and n_c of the scatterers. N_A is Avogadro's Number and ρ is the density of the scattering drops.

P(Q) is the particle scattering from factor which for measurements at 90° and for small scattering is approximately unity. S(Q) is the structure factor which is given by (20-22).

$$S(Q) = 1 + \frac{4\pi N}{Q} \int_0^\infty g(r) - 1 \;\; r \sin Qrdr \tag{5}$$

where N is the number of particles per cm^3 and g(r) is the radial distribution function. Basically S(Q) is a term, usually referred to as the structure factor, which accounts for the increase in scattering due to ordering of the particles.

For spherical microemulsion droplets,

$$M = \frac{4}{3} \pi \; R^3 \rho \; N_A \tag{6}$$

Thus,

$$R_{90} = K_o \; \frac{4}{3} \pi R^3 \; N_A \rho \; S(90) \; \phi_{H_2O} \tag{7}$$

S(90) was calculated from the Percus-Yevick hard sphere model as given by Ashcroft and Lekner (23) enabling the droplet radius R to be estimated assuming that the thickness of the surfactant layer surrounding the water cone, t, to be 0.4 nm (the size of a butan-1-ol molecule).

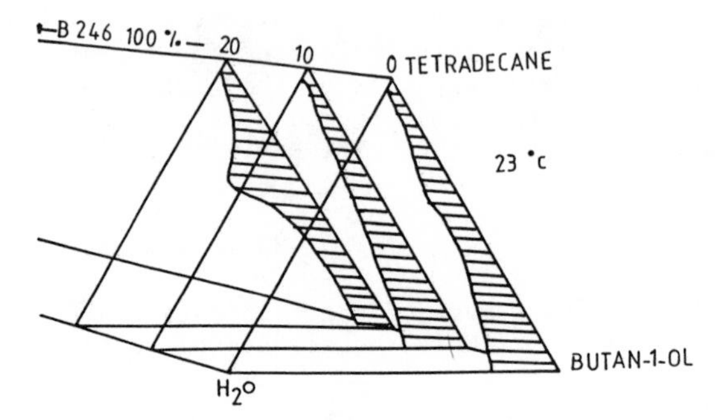

Figure 3 Transparent microemulsion region at 23°C for various B246 concentrations.

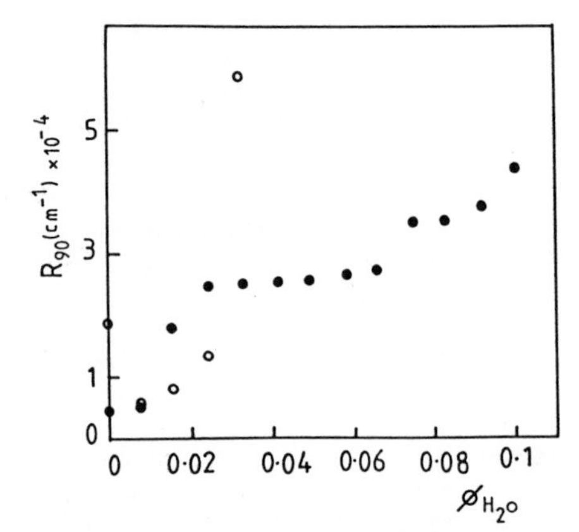

Figure 4 Scattering from microemulsions as a function of volume fraction of water at two B246 concentrations
0 10% polymer ● 20% polymer.

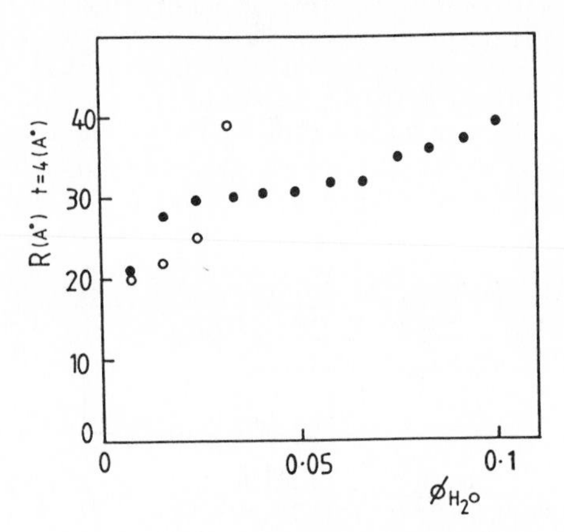

Figure 5 Water core radii, obtained from light scattering assuming the thickness of the surfactant layer to be 0.4 nm
0 10% polymer ● 20% polymer.

Figure 5 is a plot of R versus ϕ_{H_2O}. It is important to compare the results obtained in this present study with those reported in the literature. In most of the water in oil microemulsion systems studied by scattering methods, the ratio of water to surfactant was kept constant. In these cases it was found that an increase in the water volume fraction did not change the size of the droplets. Cebula et al.(19), again with a constant ratio of water to surfactant, were the first to show that the droplet size of microemulsions increases close to the phase boundary, as was noticeable in our results. Baker et al.(16) have measured the droplet size radius at a fixed surfactant concentration. It was found that the size of the microemulsion droplets increases as the volume fraction of water increases. However, at a fixed ratio of water to surfactant the droplet size was constant. The results presented here are the droplet size as a function of water concentration for a fixed surfactant concentration. The results show that the droplet size increases slowly with increasing water concentration (this is a result of the droplets swelling as more water is added to the system), in agreement with the results of Baker *et al* (16). In addition, close to the phase boundary the droplet size increases more markedly; the size of the droplets increases by 50%, as was observed by Cebula et al. (19).

Literature Cited

1. Danielsson, I; Lindman, B. *Colloids and Surfaces* 1981, *3*, 391.
2. Overbeek, J.Th.G., *Faraday Disc. Chem. Soc.*, 1978, *65*, 7.
3. *Microemulsions Theory and Practice*; Prince, L.M., Ed.; Academic: New York, 1977.
4. Shinoda, K; Friberg, S., *Adv. Colloid Interface Sci.* 1975, *4*, 281.
5. Tadros, Th.F. In *Surfactants*; Mittal, K.; Lindman, B., Eds.; Plenum: New York, 1984; Vol.3, p.1501.
6. Overbeek, J.Th.G.; de Bruyn, P.L; Verhoeckx, F. in *Surfactants*; Tadros, Th.F., Ed.; Academic: New York, 1984; p111.
7. Reiss, G.; Nervo, J.; Rogez, D. *Polym. Eng. Sci.* 1977, *17*, 8, 634.
8. Reiss, G.; Nervo, J. *Inf. Chim.*, 1977, *170*, 185.
9. Reiss, G., Nervo, J.; Rogez, D. *Polym. Prep. Am. Chem. Soc., Div. Polym. Chem.* 1977, *18*, 329.
10. Marie, P.; Gallot, Y., *C.R. Acad. Sci. (Paris) Ser C*, 1977, *327*, 284.
11. Boutillier, J.; Candau, F. *Colloid Polym. Sci.* 1979, 267, 46.
12. Candau, F.; Boutillier, J.; Tripier, F.; Wittmann, J.C. *Polymer* 1979, *20*, 1221.
13. Ballet, F.; Debeuvois, F.; Candau, F. *Colloid Polym. Sci.* 1980, *258*, 1253.
14. Barker, M.C.; Vincent, B. *Colloids and Surfaces* 1984, *8*, 297.
15. Baker, R.C.; Florence, A.T.; Tadros, Th.F.; Wood, R.M. *J. Colloid Interface Sci.* 1984, *100*, 311.
16. Baker, R.C.; Florence, A.T.; Ottewill, R.H.; Tadros, Th.F. *J. Colloid Interface Sci.* 1984, *100*, 332.
17. Carr, C.I.; Zimm, B.H. *J. Chem. Phys.* 1950, *18*, 1616.
18. Cabannes, J.; *La Diffusion Moleculaire de la Lumiere*; Les Presses Universitaires de France: Paris, 1929.

19. Cebula, D.J.; Ottewill, R.H.; Ralston, J.; Pusey, P.N. J. Chem. Soc. Faraday Trans I 1981, 72, 2585.
20. Kerker, M. The Scattering of Light; Academic: New York, 1969.
21. Temperley, H.M.V.; Trevein, D.H. Liquids and Their Properties; Ellis Horwood: Chichester, 1978, p.191.
22. Marcus, Y. Introduction to Liquid State Chemistry; Wiley: New York, 1977.
23. Ashcroft, N.W.; Lekner, J. J. Phys. Rev. 1966, 145, 83.

RECEIVED September 12, 1988

Chapter 3

Polymerization in Microemulsions

Stig E. Friberg, Guo Rong, Ch. Ch. Yang, and Yun Yang

Department of Chemistry, Clarkson University, Potsdam, NY 13676

Structural entities in water-in-oil (W/O) microemulsion at low water content are reviewed. These structures include monomers of surfactant associated with a few water and cosurfactant molecules. These small aggregates are stable in a non-polar environment in spite of their polar character and provide an interesting case of unusual interaction between polymers and microemulsion structures. Examples are provided of cases when this interaction is important for stability of microemulsions with added organic or inorganic polymers.

Microemulsions (1-3) are colloidal dispersions of water-in-hydrocarbons or vice versa stabilized by a single surfactant (4,5) or a combination of an ionic surfactant and a hydrophobic amphiphilic called a cosurfactant (6-8). The droplet size of these dispersions is far less than the wave length of light; hence they appear transparent to the eye.

They are an attractive medium for polymerization; a successful reaction of that kind retaining the structure would mean new and interesting materials. Claims to such reactions have been made in several instances from micellar structures to highly solubilized systems (9-13) with varied degree of success.

In this article we will focus the attention on some basic factors in the stability of microemulsion structures and the influence on these factors by the structural changes caused by the polymerization.

Stability of Microemulsions and Polymer Content

The stability of microemulsions has been treated in the form of a model with dispersed droplets in a continuous medium (14,15). The free energy of such a system may schematically be described as the sum of a series of terms according to Table I.

0097-6156/89/0384-0034$06.00/0

Table I. Free Energy Terms of a Microemulsion

1) Chemical Potential of Components
2) Interparticle Potentials
a) Van der Waals potential
b) Electric double layer potential
c) Repulsive potential from adsorbed molecules
3) Interfacial Free Energy
a) Stretching Component
b) Bending Component
c) Torsional Component
4) Entropic Contribution From Location of Droplets

SOURCE: Reprinted with permission from ref. 14.

The significance of the different terms varies considerably; the Van der Waals potential is usually without much influence (14), while the interfacial free energy as well as the entropic contribution from the droplets are decisive. In fact the interfacial free energy must reach a low level ($\cong 10^{-3}$ mN/m) to provide stability for the microemulsion.

It is easy to realize that polymerization in the system will severely influence these factors causing destabilization (13) because of conflict between the need for conformational freedom of the polymer and the space required for microemulsion droplets.

One of the most illustrative example possible is the change in water solubility when polyethyleneoxide was added to a W/O microemulsion stabilized by a polyethyleneglycol alkyl ether surfactant (16). The maximum water solubilized by adding tetraethyleneglycol dodecyl ether to decane could be approximated by the equation

$$W = 2.76 \cdot S \tag{1}$$

in which W is weight percentage water of total and S the corresponding measure of surfactant amount.

Replacing the water by a 5% by weight of an aqueous solution of polyethylene glycol (M = 1,000) gave the relation

$$(W/PEG)_{95/5} = 0.11 \cdot S \tag{2}$$

This is approximately 25 times reduction of water solubilizing capacity with only 5% polymer of a modest molecular weight. This result is a good illustration of the incompatability of polymers and microemulsions and the initial lack of success (13) in polymerization is a reasonable result.

Another stability problem is connected with the conditions at low water content. An observation of the W/O microemulsion phase region (6) demonstrates a minimum ratio of water to surfactant to be required to obtain solubility of the latter. At the lowest water concentrations the W/O microemulsion does not contain microemulsion droplets. The surfactant exists only in the form of premicellar aggregates or ion pairs (17,18).

Minimum Water Solubilization

The minimum solubility of water in a W/O microemulsion stabilized by an ionic surfactant/medium chain length cosurfactant combination depends on the surfactant counter ion; Figure 1 shows a typical solubility curve (19). The reason for this minimum water content is related to the water of hydration and calculations of the free energy of gaseous water/surfactant aggregates (20) are useful in order to understand the fundamental basis of the phenomenon.

In this way the free energy of the aggregate versus number of water molecules compares to the free energy of the separate components with water in liquid form and the surfactant in crystalline form according to Figure 2. At very low water content the free energy of the aggregate is greater than that of the separate components, because with few water molecules the total hydration energy per aggregate is still low, (Zone 1, Figure 2). With increased number of water molecules the pronounced hydration energy becomes significant and the aggregates free energy becomes lower than that of the separate components, Zone 2, Figure 2. For very high water contents the hydration energy is reduced because the added water molecules now are not in direct bonding contact with the surfactant ion pair, Zone 3, Figure 2.

The monomeric surfactant aggregate is no longer a stable unit and dimerization takes places as a beginning of the inverse micellization. One interesting aspect of these stability relations is the fact that the part which has monomeric aggregates in the surfactant/cosurfactant/water phase diagram also shows maximum solubility of hydrocarbon as well as polymer solutions. The relations between monomer, dimer and polymer solubility are described in the following sections.

W/O Styrene Microemulsions

Addition of styrene monomer to a W/O microemulsion base, Figures 3A-C (21), illustrates the high stability of the monomeric surfactant aggregate against dilution with a hydrocarbon. At 50 wt % of styrene the maximum water solubility is reduced to 32%, Figure 3B, and at 75% level of styrene the solubility region is limited to the original region containing monomeric surfactant associations only, Figure 3C.

The implication of this change is straightforward. Polymerization in the low water content means no conflicting space demands between the polymer and the inverse micelles because the latter do not exist. Since polystyrene is completely soluble in styrene it would be reasonable to expect polymerization to be possible in this region with no stability problems.

However, this is not the case (21). The presence of minor amounts of polymer leads to a pronounced reduction in the solubility region. Figure 4 illustrates this fact for pentanol as a cosurfactant. 10% of the styrene replaced by polystyrene (M = 1,000) reduced the solubility region to less than one third of the one for the styrene microemulsion.

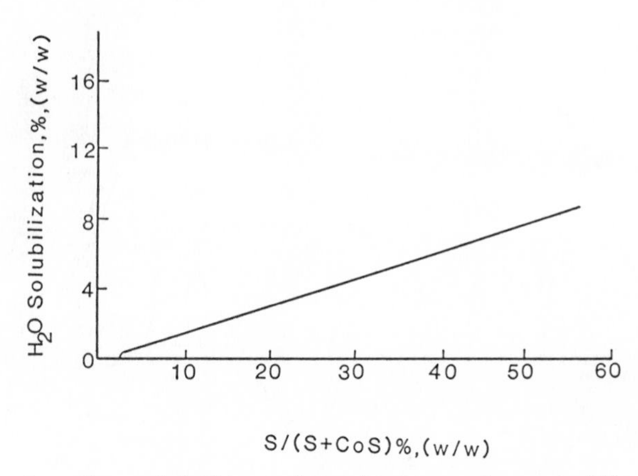

Figure 1. The solubility of an ionic surfactant (S) in a cosurfactant (CoS, pentanol) is proportional to a minimum water content.

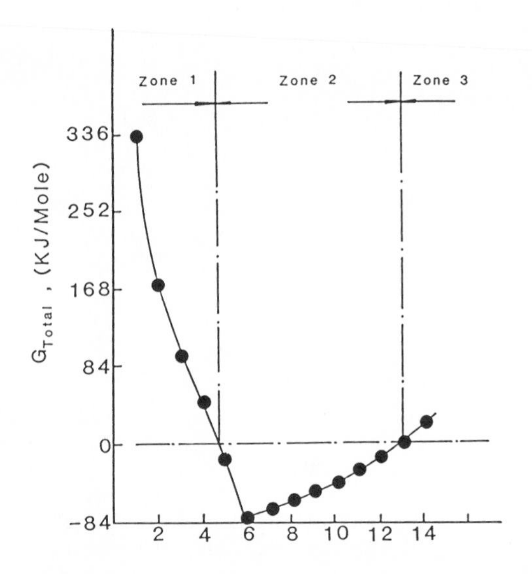

Figure 2. Free-energy difference from solid crystalline sodium octanoate and liquid water for a (mono) sodium octanoate/water molecular compound with different numbers of water molecules (Reprinted from ref. 20. Copyright 1982 American Chemical Society.)

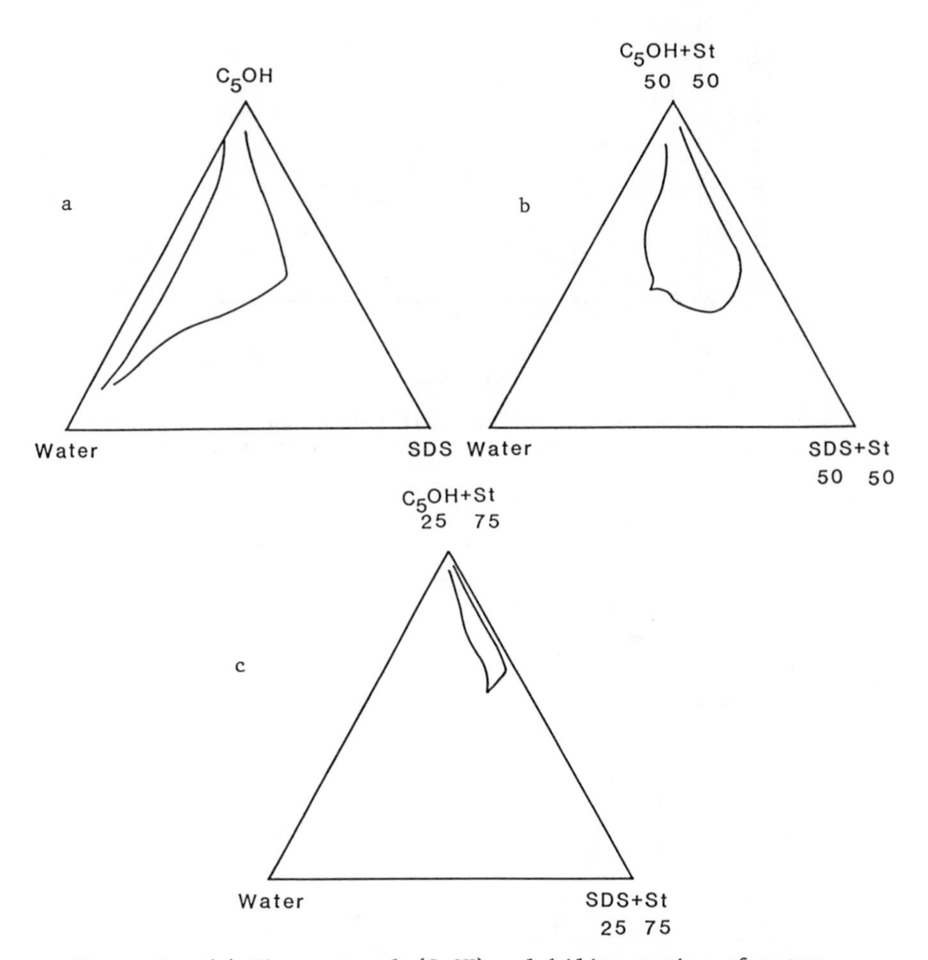

Figure 3. (a) The pentanol (C_5OH) solubility region of water reaches far toward the water corner after addition of an ionic surfactant, sodium dodecyl sulfate (SDS). (b) Addition of styrene in a 1/1 ratio to the pentanol and the surfactant gave a strong reduction in maximum water solubility. (c) With a styrene/amphiphile ratio of 3/1 only a small amount of water could be solubilized.

In fact, the surfactant monomeric aggregates and the polymer are mutually exclusive in the range of minimum water content according to the following results. The minimum water content in weight percent, Figure 4, may be expressed as

$$W = a + bS \tag{3}$$

in which W is minimum water content in weight percent, S is the surfactant content, also in weight percent while a and b are constants. For the system with pentanol and sodium dodecyl surfate a = 0.5 and b = 0.207.

This means that increased surfactant content was always accompanied by a minimum increase of water content according to Equation 3. Plotting the solubility of the polymer in the pentanol/styrene solution when pentanol is gradually replaced by the surfactant/water combination according to Equation 3 shows the reflection between polymer solubility and the amount of surfactant/water aggregate present (21). In fact, a linear reduction of polymer solubility was found with increase surfactant/water content, Figure 5. A surfactant concentration of 9.5% accompanied by water to 2.46% would result in zero solubility of the polystyrene. This means that a composition of pentanol/styrene 25/75 dissolves 42% polystyrene but that the polystyrene is completely insoluble in a composition 75% styrene, 13% pentanol, 9.5% surfactant and 2.5% water. Or, expressed in a different manner 1 molecule of polymer is removed for 1.3 molecules of surfactant and 3.3 molecules of water added.

The results clearly demonstrate that the conflicting space demands from conformational entropy component of the polymer free energy are not the only problems in the incompatibility phenomena of polymers and microemulsions. Our investigations on oligomer microemulsions (22) demonstrated the oligomers, even the dimers, to have significant influence on the solubility region. The dimer has no significant conformational entropic component in its free energy and the conclusion must be that other factors also must be important.

A comparison between the solubility region of water in W/O microemulsions of styrene and of its dimer, Figure 6 (22), reveals a significant difference. The maximum solubility of water was reduced from 29 to 11% by weight and the maximum surfactant concentration declined by 30%. In fact, the solubility region of the dimer was only 20% that of the monomer.

One interesting feature is that the minimum water content to dissolve the surfactant was unchanged, but the maximum increase of water content with added surfactant was reduced.

With styrene the maximum water solubility may be described as

$$W = 0.256 + 1.07\ F_S \tag{4}$$

in which W is the water weight fraction of total and F_S is the weight fraction surfactant/(surfactant + cosurfactant). The solubility of water in the dimer microemulsion is described by

$$W = 0.256 + 0.48 \cdot F_S \tag{5}$$

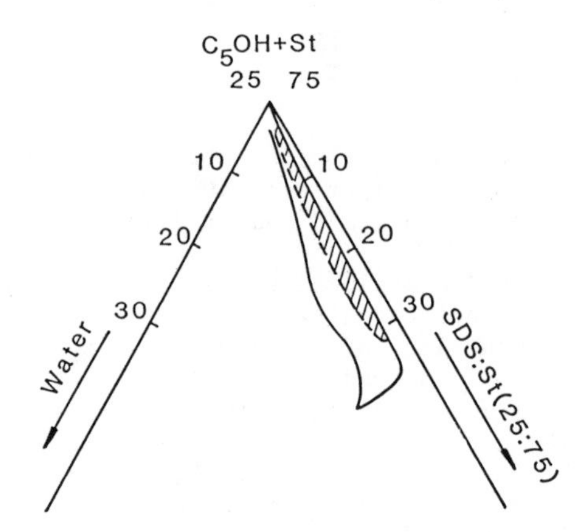

Figure 4. Replacing part of the styrene in Figure 3C with polystyrene gave further a reduction in the solubility region: —— Amphiphile/Styrene 1/3; - - Amphiphile/Styrene/Polystrene 1/2.7/0.3.

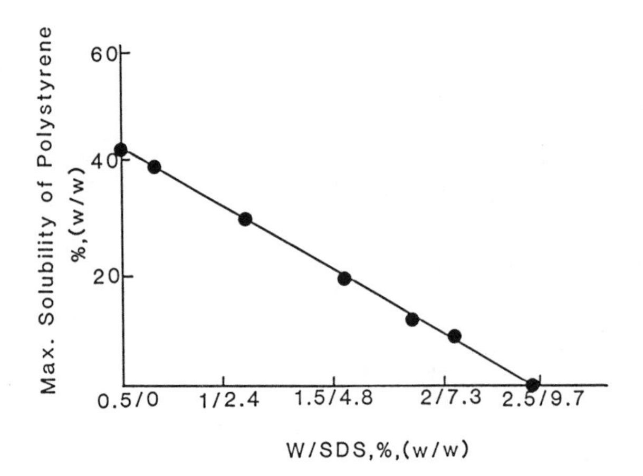

Figure 5. The solubility of polystyrene in a pentanol/styrene (25/75 wt %) solution was reduced when the pentanol was partly replaced by water plus surfactant in the amount of the abscissa.

The reduction in the coefficient in front of F_s illustrates the reduction in water solubility in a very illustrative manner. Changing the hydrocarbon phase from styrene monomer to dimer reduced the maximum water solubilization by a factor of 2.2.

It is obvious that a different factor must be found to explain the sensitivity to dimer formation for maximum water content in the monomeric surfactant aggregates because this phenomenon is entirely outside the realm of polymers and their entropic conformational demands. One such factor is the interaction between aromatic compounds and the polar group of a surfactant. Two examples of such interaction are described in the next section.

Aromatic Hydrocarbon/Polar Group Interaction

The direct manifestation of this interaction was observed a long time ago in the water concentration at which a change from monomeric aggregates to inverse micelles took place in a W/O microemulsion (23). Adding an aromatic compound has a pronounced influence on the water content at which micellization begins.

Figure 7 (23) illustrates this fact. Without hydrocarbon the onset of micellization as indicated by the suddent rise in light scattering intensity is observed at 37% water in the microemulsion. Addition of an aliphatic hydrocarbon did not change this concentration significantly (33% water). An aromatic hydrocarbon, on the other hand, had a pronounced influence; 50% benzene brings the water concentration down to 20%.

This change is caused by the interaction between the aromatic hydrocarbon and the polar group of the surfactant. This phenomenon has been clarified for normal micelles of ionic surfactants (24); for inverse micelles material is available only for nonionic surfactants. Christenson and collaborators (25-27) made an extensive study using NMR and calorimetry. The results of both studies agree showing a partition coefficient for benzene between the surfactant polar group and its hydrocarbon chain of approximately 3.

These results demonstrate two facts. The aromatic hydrocarbons have a strong interaction with the polar groups of surfactants as evidenced by NMR and calorimetry investigations. In addition, the light scattering and dipole moment determination show that this interaction influences the transition from monomeric aggregates to inverse micelles.

In summary there is compelling evidence of strong interaction between aromatic nuclei and the polar group of a surfactant and a connection to the results cited in this article on microemulsion stability and polymerization appears forthright. A dimer for an aromatic monomer should be expected to have a more pronouned interaction with the polar group. A rational consequence will be the influence on the stability of the W/O microemulsion in the water poor area of surfactant/water aggregates as found experimentally (21,22).

Minimum Amount of Water and Inorganic Polymers

Inorganic polymers in microemulsion have not been investigated earlier, but the potential for sol-gel processes (28) to form ceramics is well established (29).

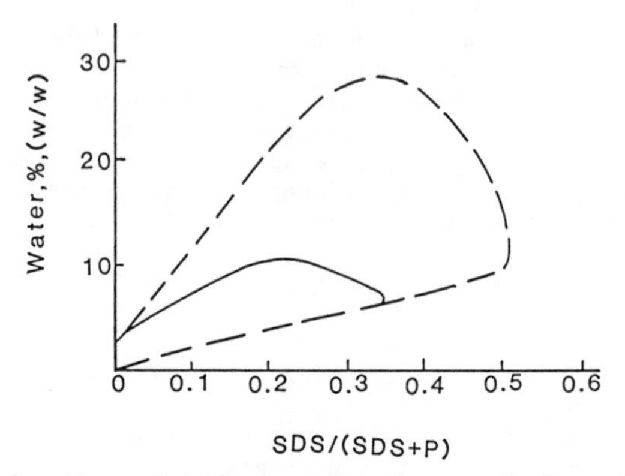

Figure 6. The solubility region for conditions in Figure 3B (- -) was considerably reduced when the styrene was replaced by its dimer (——).

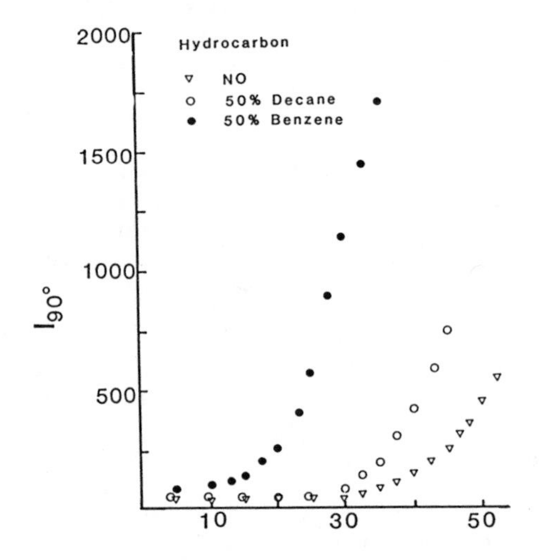

Figure 7. The onset of inverse micellization in the pentanol solution in the water, pentanol (C_5OH), potassium oleate (KOL) system (∇) was changed to a small degree at the addition of decane (o) but to a pronounced degree by benzene (•) (Reprinted with permission from ref. 23. Copyright 1978 Academic Press.)

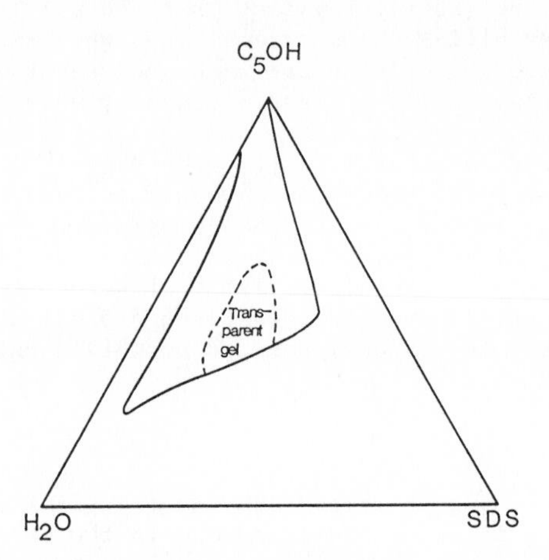

Figure 8. The transparent gel of SiO_2 is formed in a limited region of the W/O microemulsion base only.

Our preliminary investigation (30) have demonstrated that inorganic polymers of SiO_2 can be formed in a W/O microemulsion without destroying its stability. This means that inorganic polymers may be used for gelation of microemulsions; a difficult problem using the common organic polymers.

The reaction leading to the SiO_2 polymers consists of two steps. In the first one an organic silicone compound, such as silicone tetraethoxide, is hydrolyzed to some form of silicic acid. This step is immediately followed by a condensation reaction giving polymers with Si-O-Si bonds. The total reaction is

$$Si(OC_2H_5)_4 + \frac{8-x}{2} H_2O \rightarrow Si(OH)_{4-x}O_{x/2} + 4C_2H_5OH \quad (6)$$

Except for the interesting fact that the polymeric SiO_2 structures do not destabilize the microemulsion there is also an intriguing feature of the competition for water between the surfactants and the silicone tetraethoxide. Figure 1 clearly shows a minimum amount of water to be necessary in order to dissolve the surfactant and Equation 6 shows a need for water for the hydrolysis reaction. The minimum amount of water needed is 2 when x in Equation 6 is close to 4, which actually is the case in the experiment. For that case two water molecules are needed for each molecule of silicone tetraethoxide. A calculation of the lower limit of water in Figure 8 shows the increase of water concentration to be approximately four times this value. Water is obviously needed to retain the silica in the microemulsion state.

Conclusion

The interaction between polymers and the colloidal droplets in microemulsions results in space conflicts due to the entropic conformational demands by the polymer. This is a serious problem as is evident by a multitude of systems which has shown phase separation during polymerization. However this interaction is not the only one of importance for the polymer/microemulsion system. An aromatic dimer has been shown to also have a strong influence on the microemulsion region even though its conformation variation does not exist.

Its effect is instead referred to the well known interaction between an aromatic compound and a polar group. This interaction has been proven sufficiently strong to influence the initial inter-amphiphilic association during the formation of inverse micelles. It appears reasonable to conclude that dimerization would increase the interaction.

A more direct interaction is found in a W/O microemulsion system in which an inorganic polymer of SiO_2 is formed by hydrolysis of silicone tetraethoxide. The competition for water molecules to dissolve the surfactant and to feed the hydrolysis of the silicone alcoxide leads to phase separation and crystallization of the surfactant at low water content.

Acknowledgments

This research was partially supported by the CAMP program at Clarkson University.

Literature Cited

1. Microemulsions: Structure and Dynamics; Friberg, S.E.; Bothorel, P., Eds.; CRC : New York, 1986.
2. Physics of Amphiphiles: Micelles, Vesicles and Microemulsions; Digiorgio, V.; Corti, M., Eds.; North Holland: New York, 1985.
3. Microemulsion Systems; Rosano, H.L.; Clausse, M., Eds.; Marcel Dekker: New York, 1987.
4. Friberg, S.E.; Lapczynska, I. Progr. Colloid Polym. Sci. 1975, 56, 16.
5. Angel, L.R.; Evans, D.F.; Ninham, B.W. J. Phys. Chem. 1983, 87, 538.
6. Friberg, S.E.; Buraczewska, I. Progr. Colloid Polym. Sci. 1978, 63, 1.
7. Lindman, B.; Stilbs, P.; Moseley, M.E. J. Magn. Reson. 1980, 40, 401.
8. Lang, J.; Rueff, R.; Dinh-Cas, M.; Zana, R. J. Phys. 1984, 45, 257.
9. Atik, S.S.; Thomas, J.K. J. Am. Chem. Soc. 1981, 103, 4279.
10. Gan, L.M.; Chew, C.H.; Friberg, S.E. J. Macromol. Sci. Chem. 1983, A19, 739.
11. Fendler, J.H. Acc. Chem. Res. 1980, 13, 7; 1984, 17, 3.
12. Gooss, L.; Ringsdorf, H.; Schupp, H. Augen. Chem. Inc. Ed. Eng. 1980, 20, 305.
13. Stoffer, J.O.; Bone, T. J. Dispersion Sci. Technology 1980, 1, 37.
14. Ruckenstein, E. J. Dispersion Sci. Technology 1981, 2, 1.
15. Ninham, B.W.; Mitchell, D.J. J. Chem. Soc. Faraday Trans. 2 1980, 77, 601.
16. Friberg, S.E.; Lapczynska, I.; Gillberg, G. J. Colloid Interface Sci. 1976, 56, 19.
17. Clausse, M.; Rayer, R. In Colloid and Interface Science II; Kerker, M., Ed.; Academic Press: New York, 1976; p 217.
18. Eicke, H.I.; Christenson, H. J. Colloid Interface Sci. 1974, 48, 281.
19. Friberg, S.E.; Venable, R. In Encyclopedia of Emulsion Technology; Becher, P., Ed.; Marcel Dekker: New York, 1983; Vol. 1, p 287.
20. Friberg, S.E.; Flaim, T.D.; Holt, S.L.; Eds.; ACS Symposium Series No. 177; American Chemical Society: Washington, DC, 1982; p 1.
21. Gan, L.M.; Chew, C.H.; Friberg, S.E. J. Macromolecular Sci. 1983, A19, 739.
22. Gan, L.M.; Chew, C.H.; Friberg, S.E.; Higashimura, T. J. Pol. Sci. Chem. 1981, 19, 1585.
23. Sjoblom, E.; Friberg, S.E. J. Colloid Interface Sci. 1978, 67, 16.
24. Eriksson, J.C.; Gillberg, G. Acta Chem. Scand. 1966, 20, 2019.
25. Christenson, H.; Friberg, S.E. Colloid Polym. Sci. 1980, 75, 276.
26. Christenson, H.; Friberg, S.E.; Larsen, D.W. J. Phys. Chem. 1980, 84, 3633.

27. Friberg, S.E.; Christenson, H.; Bertrand, G.; Larsen, D.W. Reverse Micelles 1984, 5, 105.
28. Roy, R. J. Am. Ceram. Soc. 1986, 39, 145.
29. Roy, R. Science 1987, 238, 1664.
30. Friberg, S.E.; Yang, C.C., to be published.

RECEIVED August 28, 1988

Chapter 4

Polymerization of Water-Soluble Monomers in Microemulsions

Potential Applications

Françoise Candau

Institut Charles Sadron, Centre de Recherches sur les Macromolécules, Ecole d'Applications des Hauts Polymères, 6 rue Boussingault, 67083 Strasbourg Cédex, France

The general features of inverse microemulsion polymerization at the present state of knowledge are presented. The influence of various water-soluble monomers on the structural properties of the polymerizable microemulsions is analyzed and an optimization of the process is proposed. The best conditions correspond to the polymerization of monomers in bicontinuous microemulsions characterized by very low interfacial tensions, which subsequently leads to clear and stable microlatices of polymers with high molecular weights. The potential applications of the process are discussed.

Polymerization in microemulsion systems has recently gained some attention as a consequence of the numerous studies on microemulsions developed after the 1974 energy crisis (1,2). This new type of polymerization can be considered an extension of the well-known emulsion polymerization process (3). Microemulsions are thermodynamically stable and transparent colloidal dispersions, which have the capacity to solubilize large amounts of oil and water. Depending on the different components concentration, microemulsions can adopt various labile structural organizations - globular (w/o or o/w type), bicontinuous or even lamellar - Polymerization of monomers has been achieved in these different media (4-18).

The first experimental studies aimed at elucidating the mechanism and the characteristics of the process from an academic point of view. Relatively small amounts of monomers were dispersed in microemulsions, which stability was ensured by large surfactant concentrations. The major problem was often the onset of phase separation or polymer precipitation in the course of polymerization (7,8,15). This was due to some structural changes of these dynamic structures and to entropic effects associated with the polymerization. In some cases, these difficulties have been overcome

0097-6156/89/0384-0047$06.00/0

as described in our recent publications on the polymerization of acrylamide in AOT reverse micelles (13,14). The main conclusions were that the mechanism and reaction kinetics in inverse microemulsion polymerization are basically different from those observed in inverse emulsion. In addition, it was shown that stable uniform microlatices (d<50nm) could be formed with high molecular weight hydrosoluble polymers entrapped within water-swollen particles. However, these experiments presented industrial application limitations because of the high ratio of surfactant to monomer concentration (≃ 2.5) required. High solid contents are usually desirable in most applications. This led us to optimize the process and focus on the formulation of these systems. A matter of great importance since it controls the properties of the resultant latices.

In this paper, we review the main results obtained so far in the field of inverse microemulsion polymerization. The role played by the monomers on the structural characteristics of the microemulsion prior to polymerization will be emphasized. The main properties of the final latices will be discussed in light of both basic and applied research.

Formulation

Much effort has been devoted to find inexpensive formulations compatible with an economical process. To obtain an inverse monophasic microemulsion, special conditions have to be met. Their main parameters are as follows : surfactant concentration, HLB of the surfactant(s), temperature, nature of the organic phase and composition of the aqueous phase. A preferred class of surfactants are those forming microemulsions without any added cosurfactant (alcohol). The presence of an alcohol is indeed liable to favor chain transfer reactions, limiting the range of attainable molecular weights (19,20). In addition, the dilution procedure of these microemulsions containing alcohols is not trivial since the continuous phase consists of a solvent mixture of unknown composition.

The selection of the organic phase has a substantial effect on the minimum surfactant concentration necessary to obtain the inverse microemulsion. Nevertheless, there is a threshold for the minimum surfactant amount needed to form the small microemulsion droplets (d ≃ 5-10 nm). Calculations give a limiting value of approximately 10% of all other components ; lower concentrations of surfactant lead to conventional macroemulsions.

For nonionic surfactants, an optimization of the process was achieved by using a similar approach to the so-called Cohesive Energy Ratio (CER) concept developed by Beerbower and Hill for the stability of classical emulsions (21). Its basic assumption is that the partial solubility parameters of oil and emulsifier lipophilic tail and of water and hydrophilic head are perfectly matched. Thus, the Winsor cohesive energy ratio R_o, which determines the nature and the stability of an emulsion, is directly related to the emulsifier HLB (hydrophile-lipophile balance) by

$$R_0 = \frac{d_H}{d_L}\left(\frac{20}{HLB} - 1\right)\frac{(\delta^2_d + 0.25\,\delta^2_p + 0.25\,\delta^2_h)_L}{(\delta^2_d + 0.25\,\delta^2_p + 0.25\,\delta^2_h)_H} \qquad (1)$$

where d_L is the density of the lipophile (L) and d_H that of the hydrophile (H). The subscripts d,p and h in the partial solubility parameters δ refer to the London dispersion forces, polar and hydrogen bonding forces respectively. As the chemical match has established equivalence between oil and lipophile it is thus possible from Equation 1 to calculate the required HLB from a given oil.

These criteria have been used to select systems most appropriate to the formation of polymerization microemulsions (22). The surfactants usually consist of two nonionics. The complex resulting from their association at the w/o interface favors greater stability (23). Blending surfactants also allows the selection of an optimum HLB value by varying their composition. The monomer content in the aqueous phase can vary from 30-70% by weight but is usually in the range of 50%. Neutral monomers (acrylamide) (12,22,24), anionic (sodium acrylate) (11,25) and cationic (methacryloxyethyltrimethylammonium chloride) (26) have been investigated. Various oils have been tested to check the best match to lipophiles.

The oil structure influence on the formulation is illustrated in Figure 1. It represents the minimum percentage of emulsifiers required to induce the transition macro-microemulsion versus their HLB values for monomer-water mixtures dispersed in different oils. It can be seen that in the case of acrylamide (AM) or acrylamide-sodium acrylate (Aa) mixtures, the amount of surfactant needed to form a microemulsion is much larger for toluene or cyclohexane than for Isopar M (11,22). When methacryloxyethyltrimethylammonium chloride (MADQUAT) is the monomer, the optimal conditions are obtained in cyclohexane. These results closely follow the differences calculated for the solubility parameters between oils and lipophiles as shown in Table I.

Table I. Solubility Parameters of Oils and Surfactants

Component	$\delta = (\delta^2_d+\delta^2_p+\delta^2_h)^{1/2}$ (Hildebrand)
Oil	
Isopar M	7.79
Dodecane	7.60
Decane	7.52
Heptane	7.30
Cyclohexane	8.20
Toluene	8.88
Surfactant	
Lipophile tail G 1086	7.87
Lipophile tail Arlacel 83	7.87
Lipophile tail Tween 80	7.87

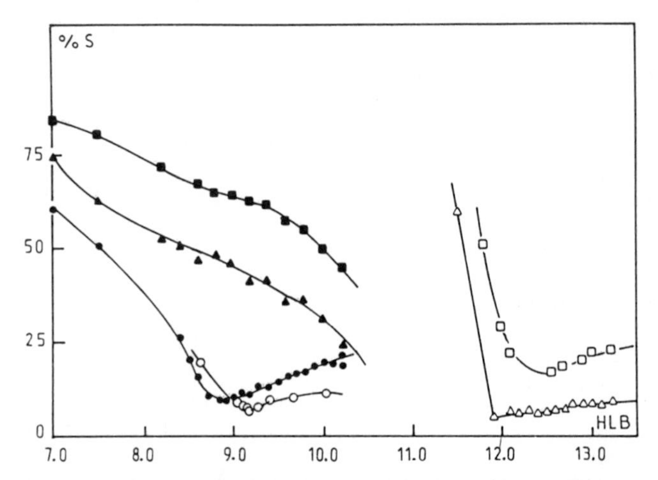

Figure 1. Effect of the nature of the oil and monomer on the percentage of surfactant(s) necessary for the formation of microemulsions as a function of HLB (oil to aqueous phase weight ratio ≃ 1).
Aqueous phase : ●, ▲,■ (water 55%, AM 40%, sodium acetate 5%) T = 20°C, oil : ● Isopar M, ▲ Cyclohexane,■ Toluene
O : (water 50%, AM 39%, sodium acrylate 11%) oil : Isopar M, T = 24.6°C.
Δ,□ (water 50%, MADQUAT 50%) T = 20°C
oil : Δ cyclohexane,□ heptane
Surfactant(s) : ●, Δ, ■, O : (G 1086 + Arlacel 83)
Δ,□ : (Tween 80 + Arlacel 83).

Microemulsions form in an HLB range of 7-10 (acrylamide or acrylamide-sodium acrylate) or ≃ 11-14 (MADQUAT) depending on the monomer structure. In most cases, the curves exhibit a more or less marked minimum for an optimum HLB_{opt} value corresponding to the best conditions for the formation of monophasic thermodynamically stable systems. The values found for the HLB_{opt} (up to ≃ 13) are much higher than those calculated from Equation 1 (HLB ≃ 6) or those used in practice in inverse emulsions (HLB ≃ 4-6). This unusual behavior is the result of an intricate combination of several effects linked to the initial structure of the microemulsions. In this respect, one must notice that HLB values of about 8-11 are indicative of systems located in a phase inversion region (21). This is confirmed by the value of the correponding CER calculated from Equation 1 which is of the order of R_0 ≃ 1. In this part of the phase diagram, where the proportions of oil and water are comparable, the microemulsions have a bicontinuous character usually described as a sponge-like structure with randomly connected oil and water domains (27,28).

The formation of bicontinuous microemulsions is conditioned by the nature of the monomer which is present in large amounts (up to 25% by weight) and which exerts a great effect on the HLB and interfacial properties of the systems. Furthermore as the polymerizable microemulsions contain a fairly large concentration of surfactant(s), interactions between surfactants and monomers cannot be neglected, especially when the latter are electrolytes.

The Monomer Role as a Cosurfactant. A common feature to all the monomers investigated is that the addition of the polymerizable component to the water-oil-surfactant(s) system causes a considerable increase in the microemulsion domain of the phase diagram (11,14,22,29). A typical example of a phase diagram determined in the absence (pure water) and in the presence of monomer (acrylamide-sodium acrylate) is given in Figure 2. It is obvious the monomer acts as a cosurfactant and contributes to the microemulsion stabilization by interacting with the different components. Note the high solubilizing capacity of the nonionic surfactant blend which forms microemulsions with monomer concentrations as large as 43% and surfactant contents as low as 10%.

The surface-active properties of the monomers were also confirmed by surface-tension experiments. Figure 3 shows the variation of the surface tension γ with the concentration of a cationic monomer (MADQUAT) (29) and a neutral one (AM) (Graillat, C.; Pichot, C. unpublished results). The surface-tension drops for example from 70 dyn/cm to 40 dyn/cm when the MADQUAT concentration varies from 10^{-3} to about 3M (≃ 3M is the monomer concentration based on the aqueous phase used in most polymerization reactions). However, the amphiphilic character of monomers is not sufficiently pronounced to give rise to micellization since no sharp transition corresponding to the C.M.C. is detectable.

The amount of monomers which is located at the water-oil interface has been estimated for acrylamide bicontinuous microemulsions and has been found about 30% (wt/wt) (22).

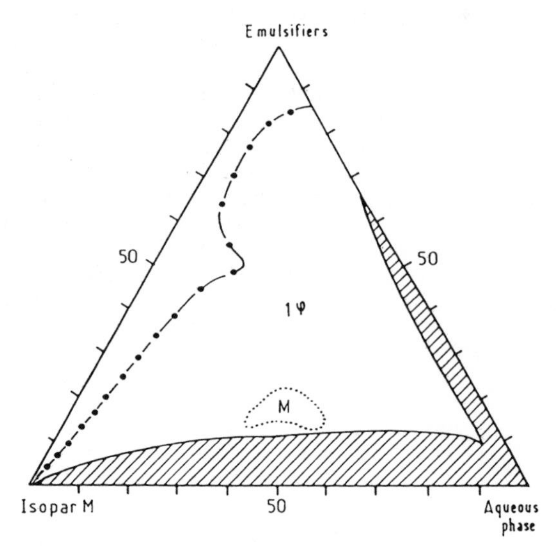

Figure 2. Pseudo-ternary phase diagram (wt/%). Hatched area: Emulsion domain (-•-•-) in the absence of monomer, (T = 20°C). (—) in the presence of monomer (T = 24.6°C). Aqueous phase : acrylamide-sodium acrylate-water (39: 16.5 : 44.5).
Surfactant(s) : G 1086 + Arlacel 83, HLB = 9.25.
Systems have been polymerized in the area M. (Reproduced with permission from Ref.11. Copyright 1986 Academic Press).

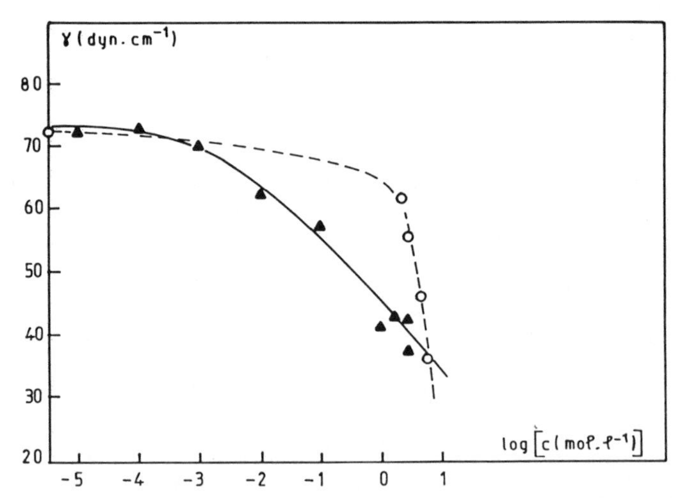

Figure 3. The surface tension against monomer concentration.
▲ MADQUAT ; O Acrylamide.

The Monomer Role as an Electrolyte. Addition of electrolytes to aqueous solutions of ethoxylated surfactants has long been known to modify their cloud points (30-32). Two behaviors can be observed, depending on the nature of the salt : a cloud point decrease due to a salt dehydration of the emulsifier hydrophilic moiety (salting out) or an increase showing an enhanced solubility of the surfactant in water (salting in).

The electrolyte effect of some water-soluble monomers on the cloud point of ethoxylated surfactants is illustrated in Figure 4. In the absence of salt, the cloud point of the emulsifier blend (G 1086 + Arlacel 83, HLB = 9.3) is equal to 64°7 (12). Three monomers-sodium acrylate, MADQUAT - ADQUAT (acryloxyethyltrimethylammonium chloride) - salt the surfactant blend out, the strongest effect being observed with ADQUAT (29).

Also reported for comparison are the curves relative to two non polymerizable salts, sodium acetate and sodium chloride which cause a salting out of the surfactant. The role of electrolytes in the stabilization of the polymerized systems will be discussed below. The cloud point shift values, for the surfactant blend, measured after addition of a unimolal electrolyte solution are listed in Table II.

Table II. Cloud Point Shift Values for the Surfactant Blend (Arlacel 83 + G 1086, HLB = 9.3, C = 83%) from Various Electrolytes at Molal Strength of 2

Electrolyte	ΔT_p °C
Sodium chloride	- 19°8
MADQUAT	- 20°4
Sodium acrylate	- 22°3
Sodium acetate	- 28°0
ADQUAT	- 40°5

The contributions of the individual ions constituting the electrolyte have been separated following a method by Schott et al. (32). The values thus calculated indicate that the salting effects observed on the nonionic surfactants are mainly due to the prominent influence of the anions as compared to those of the cations (Holtzscherer, C.; Candau, F. J. Colloid Interface Sci., in press).

These salting-out effects are partly responsible for the high HLB_{opt} values found experimentally (see Figure 1). At a given temperature, the HLB of the blend is made more lipophilic on addition of an electrolyte monomer. The HLB_{opt} has to be shifted to higher values in order to counterbalance the solubility decrease in water. Thus, the HLB_{opt} measured is an apparent value, the effective one being that obtained in the absence of electrolyte. Figure 5 shows the increase in the HLB_{opt} value observed when the ionic monomer content in an acrylamide-sodium acrylate feed stock increases from 0 to 65% (11). When the monomer used is not an electrolyte (acrylamide), a similar curve is obtained by addition of a non polymerizable salt such as sodium acetate.

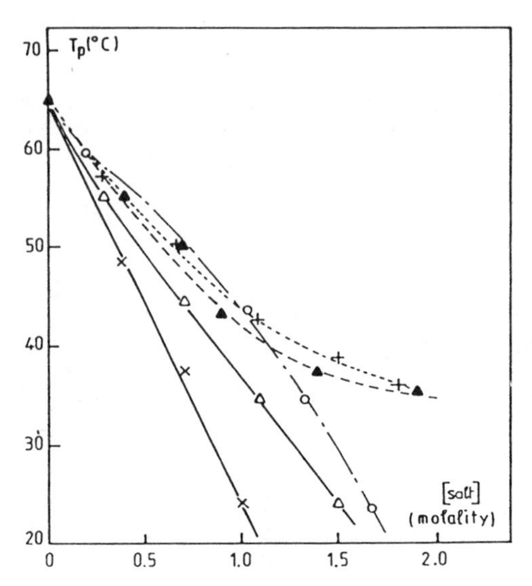

Figure 4. Variation of the cloud point of the surfactant blend (Arlacel 83 + G 1086, HLB = 9.3) in aqueous solution (c = 83%) with salt molality.
+ NaCl ; ▲ Sodium acrylate ; O MADQUAT ; Δ Sodium acetate ; x ADQUAT.

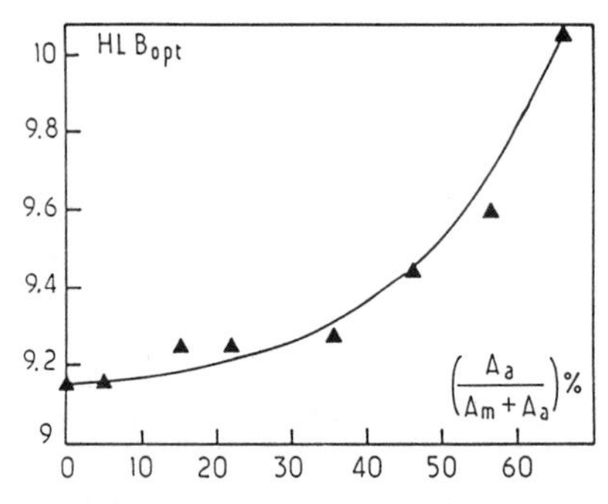

Figure 5. Variation of HLB_{opt} with the sodium acrylate content (T = 24.6°C). (Reproduced with permission from Ref.11. Copyright 1986 Academic Press).

Addition of salting-out type electrolytes to oil-water-surfactant(s) systems has also a strong influence on their phase equilibria and interfacial properties. This addition produces a dehydration of the surfactant and its progressive transfer to the oil phase (2). At low salinity, a water-continuous microemulsion is observed in equilibrium with an organic phase. At high salinity an oil-continuous microemulsion is in equilibrium with an aqueous phase. At intermediate salinity, a middle phase microemulsion with a bicontinuous structure coexists with pure aqueous and organic phases. These equilibria were referred by Winsor as Types I,II and III (33).

The influence of sodium acetate on the phase equilibria of acrylamide microemulsions has been investigated (Holtzscherer, C.; Candau, F. J. Colloid Interface Sci., in press). The interfacial tensions of the systems preequilibrated are reported versus the salt concentration in Figure 6. It can be seen that addition of sodium acetate induces a phase transition WI → W III which occurs for $S^* = 1.2M$. The intercept of the two curves which occurs in the Winsor III domain defines an optimal salinity for the formation of bicontinuous microemulsions.

In summary, these results show that the role of the monomer is twofold : as a cosurfactant, it increases the flexibility and the fluidity of the interface, which favors the formation of a bicontinuous microemulsion. An an electrolyte, it induces the latter structure. These two conditions are imperative. Similar water-oil-surfactant(s) systems do not lead to bicontinuous microemulsions in the absence of monomers or salts (when the monomer is not itself an electrolyte).

Polymerization

The conditions which have been defined for the formation of effective microemulsions (nature of the oil, R_0 and HLB values) are also required for obtaining clear and stable microlatices after polymerization. Various water-soluble monomers have been polymerized by a free radical process in anionic and nonionic microemulsions, either under U.V. irradiation or thermally with AIBN as the initiator (11,14,22,29). Total conversion to polymer was achieved in less than 20 minutes (a few minutes in some cases). Series of experiments have been performed in various oils. Table III summarizes some of the results and emphasizes the importance of the formulation. A good chemical matching between oils and emulsifiers (G 1086 + Arlacel 83, Isopar M) leads to stable latices, a poor matching (G 1086 + Arlacel 83, heptane) leads to unstable latices which settle within a few hours to a few days (22).

Similarly, electrolytes play a major role in the stability of the final latices. Polymerization of AM microemulsions in the absence of salt gives unstable latices, even at high surfactant concentrations. When salts with high salting-out efficiency (sodium acetate) are added to the systems, stable and clear microlatices are produced. Polymerization of microemulsions containing a polymerizable salt leads, under appropriate conditions, to the formation of stable microlatices (11,29).

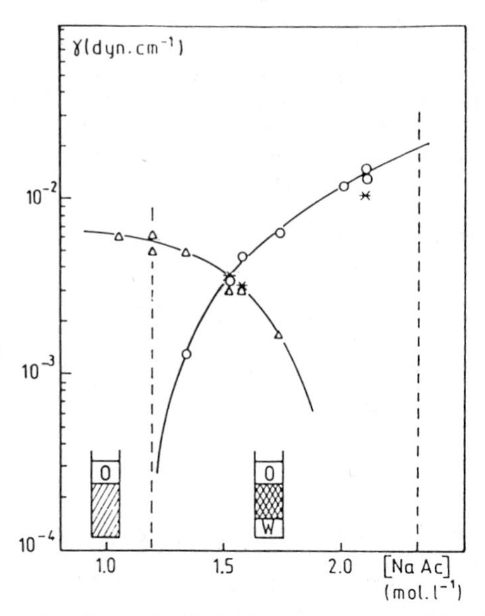

Figure 6. The interfacial tension as a function of sodium acetate concentration. Composition of the systems. (G 1086 + Arlacel 83, HLB = 9.3) 8% ; aqueous phase (water + sodium acetate 60%, AM 40%) 45.6% ; Isopar M : 46.4%.

Table III. Effect of Oil Structure on Latex Stability

Oil	(co)Polymer	(wt %)	Latex Stability	T%	d(nm)
Isopar M [1]	(AM-Aa)	25	stable	90	86.6
Isopar M [1]	AM	20	stable	99	121.0
Isopar M [1]	AM	14.5	stable	99	84.6
Dodecane [1]	AM	14.5	stable	99	90.2
Decane [1]	AM	14.5	stable	80	90.2
Heptane [1]	AM	14.5	unstable(hours)	75	-
Cyclohexane [2]	MADQUAT	25	stable	80	122.0

Surfactant Blend : [1] G 1086 + Arlacel 83 [2] Tween 80 + Arlacel 83

These results clearly show that the most effective polymerizable microemulsions for industrial applications (minimum amount of surfactant) are bicontinuous. It is interesting to note that addition of salting-in-type electrolytes (inducing no W I → W III transition) to AM microemulsions gives unstable and turbid latices.

In the course of polymerization, a progressive transformation of the random bicontinuous structure toward a concentrated stable dispersion of spherical latex particles, is observed as shown by electron microscopy and quasi-elastic light scattering (QELS) experiments (11,12,29). The diameters of the latex particles d have been determined by QELS and their optical transmissions T have been measured by turbidimetry (Table III). A good chemical matching between oil and lipophile leads to latices with a transmission close to 100%. The sizes of the latex particles range from 50 to 120 nm depending upon the experimental conditions. They increase with monomer concentration but decrease with increasing surfactant content. The variance of the auto-correlation function of the scattered intensity, which provides an estimate of the particle size distribution is rather low and around 4-5% (11,14,24,29).

The intrinsic viscosities of the resultant (co)polymers in aqueous solutions are high (up to 3700 $cm^3 g^{-1}$) and their molecular weights reach a few millions (24,29,34). By combining the molecular weight and the size of the particles, it can be inferred that the number of polymer molecules in a particle is very low, possibly as low as one (case of AOT systems) (24,35).

The microstructure of acrylamide-sodium acrylate copolymers was determined by ^{13}C NMR (36). The monomer sequence distribution was found to conform to Bernouillian statistics and the reactivity ratios of both monomers were close to unity. These results which differ from those obtained for copolymers prepared in solution or emulsion (37) confirmed a polymerization process by nucleation and interparticular collisions.

Basic Features of the Process

The novel process of inverse microemulsion polymerization allows one to prepare microlatices with characteristics significantly different from those prepared in inverse emulsions (38,39). These main characteristics are the following :

- low dimensions (d < 100 nm) and low particle size distribution.
- high stability and optical transparency.
- interesting rheological properties (high fluidity even at large volume fractions).
- low number of high molecular weight polymer chains confined within small particles.
- sequential distribution of poly(acrylamide-co-acrylates) obeying Bernouillian statistics.

Applications of the Process

The fields of applications are broad and numerous possibilities have not yet been explored. One of the major applications concerns the use of the inverse microlatices in the preparation of thickened aqueous solutions in enhanced oil recovery processes (40-42).

The method classically used consists of floodind the oil field by means of an injection of salt water in order to force petroleum out of the porous rocks where it is trapped. However, the difference in mobility between oil and water reduces the efficiency of the method and it is common to thicken the water in the hydrosoluble polymers such as partially hydrolyzed polyacrylamides or acrylamide-sodium acrylate copolymers. The polymers are usually supplied as water-in-oil emulsions, which have such advantages as easier handling, storage and dissolution as compared to solid powders whose moisture absorption causes formation of lumps and agglomerates. However, inverse latices prepared by the conventional emulsion method suffer disadvantages, particularly an unstability resulting in rapid settlement and in the requirement of delicate shearing during their dissolution in aqueous phase (i.e. during their inversion).

The microlatices prepared by inverse microemulsion polymerization are more advantageous, resulting from their lower particle size and polydispersity index, and their great stability. They are better scavengers of oil formations. Tests conducted in the laboratory have proven their efficiency. They are also self-inverting and additional surfactants are not usually needed to promote their inversion, as in some previously described methods.

In other techniques of oil production, the microlatices can be usefully employed for ground consolidation, manufacture of drilling muds and as completion or fracturation fluids. Another use concerns the prevention of water inflows into production wells. The method consists injecting from the production well into the part in the field to be treated, an aqueous solution of polymer prepared by inverse microlatex dissolution in water. The polymer is adsorbed on the walls of the formation surrounding the well. When the latter is brought in production, the oil and/or the gas selectively traverse the treated zone whereas the passage of water is inhibited.

In other uses, the water-soluble polymers prepared by microemulsion polymerization can serve as coagulants to separate solids suspended in a liquid. The polymer, more finely dispersed than when obtained by conventional emulsion polymerization has a higher activity on muds from sources such as steel works rolling-mills, coal and potash.

Alternatively, they can act as flotation and draining adjuvants (26), for example in the manufacture of paper : the addition of the polymer in stable microlatex improves water draining from the paper sheets.

They can also be used as flocculants for many substrates, such as cellulose fibers and in water treatment. The water-in-oil microlatices are destabilized and inverted by adding excess water, which releases the water-swollen polymer particle to the water phase. The efficiency of the usual polymers is however altered by rapid temperature changes, especially in winter, which favor the aggregation of the fine polymer particles, thus strongly decreasing their flocculation effect. The polyacrylamides prepared by microemulsion polymerization are entrapped within low size and low polydispersity particles and have better properties in this respect.

Another possibility is the use of the microlatices for assembling glass fibers since they have a low viscosity and good stability, as well as in the leather industry (finishing), in photographic emulsions and in paints.

An interesting characteristic of the process for industrial applications is the very high rate of polymerization which constitutes an important economical factor.

Finally, by adjunction of a cross-linking compound in the disperse phase of the microemulsion it is possible to prepare microgels which can be utilized as water-retention agents (13,26).

Conclusions

The main features of inverse microemulsion polymerization process have been reviewed with emphasis given to a search for an optimal formulation of the systems prior to polymerization. By using cohesive energy ratio and HLB concepts, simples rules of selection for a good chemical match between oils and surfactants have been established ; this allows one to predict the factors which control the stability of the resultant latices. The method leads to stable uniform inverse microlatices of water-soluble polymers with high molecular weights. These materials can be useful in many applications.

In addition to the practical interest, the process presents challenges encouraging further fundamental exploration. A thorough study not reported here, has been performed on the mechanism and kinetics of the polymerization of acrylamide in AOT/water/toluene microemulsions (Carver, M.T.; Dreyer, U.; Knoesel, R.; Candau, F.; Fitch, R.M. J. Polym. Sci. Polym. Chem. Ed., in press. Carver, M.T.; Candau, F.; Fitch, R.M. J. Polym. Sci. Polym. Chem. Ed., in press). The termination reaction of the polymerization was found to be first order in radical concentration, i.e. a monoradical reaction instead of the classical biradical reaction. Another major conclusion was that the nucleation of particles is continuous all throughout the polymerization in contrast to conventional emulsion polymerization where particle nucleation only occurs in the very early stages of polymerization. These studies deserve further investigations and should be extended to other systems in order to confirm the unique character of the process.

Literature Cited

1. Shinoda, K.; Friberg, S. Adv. Colloid Interface Sci. 1975, 4, 281.
2. Bellocq, A. M.; Biais, J.; Bothorel, P.; Clin, B.; Fourche, G.; Lalanne, P.; Lemaire, B.; Lemanceau, B.; Roux, D. Adv. Colloid Interface Sci. 1984, 20, 167.
3. Polymer Colloids II ; Fitch, R. M., Ed.; Plenum: New-York, 1980.
4. Schauber, C. Ph.D Thesis, Mulhouse University, Mulhouse, 1979.
5. Atik, S. S.; Thomas, K. J. J. Am. Chem. Soc. 1982, 104, 5868.
6. Tang, H. I.; Johnson, P. L.; Gulari, E. Polymer 1984, 25, 135.
7. Sayakrishnan, A.; Shah, D. O. J. Polym. Sci. Polym. Lett. Ed. 1984, 22, 31.
8. Stoffer, J. O.; Bone, T. J. Dispersion Sci. Technol. 1980, 1, 37.
9. Gan, L. M.; Chew, C. H.; Friberg, S. E. J. Macromol. Sci. Chem. 1983, A19, 739.
10. Kuo, P. L.; Turro, N. J.; Tseng, C. M.; El Aasser, M.; Vanderhoff, J. W. Macromolecules 1987, 20, 1216.
11. Candau, F.; Zekhnini, Z.; Durand, J. P. J. Colloid Interface Sci. 1986, 114, 398.
12. Holtzscherer, C. Ph.D Thesis, Louis Pasteur University, Strasbourg, 1986.
13. Leong, Y. S.; Candau, F. J. Phys. Chem. 1982, 86, 2269.
14. Candau, F.; Leong, Y. S.; Pouyet, G.; Candau, S. J. J. Colloid Interface Sci. 1984, 101, 167.
15. Grätzel, C. K.; Tirousek, M.; Grätzel, M. Langmuir 1986, 2, 292.
16. Lianos, P. J. Phys. Chem. 1982, 86, 1935.
17. Fouassier, J. P.; Lougnot, D. J.; Zuchowicz, I. Eur. Polymer J. 1986, 22, 933.
18. Candau, F. In Encyclopedia of Polymer Science and Engineering; Mark, H.; Bikales, N.; Overberger, C. G.; Menges, G. Eds.; 2nd edn, Wiley, New-York, 1987; 9, p 718.
19. Leong, Y. S.; Riess, G.; Candau, F. J. Chim. Phys. Phys.Chim. Biol. 1981, 78, 279.
20. Piirma, I.; Wu, M. Presented at 57th Colloid Surface Science Symposium, 1983, Toronto, Canada, June 12-15.
21. Beerbower, A.; Hill, M. W. In Mc Cutcheon's Detergents and Emulsifiers Annual; Allured: Ridgewood, NJ, 1971; p 223.
22. Holtzscherer, C.; Candau, F. Colloids and Surfaces 1988, 29, 411.
23. Boyd, J.; Parkinson, C.; Sherman, P. J. Colloid Interface Sci. 1972, 41, 359.
24. Holtzscherer, C.; Durand, J. P.; Candau, F. Colloid Polym. Sci. 1987, 265, 1074.
25. Candau, F.; Zekhnini, Z.; Durand, J. P. Progress Colloid Polym. Sci. 1987, 73, 33.
26. Candau, F.; Buchert, P. French Patent (to Soc. Chim. Charb.) 87 08 925, 1987.
27. Scriven, L. E. Nature (London) 1976, 263, 123.
28. Friberg, S.; Lapczynska, I.; Gillberg, G. J. Colloid Interface Sci. 1976, 56, 19.

29. Buchert, P. Ph.D Thesis, Louis Pasteur University, Strasbourg, 1988.
30. Schick, M. J. J. Colloid Interface Sci. 1962, 17, 801.
31. Shinoda, K.; Takeda, H. J. Colloid Interface Sci. 1970, 32, 642.
32. Schott, H.; Royce, A. E.; Han, S. K. J. Colloid Interface Sci. 1984, 98, 196.
33. Winsor, P. A. Trans. Far. Soc. 1948, 44, 376.
34. Candau, F.; Zekhnini, Z.; Heatley, F.; Franta, E. Colloid Polym. Sci. 1986, 264, 676.
35. Candau, F.; Leong, Y. S.; Fitch, R. M. J. Polym. Sci. Polym. Chem. Ed. 1985, 23, 193.
36. Candau, F.; Zekhnini, Z.; Heatley, F. Macromolecules 1986, 19, 1895.
37. Pichot, C.; Graillat, C.; Glukikh, V.; Llauro, M. F. In Polymer Latex II ; Plastics Rubber Institute: London, 1985; p 11/1.
38. Graillat, C.; Pichot, C.; Guyot, A.; El Aasser, M. J. Polym. Sci. Polym. Chem. Ed. 1986, 24, 427.
39. Vanderhoff, J.; di Stefano, F. V.; El Aasser, M.; O'Leary, R.; Schaffer, O. M.; Visioli, D. L. J. Dispersion Sci. Technol. 1984, 5, 323.
40. Candau, F.; Leong, Y.S.; Kohler, N.; Dawans, F. French Patent (to CNRS-IFP) 2 524 895, 1984.
41. Durand, J. P.; Nicolas, D.; Kohler, N.; Dawans, F.; Candau, F. French Patents (to IFP) 2 565 623 and 2 565 592, 1987.
42. Durand, J. P.; Nicolas, D.; Candau, F. French Patent (to IFP) 2 567 525, 1987.

RECEIVED September 12, 1988

MIDDLE-PHASE MICROEMULSIONS

Chapter 5

Preparation and Characterization of Porous Polymers from Microemulsions

S. Qutubuddin[1], E. Haque[1], W. J. Benton[2], and E. J. Fendler[2]

[1]Chemical Engineering Department, Case Western Reserve University, Cleveland, OH 44106
[2]Research and Development, BP America, Inc., 4440 Warrensville Center Road, Cleveland, OH 44128

Polymerization of microemulsions containing styrene monomer have produced porous solid materials. The morphology and porosity are affected by the type and concentration of surfactant and cosurfactant. The middle phase microemulsions yield maximum porosity as observed using scanning electron microscopy. Permeability measurements were done to obtain the diffusion coefficient of gases through the porous solid. The polymerization process was investigated using a polarized light screen and enhanced video microscopy in order to follow the dynamics and evaluate the role of the initial microstructure. Differential scanning calorimetry was used to study the glass transition temperature and reaction kinetics. In addition to being porous, these materials exhibit interesting thermal properties. The anionic microemulsion systems containing sodium dodecyl sulfate yield solids with a glass transition temperature (Tg) higher than observed in bulk polystyrene. In contrast, the nonionic microemulsion produces solids with lower Tg. This is due to different interactions between the surfactant and polystyrene chains in the two systems. X-ray diffractometry, FTIR and NMR were used to investigate the structure of the solid polymers. The weight average molecular weight as obtained by gel permeation chromatography varied from 0.2×10^6 to 1×10^6, depending on the initial microemulsion composition. Potential applications of the new materials include ultrafiltration, conductive polymers, polymer blends or composites, etc.

In the last decade, polymers with unprecedented qualities have emerged from advances in synthesis and novel polymerization techniques. The area of polymerization reactions in organized systems or associated structures has recently become an important focus of research.

A microemulsion may be defined as a thermodynamically stable isotropic solution of two immiscible fluids, generally oil and water, containing one or more surface active species (1a,1b). Microemulsions can be lower phase (water–continuous or oil–in–water type), upper phase (oil–continuous or

0097–6156/89/0384–0064$06.00/0

water–in–oil type) and middle phase. The middle phase microemulsions can be bicontinuous in microstructure (2). In bicontinuous microemulsions both the organic and aqueous phases coexist in interconnected domains with surfactant molecules located at the interface. The domain size in microemulsions is in the range of 100 to 1000 Å. A microemulsion is an example of the nonrandom systems which have been termed "organized solutions" by Shinoda (3). The surfactant molecules are mostly confined at the oil/water interface. The water, oil, and cosurfactant (usually a short–chain alcohol) in bicontinuous microemulsion systems diffuse at rates that are comparable to those of the neat components.

Microemulsions differ from macroemulsions and miniemulsions. Macroemulsions, conventionally known as emulsions, are thermodynamically unstable mixtures of two immiscible liquids, one of them being dispersed in the form of fine droplets with diameter greater than 0.1 μm in the other liquid. Macroemulsions are turbid, usually milky white in color. The concentration of surfactants is usually very low ($<$1.0%) in macroemulsions whereas it is high (~10.0%) in microemulsions which are thermodynamically stable. On the other hand, miniemulsions are stable oil–in–water emulsions containing two immiscible fluids, prepared using a mixture of ionic surfactant and a cosurfactant such as a long chain fatty–alcohol or n–alkane (4). The surfactant concentration is usually between 0.5 to 5.0% with average droplet size of 100 to 400 nm. Miniemulsions are opaque and also thermodynamically unstable.

Polymerization in microemulsions has gained some attention in the past few years as evident from the papers in this volume. The significance of polymerization in organized media is primarily attributable to its resemblance to biological reactions, which usually occur in organized systems. The polymerization of such structures include liquid crystal polymers and polymeric surfactants, which, depending on their molecular structure, may form intra–molecular and/or inter–molecular micelles or polymerized vesicles (5). On the other hand, microemulsions have long been investigated for tertiary oil recovery (6,7). The tertiary oil recovery process contains important ramifications of the interactions between microemulsion systems and polymer solutions due to the use of polymer solutions for mobility control. Polymers as such have also been used as stabilizers for solubilized systems. Microemulsions provide suitable media for polymerization because by introducing the monomer in either the dispersed or the continuous phase, or alternatively, solubilizing the monomers in both the oil and water phases, the outcome of the polymerization process can be altered.

In this section, a brief review of various studies of microemulsion polymerization will be presented. Most of the published work in this field has dealt with the formation of microlatex particles. The surfactants used in most cases were ionic, with a few studies involving nonionic surfactants. Jayakrishnan and Shah (8) studied the polymerization of o/w microemulsion systems containing styrene and methylmethacrylate using oil–soluble initiators, azobisisobutyronitrile (AIBN) and benzoyl peroxide. They used a nonionic surfactant (Pluronic L–31) and were able to obtain latex particles. However, the polymerization rate was not rapid and there was problem with stability during polymerization. Atik and Thomas (9) reported o/w microemulsion polymerization of styrene with AIBN as the initiator and γ–irradiation as an initiating source. They were able to achieve monodisperse latices with particle diameters of 20nm and 35 nm. Tang et al. (10) also studied the polymerization of an o/w microemulsion containing styrene. The polymerization duration was extremely long (8 days at 50^{0}C in N_2 environment). The polymerized system was characterized using photon correlation spectroscopy and time–average intensity measurements. The observed bimodal distribution of polystyrene latex size in their system was explained by assuming two mechanisms of initiation and particle growth. Johnson & Gulari (11) observed that in the case of the oil–soluble initiator the size of the latex can be correlated to the droplet size in the microemulsion, whereas in the case of the water–soluble initiator there is a weak dependence. Gratzel et al. (12) studied photoredox–induced polymerization of o/w styrene microemulsions. The polymerization rate curves were found to be similar to those observed for a radical initiated emulsion polymerization process. Kuo et al. (13) reported that photoinitiated polymerization of styrene in o/w microemulsions resulted in the formation of uniform latexes in the nanometer range (30–60nm). The molecular weights of polymers produced were on the order of 10^5. Kuo et al. (13) discussed the mechanism of polymerization in microemulsions on the basis of observed polymerization rate and particle size. They found the initiation mechanism in microemulsion polymerization to be similar to miniemulsion polymerization, as opposed to similarity to macroemulsion polymerization observed by Gratzel et al. (12).

Rabagliati et al. (14) studied the polymerization of styrene in a three phase system containing an anionic–nonionic surfactant mixture and brine. Both AIBN and potassium persulfate initiators were used. The system was reported to be microemulsion continuous and even multicontinuous. (14). No autoacceleration was observed and the authors concluded that the polymerization exhibits an inverse dependence of the degree of polymerization on initiator concentration, similar to bulk solution polymerization.

Stoffer and Bone (15,16) studied the phase behavior and morphology of polymers obtained by polymerization of w/o microemulsions. They observed that the cosurfactant pentanol acts as a chain transfer agent. They also found that the polymer formed in a microemulsion is as large as the droplet size in the macroemulsion, thus explaining problems encountered with phase separation. Gan, Chew and Friberg (17) studied the stability behavior of w/o microemulsions containing styrene and polystyrene with particular reference to the effects of pentanol and butylcellosolve.

Leong and Candau (18) obtained inverse latices of small size (<50nm) via photopolymerization of acrylamide in a microemulsion system of acrylamide, water, toluene and Aerosol OT. They observed that rapid polymerization and total conversion was achieved in less than 30 minutes. The microemulsions remained transparent and stable during polymerization. Candau et al. (19) also reported the results of a kinetic study of the polymerization of acrylamide in inverse microemulsions. Both oil soluble AIBN and water soluble potassium persulfate initiators were used. The rate was found to depend on the type of initiator, but in both cases neither autoacceleration nor dependence on initiator concentration was observed. An excellent review of microemulsion polymerization was published recently by Candau (20).

The preparation and characterization of novel porous solid polymers by the polymerization of different types of microemulsions are discussed in this paper.

MATERIALS

Styrene (99% pure) was obtained from Fisher Scientific and Aldrich Chemical. It was used either as supplied or after purification to remove the inhibitor. To remove the inhibitor, styrene (Aldrich) was washed three time with 10% sodium hydroxide and dried over phosphorus pentoxide in a desiccator after repeated washing with water. The purified sample was stored under nitrogen at 0^0C. Potassium persulfate and AIBN initiators were obtained from Fisher and DuPont, respectively, and were used without further purification. Sodium dodecyl sulfate (99% electrophoresis purity) was obtained from Bio–Rad Laboratories and used as received. The nonionic surfactants, Neodol 91–5 (Shell) and Emsorb 6916 (Emery), were 99% pure and used without further purification. The cosurfactants used in this study were 2–pentanol and butyl cellosolve. They were at least 99% pure and used as received. Water was deionized and doubly distilled. Benzene, methanol, and tetrahydrofuran (Fisher 99% pure) were used without further purification to dissolve the solid polymers.

METHODS

Microemulsions were prepared in Teflon–capped glass tubes cleaned with chromic acid and thoroughly rinsed in double–distilled water. Aqueous solutions were prepared by contacting required amounts of the surfactant, cosurfactant, and water. The solutions were gently heated to 60°C and mixed, first using a vortex mixer and then by ultrasonication. The aqueous solutions were contacted with the required amount of the monomer styrene. The mixtures were then thoroughly agitated by ultrasonication for at least five minutes. The samples were then kept at a constant temperature (22°C) in an environmental room to allow equilibriation. The phase behavior was observed using the polarized light screen (21). Equilibrium was assumed to have been achieved when the volumes and the appearance of the different phases did not show any sign of change with time as observed under polarized light. The equilibriation time ranged from two to three days for anionic surfactant systems, and three to four weeks for nonionic surfactant systems. After equilibriation, conductivity measurements of selected microemulsions were made using an Orion 1011 conductivity meter.

Polymerization of the microemulsion systems was performed under nitrogen environment using both purified styrene and unpurified styrene containing inhibitor. The polymerization was initiated by the thermal decomposition of potassium persulfate or AIBN at 60°C. The polymerization duration for the purified styrene system under nitrogen environment was 18 hours, whereas it took 36 hours for the styrene with inhibitor. The mode and dynamics of polymerization were observed using both polarized light and enhanced video microscopy (22).

The morphology of the solid polymer was observed using a scanning electron microscope (ISI–SX–30) connected to an EDS analyzer (Princeton Gamma–Tech). The samples were mechanically fractured in flexure and sputter coated with gold for 120 seconds before observation by SEM. Thermochemical measurements of microemulsion polymerization and thermal properties of the polymers were made on a differential scanning calorimeter (Perkin–Elmer, DSC–7). The glass transition temperature was measured at a scan rate of 10°C/min. In all cases the sample was heated from 40°C to 160°C at 10°C/min. The sample was kept at 160°C for one minute before cooling at 30°C/min. to 40°C. The sample was kept at 40°C for one minute before heating again to 160°C at 10°C/min. The Tg reported here is that obtained on the first scan. The Tg remained almost the same in the second scan within experimental error. However, the intensity of the change of the slope (ΔCp) decreased in the second scan.

Reprecipitation of the polymers was done by first dissolving the polymer in benzene so as to form a dilute solution (≃5%). Methanol was then added to the system (20 to 25 times the volume of the solution) to precipitate the polystyrene. The reprecipitated solid was then slowly filtered and dried under vacuum for 24 hours.

X–ray diffraction work was performed using a Phillips X–ray

Diffractometer (Model PW 1710), using CuKα radiation. A thin film (≃1mm) of 1.5 by 1.5 cm square sample was used for X–ray studies. Both water and gas phase permeability measurements were done with a machined 1 cm diameter thin sample (≃1mm). The molecular weight distribution was obtained using a gel permeation chromatograph (Waters) with a styragel column and using tetrahydrofuran as the solvent. Attenuated total reflectance (ATR) and KBr pellet techniques were used for FTIR. Solid state NMR was done on precipitated powdered samples. A Rayonet RPR–100 reactor at 35^{0}C was used for UV irradiation of the samples.

PHASE BEHAVIOR OF MICROEMULSIONS

Microemulsions were obtained using different types and concentrations of surfactant, cosurfactant, and styrene. An anionic surfactant, sodium dodecyl sulfate (C12H250SO3Na), and two types of nonionic surfactants, Emsorb 6916 (sorbitan monolaurate) and Neodol 91–5 (ethoxylated alcohol), were used. The surfactant concentration was varied between 5 to 10% (w/w) for the anionic system and between 5 to 15% (w/w) for the nonionic systems. Either 2–pentanol or ethylene glycol monobutyl ether (butyl cellosolve, C4H9OCH2CH2OH) was used as the cosolvent with the anionic surfactant. The amount of cosurfactant used depended on the anionic surfactant concentration and varied form 12.5 to 25% (w/w).

Phase behavior of the microemulsions was studied as functions of the type and concentration of surfactant and cosurfactant, and also temperature. Figure (1) shows the phase behavior at 22^{0}C of the anionic system containing sodium dodecyl sulfate (SDS) and 2–pentanol. The ratio of SDS/2–pentanol was kept constant at 0.4. Figure (2) illustrates the phase behavior of the same system after it has been equilibriated for about 30 hours at 60^{0}C (polymerization temperature). As shown in Figure (2), there is a phase transition around the binodal region at 60^{0}C giving rise to a three phase system consisting of a middle phase microemulsion in equilibrium with excess water and organic phases. The right hand corner in both Figures (1) and (2) shows the coarse emulsion region. Also, as evident from Figure (2), the single phase region at 60^{0}C is narrow compared to that at 22^{0}C. This is due to thermally induced phase separation of the initial microemulsion systems. Phase behavior of systems containing nonionic surfactants was also partially studied. The equilibriation time for systems containing nonionic surfactant is longer than the anionic system. Phase behavior of SDS microemulsions with butyl cellosolve was also partially studied. However, the work was suspended due to the carcinogenic nature of butyl cellosolve.

Figure (3) shows the solubilization parameters as functions of water concentration for SDS/2–pentanol ratios of 0.25 and 0.40 at 25^{0}C. The solubilization parameters are defined as Vo/Vs and Vw/Vs, where Vo, Vs and Vw are the volumes of organic phase, surfactant and aqueous phase in the microemulsions. The parameters are related to the drop size and also interfacial tensions (7,23). The bicontinuous phase is located around the composition range corresponding to equal values of solubilization parameters. The solubilization parameters are dependent on the initial surfactant and/or cosurfactant concentration. Similar dependence has been observed in other systems as a function of salinity and pH (7,23). Conductivity measurements performed as a function of water content indicate an S–shaped curve as shown in Figure (4). This is typical of microemulsions showing transition from oil–continuous to bicontinuous to water–continuous microstructure with increasing water content.

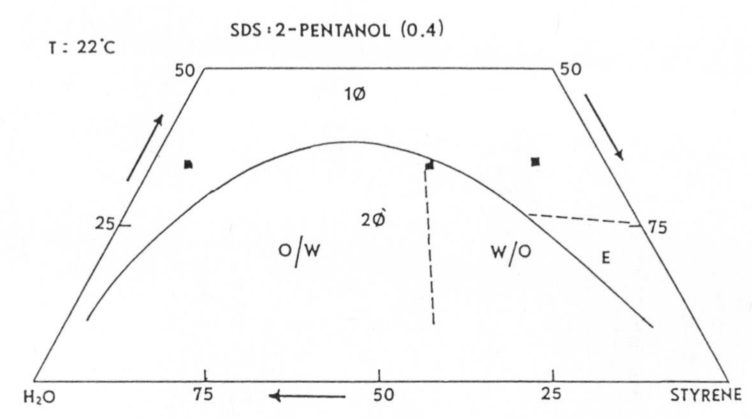

Figure 1: Partial phase behavior of SDS/2–pentanol/Styrene microemulsion system at 22°C. Squares on the diagram refer to compositions noted in Table I. Reproduced with permission from Ref. 25, Figure l. Copyright 1988 Society of Plastics Engineers.

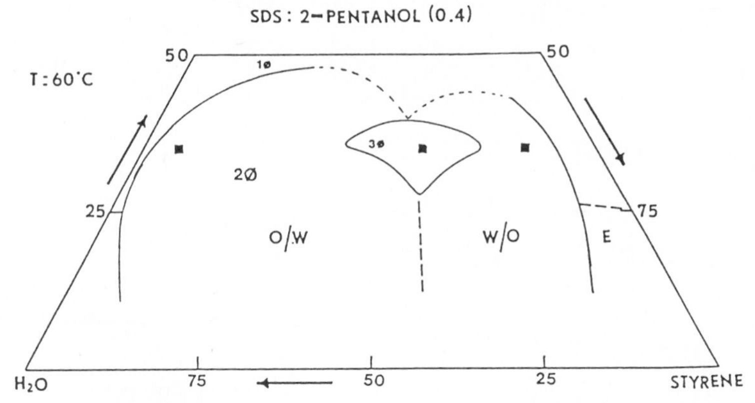

Figure 2: Approximate phase diagram of SDS/2–pentanol/Styrene microemulsion systems (same as in Figure 1) at 60°C. Squares on the diagram refer to compositions noted in Table I. Reproduced with permission from Ref 25, Figure 2. Copyright 1988 Society of Plactics Engineers.

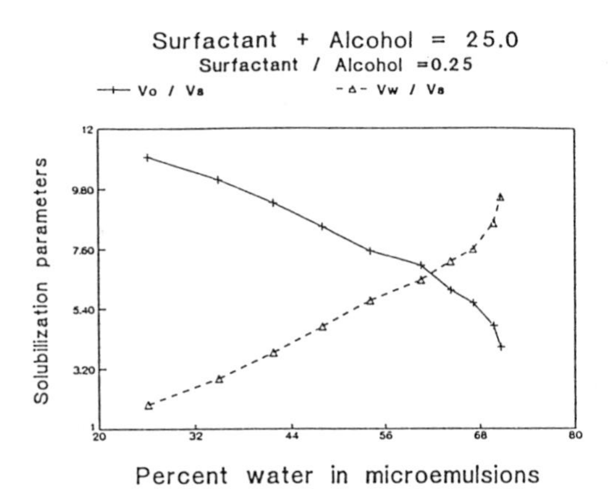

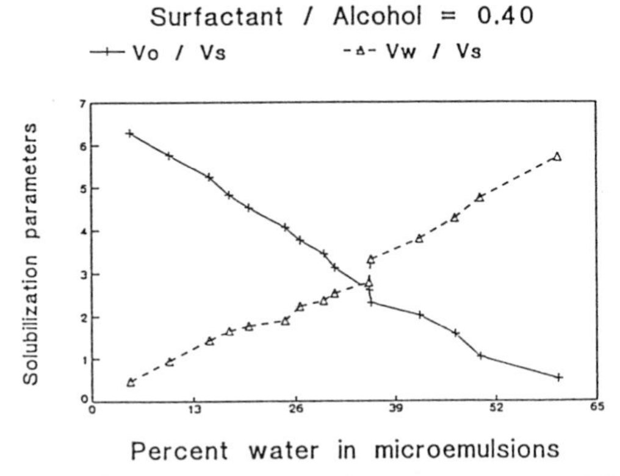

Figure 3: Solubilization parameters for SDS/2–pentanol/Styrene microemulsion system at 25°C.

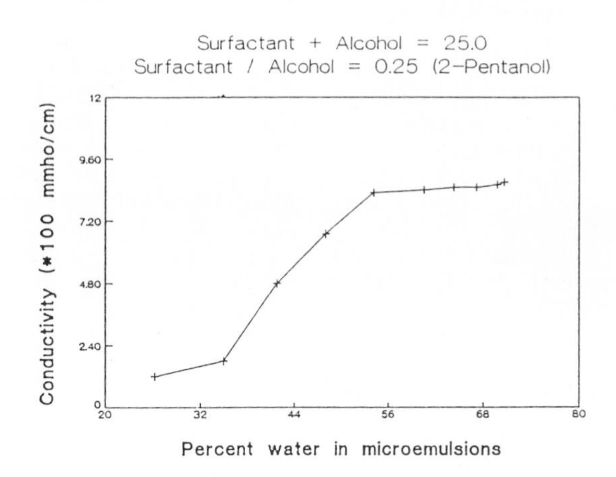

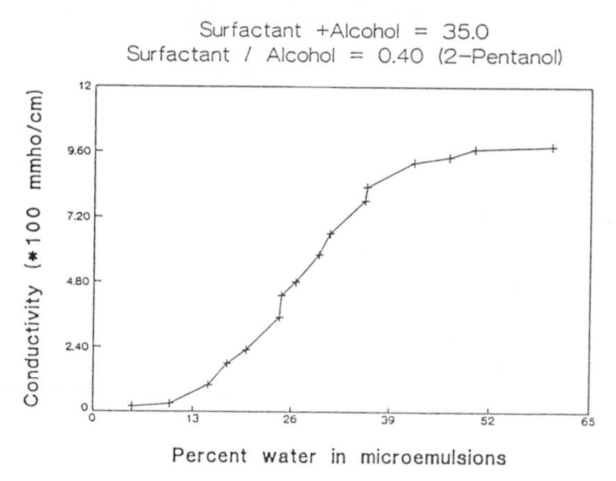

Figure 4: Conductivity plots for SDS/2-pentanol/Styrene microemulsion system at 25°C.

POLYMERIZATION OF MICROEMULSIONS

Polymerization of styrene in each of the three types of microemulsions was performed using a water soluble initiator, potassium persulfate (K2S2O8), as well as an oil–soluble initiator, AIBN. As desired, solid polymeric materials were obtained instead of latex particles. In the anionic system, the cosolvent 2–pentanol or butyl cellosolve separates out during polymerization. Three phases are always obtained after polymerization. The solid polymer was obtained in the middle with excess phases at the top and bottom. GC analysis of the upper phase indicates more than 80% 2–pentanol, while Karl–Fisher analysis indicated more than 94% water in the lower phase. Some of the initial microemulsion systems have either an excess organic phase on top or an excess water phase as the bottom layer. GC analysis showed the organic phase to be rich in 2–pentanol. However, the volume of the excess phase is much less in the initial system than in the polymerized system.

The use of butyl cellosolve as cosurfactant instead of 2–pentanol increases the stability of microemulsions during polymerization. This is partly due to the fact that the solubility of butyl cellosolve in polystyrene is higher than the solubility of 2–pentanol in polystyrene as experimentally verified by Gan and Chew (17).

The mode and dynamics of polymerization as observed under a polarized light screen and enhanced video microscopy (EVM) were found to be different for anionic and nonionic systems. In the anionic system gelation takes place during polymerization indicating cross–linking of the system. On the contrary, in the nonionic system the microemulsion first becomes a gel and then polymerization takes place. In the anionic system, as observed using EVM, the initial isotropic microemulsion showed no tendency of nucleation and growth until it reached 58–60°C. The system started to phase separate after reaching 60°C. The 2–pentanol rich phase nucleated out first and then the water separated. Initially the growth of such nucleating bubbles is mainly by diffusion. After about 20 hours, the growth is by coalescence. The solidification of inhibitor containing styrene starts around 26 hours and is essentially complete after 36 hours.

In the nonionic system observed under EVM, the initial microemulsion showed no tendency of gelation until it reached 60°C. After reaching 60°C, the system gels and starts to polymerize after 10–12 hours. As polymerization proceeds, the water separates out. After about 20–24 hours, the gel starts to become a solid with an excess emulsion phase formed at the bottom. The polymerization is essentially complete after 36 hours. Due to different modes of polymerization in the anionic and nonionic surfactant systems, the mechanical properties of the solid are different. The polymers obtained from anionic microemulsions are brittle, while those obtained from nonionic microemulsions are ductile.

MORPHOLOGY AND PROPERTIES

Styrene polymerization in microemulsions yields a different morphology compared to bulk polymerization. Preliminary results have been reported earlier (24,25). The solids obtained by polymerization of microemulsions are opaque, as opposed to transparent polystyrene obtained from bulk polymerization. The opacity is due to the presence of the surfactant. SEM micrographs indicate porous structures in the solid materials obtained by polymerizing microemulsions. Figure 5(a) shows the structure of styrene polymerized in bulk. As expected, it

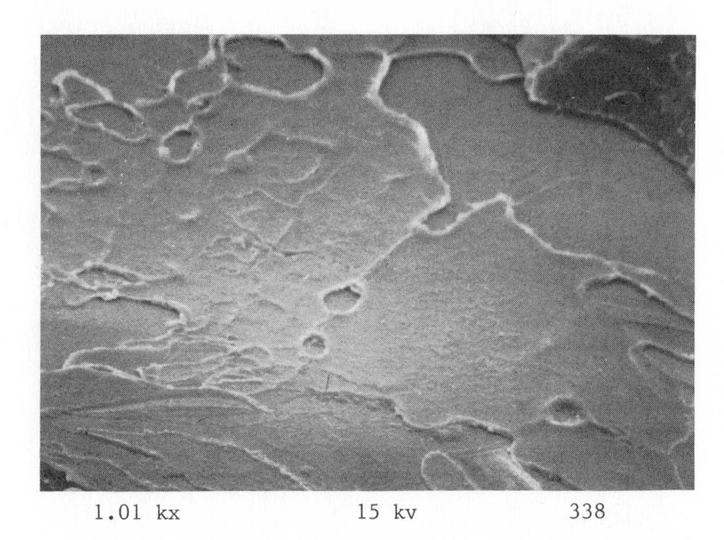

Figure 5(a): Electron micrograph of styrene polymerized in bulk (x1000). Reproduced with permission from Ref 25, Figure 4a. Copyright 1988 Society of Plastics Engineers.

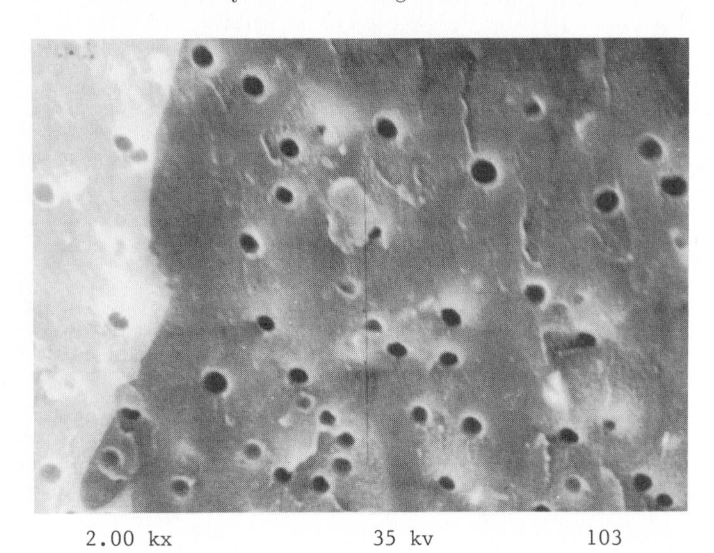

Figure 5(b): Electron micrograph of inhibitor free styrene polymerized in a oil–continuous microemulsion system containing SDS and 2–pentanol (x2000).

Continued on next page

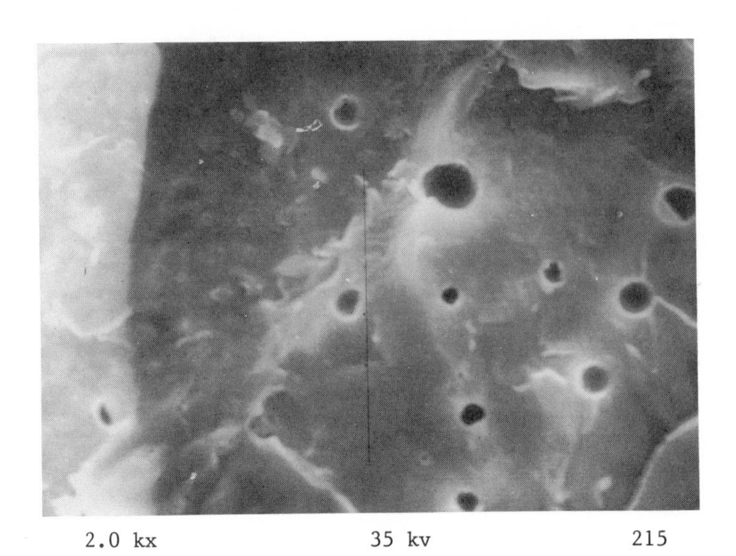

Figure 5(c): Electron micrograph of inhibitor free styrene polymerized in a water–continuous microemulsion system containing SDS and 2–pentanol (x2000).

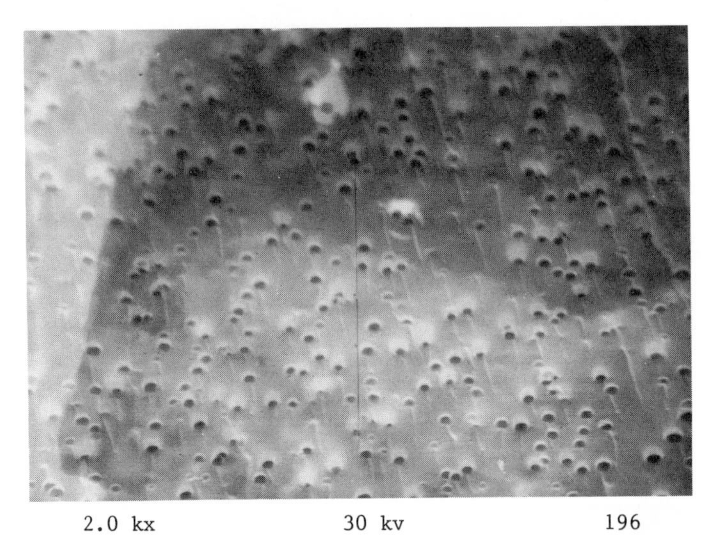

Figure 5(d): Electron micrograph of inhibitor free styrene polymerized in a middle phase microemulsion system containing SDS and 2–pentanol (x2000).

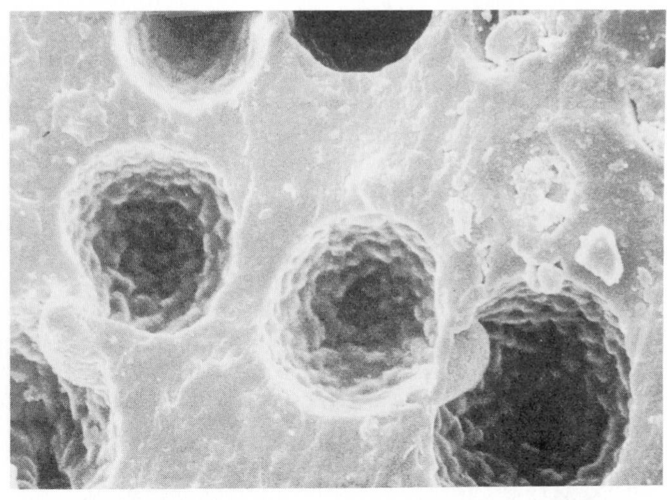

1.00 kx 15 kv 450

Figure 5(e): Electron micrographs of inhibited styrene polymerized in a middle phase microemulsion containing SDS and 2–pentanol (x1000). Reproduced with permission from Ref. 24, Figure 1. Copyright 1988 John Wiley & Sons.

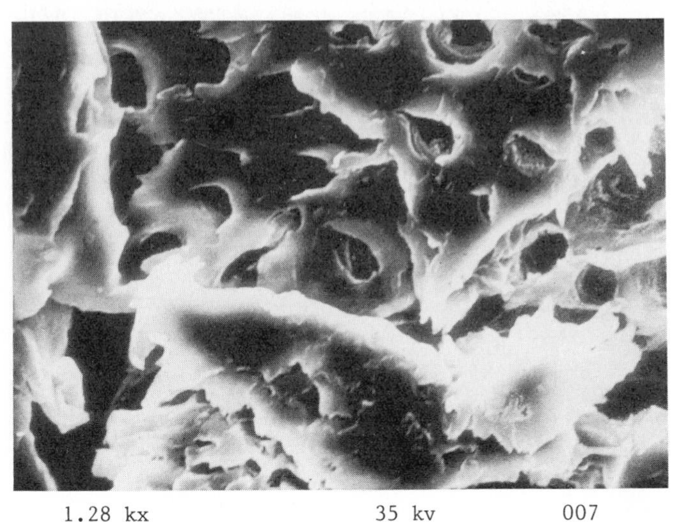

1.28 kx 35 kv 007

Figure 5(f): Electron micrograph of inhibited styrene polymerized in a non–ionic (Neodol 91–5) microemulsion system (x1280). Reproduced with permission from Ref. 24, Figure 2. Copyright 1988 John Wiley & Sons.

Continued on next page

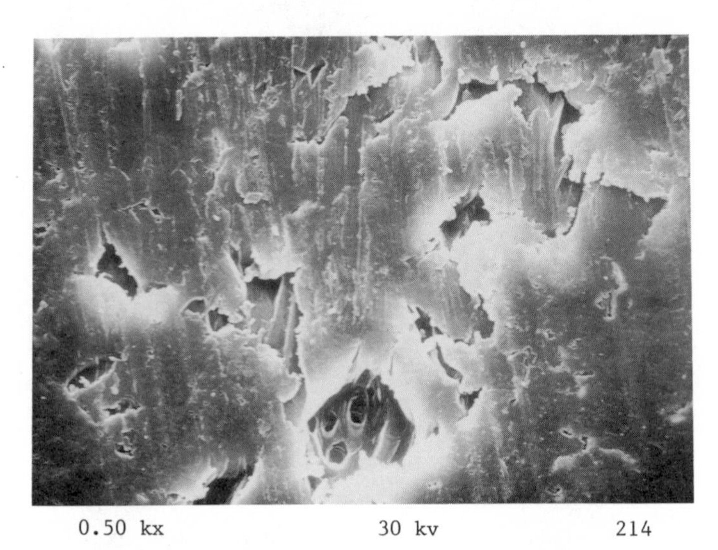

Figure 5(g): Electron micrograph of inhibited styrene polymerized in a non–ionic (Emsorb 6916) microemulsion system (x500).

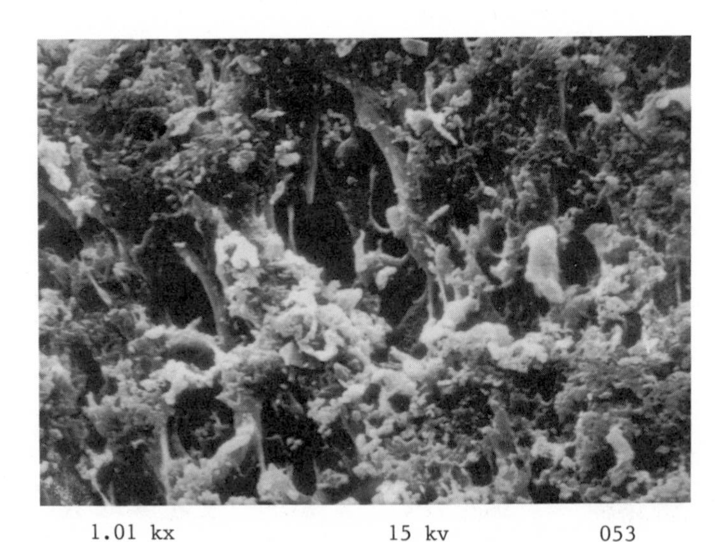

Figure 5(h): Electron micrograph of inhibited styrene polymerized in a middle phase microemulsion containing SDS and butyl cellosolve (x1010).

has a typical glassy structure (amorphous) with no pores. Figures 5(b), 5(c) and 5(d) show the porous structure when purified styrene is polymerized in microemulsions which are oil–continuous, water–continuous and middle phase, respectively, at room temperature. The porosity of polystyrene solid obtained in middle phase microemulsions is greater than that obtained with either water–continuous or oil–continuous microemulsions. This is related to the fact that middle phase microemulsions contain interconnected domains of both water and organic phases. When styrene is not purified, the solid material contains pores in both micron and submicron ranges, whereas the pores are always less than one micron in polystyrene solid from pure monomer. This is due to a faster polymerization rate and a shorter induction period. Figure 5(e) shows the structure of polystyrene when obtained from middle phase microemulsions containing inhibited styrene. Also it is evident from Figures 5(b)–5(d) that the pores are smallest in size in the middle phase system compared to water–continuous and oil–continuous systems.

The polymeric material based on nonionic surfactant (Neodol 91–5) has diferent pore morphology (Figure5(f)) as compared to the anionic system. Even the polymeric material based on the anionic system with a different cosolvent, butyl cellosolve, shows a different morphology (Figure5(g)). Figure5(h) illustrates the structure of polymers obtained from a microemulsion using a different nonionic surfactant (Emsorb 6916). Thus, the pore morphology depends on the initial microstructure of the microemulsion as determined by the type of surfactant and cosolvent in addition to composition and polymerization conditions.

Differential scanning calorimetry (DSC) was used to determine the kinetics of polymerization and the glass transition temperature of the solid polymer. Preliminary results indicate the dependence of kinetics on the microstructure as determined using Borchardt and Daniels method (26). The reaction order, rate constant, and conversion were observed to be dependent on the initial microstructure of the microemulsions. The apparent glass transition temperature (Tg) of polystyrene obtained from anionic surfactant (SDS) microemulsions is significantly higher than the Tg of normal bulk polystyrene. In contrast, polymers from nonionic microemulsions show a decrease in Tg. Some representative values of Tg are shown in Table I.

Several factors may contribute to the observed increase in Tg values in anionic surfactant based systems. There is a strong ionic interaction between polystyrene and SDS. In the presence of water, the surfactant dissociates and apparently interacts strongly with the π electrons of polystyrene. This results in decreased segmental mobility (i.e. increased chain stiffness) which increases the Tg. Chain stiffness also leads to increased physical cross–linking. As shown in Table I, when SDS is physically mixed with styrene and then polymerized, the Tg does not change. In order to find out if the SDS is only physically trapped in the system or whether there is some chemical interaction, the virgin solid was dissolved in benzene and then precipitated from methanol. The Tg of the precipitated polystyrene was then measured. As indicated in Table I, the Tg decreased from that of the virgin solid but it is still higher then the Tg of normal bulk polystyrene.

The decrease in Tg of nonionic microemulsion systems is due to the absence of electrostatic interactions between the surfactant and polystyrene. The nonionic surfactant behaves as a low molecular weight additive (i.e. plasticizer) which then lowers the Tg. In order to determine whether the cosurfactant 2–pentanol has any effect on Tg, a nonionic system was prepared containing 2–pentanol. The Tg of the polymerized solid was then determined. The Tg remained the same as in the nonionic system containing no cosurfactant.

TABLE I: Glass Transition Temperature and Molecular Weight Distribution of Polymers obtained from Anionic Microemulsions

SYSTEM	Composition Before Polymerization	Tg (°C) Before Reprecipitation	M_W	$\frac{M_W}{M_n}$	Tg (°C) After Reprecipitation
Pure Polystyrene	Styrene:100%	99 ± 1.0	548,500 ± 500	4.53 ± 0.40	–
Polystyrene + Sodium Dodecyl Sulfate (SDS)	Styrene:90% SDS:10%	98 ± 1.0	963,500 ± 1500	–	–
Polystyrene (Oil–continuous Microemulsion)	SDS:10% 2–Pentanol:25% Styrene:55% Water:10%	133 ± 2.0	709,000 ± 7000	5.41 ± 0.05	121 ± 2
Polystyrene (Middle Phase Microemulsion)	SDS:10% 2–Pentanol:25% Styrene:40% Water:25%	128 ± 1.2	274,000 ± 1000	5.71 ± 0.05	114 ± 2
Polystyrene (Water–contiunous Microemulsion)	SDS:10% 2–Pentanol:25% Styrene:5% Water:60%	118 ± 2.0	284,500 ± 1500	5.96 ± 0.06	105 ± 1

FTIR and X-ray diffraction (XRD) studies were performed on samples which were very finely ground and repeatedly washed with water and methanol. XRD results indicate low angle ordering. FTIR spectra of the solid polymers were obtained for solids before and after reprecipitation. The FTIR spectrum of the virgin solid shows that there is alteration in the structure of polymers obtained from microemulsions as compared to bulk polystyrene. There is evidence of free SDS in the solid. The spectrum of the reprecipitated solid shows the structure to be similar to high molecular weight atactic polystyrene. There is no evidence of free SDS in the structure but unexpected bands were seen around wave numbers of 1200, 1300 and 1580 cm^{-1} which were not observed in the virgin solid. These bands are not present in either pure polystyrene or pure SDS. NMR studies on reprecipitated solid also show no evidence of SDS in the structure, but at around 3.8 and 5.2 ppm there is an indication of an unidentified component in the structure. SDS tends to hydrolyze to form dodecanol and sodium hydrogen sulfate, especially under acidic conditions and elevated temperature (27). Work is now currently in progress to check for the residual presence of possible degradation products.

The weight average molecular weights of the solid polymers as determined using gel permeation chromatography are also shown in Table I. The molecular weight of the polystyrene obtained from microemulsions ranged between 270,000 to 700,000. The results indicate the dependence of molecular weight on the initial microstructure of the microemulsions. The high ratio of Mw/Mn is typical of vinyl polymerization with high conversion.

Preliminary conductivity measurements indicate that the polymers based on the anionic system are ionically conductive, whereas the nonionic based polymers are non-conductive. AC impedance tests were done on a thick film ($\simeq$1mm thick) using sodium sulfate as the electrolyte in a specially designed closed cell. The resistivity of polystyrene obtained from middle phase microemulsions was found to be in the range of 10^2–10^3 ohm–cm, compared to 10^{20} – 10^{22} ohm–cm for bulk polystyrene. A thin film of the polymer was also obtained on graphite electrodes by UV irradiation. Electrochemical measurements using such polymer coated electrodes also suggest that the film is conductive. SEM micrographs before and after the electrochemical measurements indicate that the polymeric film is stable and porous.

Permeability measurements were done on the porous polymers obtained by polymerization of the middle phase microemulsion containing SDS. As shown in Figure 6 and Table II, the permeability coefficients in both N2 and O2 gas are higher than in polystyrene. As expected, the diffusivity of nitrogen is higher than that of oxygen. Values of the solubility coefficient (ratio of permeability to diffusivity) are also listed in Table II.

APPLICATIONS

The novel porous materials have many potential applications including ultrafiltration, conductive polymers, polymer composites, etc. As reported in the previous section, preliminary experiments on characterization of deposited film on graphite electrode shows promising results. Work is currently in progress to characterize such films and study the effect of the incorporated surfactant on different electrochemical reactions on a variety of electrodes.

It is possible to develop novel polymer–polymer composites or blends by simultaneous or sequential polymerization of hydrophilic and hydrophobic monomers. Preliminary work on microemulsions containing styrene/acrylamide

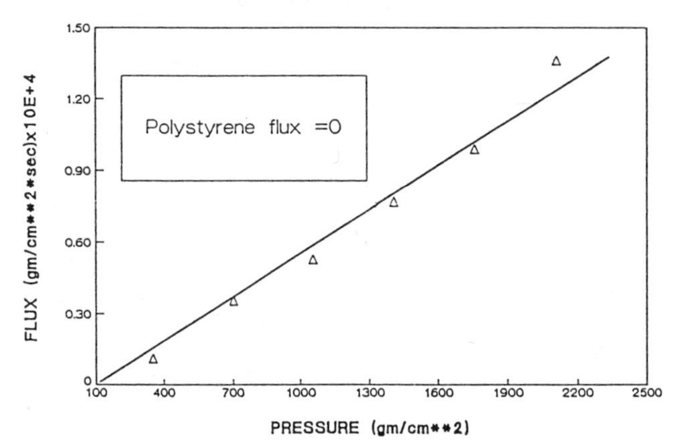

Figure 6: Water permeability of solid obtained from a middle phase microemulsion system containing SDS and 2–pentanol.

TABLE II: Transport Properties of Polymers Obtained from SDS Microemulsions

SYSTEM	N_2 Gas			O_2 Gas		
	Permeability $\left[\frac{cc(stp)cm}{cm^2 cmHg.sec}\right]$	Diffusivity $\left[\frac{cm^2}{sec}\right]$	Solubility $\left[\frac{cc(stp)}{cc.cmHg}\right]$	Permeability $\left[\frac{cc(stp)cm}{cm^2 cmHg.sec}\right]$	Diffusivity $\left[\frac{cm^2}{sec}\right]$	Solubility $\left[\frac{cc(stp)}{cc.cmHg}\right]$
POLYSTYRENE	0.79×10^{-10}	0.17×10^{-9}	0.46	8.07×10^{-11}	0.11×10^{-6}	13×10^{-3}
POLYSTYRENE OBTAINED FROM MIDDLE PHASE MICRO-EMULSION (SDS)	0.60×10^{-7}	0.37×10^{-6}	0.16	7.48×10^{-10}	1.30×10^{-6}	5.74×10^{-4}

indicate interesting thermal and mechanical properties. The characterization of such solid composites and the effect of surfactant as compatabilizing agents will be discussed in another paper (28).

SUMMARY

Polymerization of styrene in microemulsions has produced porous solid materials with interesting morphology and thermal properties. The morphology, porosity and thermal properties are affected by the type and concentration of surfactant and cosurfactant. The polymers obtained from anionic microemulsions exhibit Tg higher than normal polystyrene, whereas the polymers from nonionic microemulsions exhibit a lower Tg. This is due to the role of electrostatic interactions between the SDS ions and polystyrene. Transport properties of the polymers obtained from microemulsions were also determined. Gas phase permeability and diffusion coefficients of different gases in the polymers are reported. The polymers exhibit some ionic conductivity.

ACKNOWLEDGMENTS

This work was supported in part by Edison Polymer Innovation Corporation, GenCorp and National Science Foundation (Presidential Young Investigator program). Thanks are due to Dr. Jerry Coffee and Dr. Larry Bowa at BP America, Inc. for their valuable suggestions. The assistance of Dr. E. Dayalan in the electrochemical experiments is acknowledged. Thanks are also due to Mr. J. Gatto for some phase behavior experiments.

LITERATURE CITED

1a. Prince, L.M. Microemulsion, Theory and Practice; Academic : New York, 1977.
1b. Friberg, S.E. J. Dispersion Sci. Technol. 1985, 6, 317.
2. Scriven, L.E. Nature. 1976, 263, 123.
3. Shinoda, K.J. Phys. Chem. 1985, 89, 2429.
4. Choi, Y.T.; El–Aasser, M.S.; Sudol, E.D, Vanderhoff, J.W. J. Polymer Sci, Polymer Chem. Ed. 1985, 23, 2973.
5. Paleous, C.M. Chem. Soc. Review. 1985, 14 (1), 45.
6. Surface Phenomena in Enhanced Oil Recovery; Shah, D.O., Ed, Plenum: New York, 1981.
7. Miller, C.A.; Qutubuddin, S. Interfacial Phenomena in Non–Aqueous Media, Marcel Dekker: New York, 1986, p. 117 – 184.
8. Jayakrishnan, A.; Shah, D.O. J. Polymer Sci., Polymer Letter Ed. 1984, 22, 31.
9. Atik, S.; Thomas, J.K. J. Am. Chem. Soc. 1981, 103, 4279.
10. Tang, E.I.; Johnson, P.L.; Gulari, E. Polymer. 1984, 25, 1357.
11. Johnson, P.L.; Gulari, E. J. of Polymer Sci., Polymer Chem. Ed. 1984, 22, 3967.
12. Gratzel, C.K.; Jirousek, M.; Gratzel, M. Langmuir 1986, 2, 292.
13. Kuo, P.K.; Turro, N.J.; Tseng, C.M.; El–Aasser, M.S.; Vanderhoff, J.W. Macromolecules 1987, 20, 1216.
14. Rabagliato, F.M.; Falcon, A.C.; Gonzalez, D.A.; Martin, C. J. Disp. Sci. & Tech. 1986, 7 (2), 2245.
15. Stoffer, J.O.; Bone, T. J. Polymer Sci., Polymer Chem. Ed. 1980, 18, 2641.

16. Stoffer, J.O.; Bone, T. J. Disp. Sci. & Tech. 1980, 1, 4, 393.
17. Gan, L. M. Chew, C. H.; Friberg, S.E. J. Macromolecular Sci. - Chem. 1983, A19 (5), 739.
18. a. Leong, Y.S.; Candau, F. J. Phy. Chem. 1982, 86, 13, 2269.
b. Candau, F; Leong, Y.S.; Pouyet, G.; Candau, S. J. Colloid Int. Sci. 1984, 101, 1, 167.
19. Candau, F.; Leong, Y.S.; Fitch, R.M. J. Polym. Sci. Polym Letter Ed. 1985, 23, 193.
20. Candau, F. In Encyclopedia of Polymer Science Engineering; Wiley: New York 1987;vol 9,p. 718.
21. Benton, W.J.; Natoli, J.; Qutubuddin, S.; Mukherjee, S.; Miller, C.A.; Fort, T. Jr. Soc Pet. Eng. J. 1982, 22, 53.
22. Raney, K.H.; Benton, W.J.; Miller, C.A. Macro–and Microemulsions: Theory and Practice; Shah, D.O., Ed.; ACS Symposium Series No. 272; American Chemical Society: Washington, DC, 1985.
23. Qutubuddin, S.; Miller, C.A.; Fort, T. Jr. J. Colloid Int. Sci, 1984, 101, 46.
24. Haque, E.; Qutubuddin, S. J. Polymer Sci., Poly Letters Ed. 1988, 26, 429 .
25. Haque, E.; Qutubuddin, S. Proc 46th Ann. Tech. Conf. Soc. Plastics Eng. 1988, p. 1032.
26. Borchardt, H.J.; Daniels, F. J. Amer. Chem. Soc. 1956, 79, 41.
27. Kekicheff, P.; Cabone, B.; Rawiso, M. J. Colloid Int. Sci. 1984, 102, 51.
28. Haque, E.; Qutubuddin, S. Polymeric Blends from Microemulsions. To be submitted.

RECEIVED September 16, 1988

OIL-IN-WATER MICROEMULSIONS

Chapter 6

Microemulsion Polymerization of Styrene

J. S. Guo, M. S. El-Aasser, and J. W. Vanderhoff

Emulsion Polymers Institute and Departments of Chemical Engineering and Chemistry, Lehigh University, Bethlehem, PA 18015

The polymerization of styrene microemulsions prepared from water, sodium dodecyl sulfate, and 1-pentanol was carried out using water-soluble potassium persulfate or oil-soluble 2,2'-azobis(2-methyl butyronitrile) initiator at 70°C. The latexes were stable, bluish, and less translucent than the microemulsions. The polymerization rates measured by dilatometry increased to a maximum and then decreased (only two intervals). The maximum polymerization rate and number of particles varied with the 0.47 and 0.40 powers of potassium persulfate concentration, and the 0.39 and 0.38 powers of 2,2'-azobis(2-methyl butyronitrile) concentration, respectively. The small average latex particle sizes (20-30 nm) and high polymer molecular weights ($1\text{-}2 \times 10^6$) showed that each latex particle comprised only 2-3 polystyrene molecules. The number of particles remained unchanged when the styrene was diluted with toluene at constant oil-phase volume. The mechanism proposed for both water-soluble and oil-soluble initiators comprised nucleation in the microemulsion droplets by radical entry from the aqueous phase, with the droplets which did not capture radicals serving as reservoirs to supply monomer to the polymer particles (homogeneous nucleation was not ruled out, however). This mechanism was compared with those proposed for conventional emulsion polymerization and miniemulsion polymerization.

Many workers have studied microemulsions since the concept was introduced in 1943 by Hoar and Schulman (1), who showed that mixtures of oil, water, and alkali-metal soaps with certain alcohols or amines formed transparent dispersions, which they called "oleopathic hydromicelles." The research then continued desultorily for 30 years, but accelerated when microemulsions were found to be effective in enhanced oil recovery (2-4). The literature on microemulsions is now extensive and includes several books (5-12) published since 1977.

0097-6156/89/0384-0086$06.00/0

In contrast to the opaque, milky conventional emulsions and miniemulsions, microemulsions are isotropic, transparent or translucent, and thermodynamically stable. They form spontaneously when oil and water are mixed with surfactant and cosurfactant (usually 1-pentanol or 1-hexanol). Vigorous agitation, homogenization, or ultrasonification are not needed. Microemulsions are postulated to comprise dispersions of droplets of size smaller than 100 nm or bicontinuous lamellar layers. Both structures are consistent with their transparency or translucency. Which structure is more applicable is the subject of some controversy, a discussion of which is beyond the scope of this paper.

Microemulsion polymerization, developed about 1980, is new compared to conventional emulsion polymerization. Atik and Thomas (13) polymerized styrene oil-in-water microemulsions using 2,2'-azobis-(isobutyronitrile) initiator or gamma-ray initiation. Johnson and Gulari (14) polymerized diluted styrene oil-in-water microemulsions using potassium persulfate or 2,2'-azobis(isobutyronitrile) initiators, and measured the size of the microemulsion droplets and latex particles by photon correlation spectroscopy. Jayakrishnan and Shah (15) polymerized styrene and methyl methacrylate oil-in-water microemulsions using 2,2'-azobis(isobutyronitrile) or benzoyl peroxide initiators. Kuo et al. (16) polymerized styrene-toluene oil-in-water microemulsions photochemically using dibenzylketone photoinitiator and an ultraviolet light source.

Although detailed and extensive, these works have not presented a definitive mechanism for particle nucleation and growth in styrene microemulsion polymerization. This paper describes a kinetic investigation of this system using water-soluble and oil-soluble initiators and a comparison of microemulsion polymerization with conventional emulsion polymerization and miniemulsion polymerization, to determine the nucleation and particle growth mechanism.

Experimental

The styrene (Polysciences) was washed with 10% aqueous sodium hydroxide to remove the inhibitor and vacuum-distilled under dry nitrogen; the 1-pentanol (Fisher Scientific) was dried over potassium carbonate and vacuum-distilled; the potassium persulfate (Fisher Scientific) was recrystallized twice from water; the 2,2'-azobis(2-methyl butyronitrile) (E. I. du Pont de Nemours) was recrystallized twice from methanol; the sodium dodecyl sulfate (Henkel) was used as received; its critical micelle concentration measured by surface tension was 5.2 mM. Distilled-deionized water was used in all experiments.

The microemulsion polymerization recipe comprised 82.25% water, 9.05% sodium dodecyl sulfate, 3.85% 1-pentanol, and 4.85% styrene by weight. The rates of polymerization at 70°C were measured dilatometrically in a 25 ml Erlenmeyer flask equipped with a 45-cm long 1-mm ID capillary. The microemulsions containing initiator were degassed, loaded into the dilatometer, and polymerized in a thermostated water bath. This polymerization procedure was described in more detail earlier (17). The latex particle size distributions were determined using the Philips 400 transmission electron microscope with phosphotungstic acid negative-staining. After polymerization, the latex was poured into methanol; the precipitated polymer was filtered, washed with methanol and water, and dried. The polymer molecular weight distributions were determined using the Waters Model 440 gel permeation chromotograph with tetrahydrofuran as the eluant solvent.

Results and Discussion

The latexes prepared using the water-soluble potassium persulfate or oil-soluble 2,2'-azobis(2-methyl butyronitrile) initiators were stable, bluish, and less translucent than the original microemulsions. This change in appearance was attributed to the larger size of the latex particles and the greater refractive index ratio. The effects of initiator and monomer concentrations are described below.

Effect of Potassium Persulfate Concentration. Figure 1 shows the variation of conversion and rate of polymerization with time at 70°C for 0.14-0.69 mM potassium persulfate based on water. The polymerization rate increased with increasing initiator concentration, and the polymerization rate-conversion curves showed only two intervals; neither showed a constant rate or a gel effect. Interval I of the polymerization rate-conversion curve was characterized by an increase to a maximum, Interval II, by a decrease from this maximum. The maximum polymerization rate was reached at 20-25% conversion, in contrast to the 2-15% usually observed in conventional emulsion polymerization, which indicates that the particle nucleation stage was long, similar to that of miniemulsion polymerization (18,19). The long particle nucleation stage of miniemulsion polymerization was attributed to a slower rate of radical entry into the monomer droplets, which in turn was attributed to a higher concentration of adsorbed emulsifier on the miniemulsion droplet surface, or adsorbed mixed emulsifier liquid crystals on the droplet surface, or the larger size of miniemulsion droplets compared to micelles. In this microemulsion polymerization, the long particle nucleation stage was attributed to the adsorbed layer of surfactant-cosurfactant complex on the microemulsion droplet surface, which hindered the entry of radicals and thus gave a low radical capture efficiency. The shorter nucleation stage and faster polymerization rate of the microemulsion system relative to the miniemulsion system (18) was attributed to the smaller microemulsion droplet size and higher radical capture efficiency. The slower rate of the microemulsion polymerization relative to that of the conventional emulsion polymerization was attributed to the slower rate of free radical entry into the microemulsion droplets and the dilution of the styrene inside the droplets by the 1-pentanol.

The increasing rate of polymerization in Interval I was attributed to an increasing number of polymerizing particles formed by nucleation in the microemulsion droplets, similar to the mechanism proposed for miniemulsion polymerization by Chamberlain et al. (20). Interval I ends when all microemulsion droplets have disappeared, either by capturing radicals to become polymer particles or losing monomer by diffusion to the polymer particles which have captured radicals. In Interval II, the polymerization rate decreases because of the decrease in monomer concentration in the polymer particles. No gel effect was observed; termination occurred immediately upon entry of the second radical into the small latex particles which contained a growing radical.

Figure 2 shows that the maximum polymerization rate and number of particles (calculated from the volume-average diameter D_v) varied

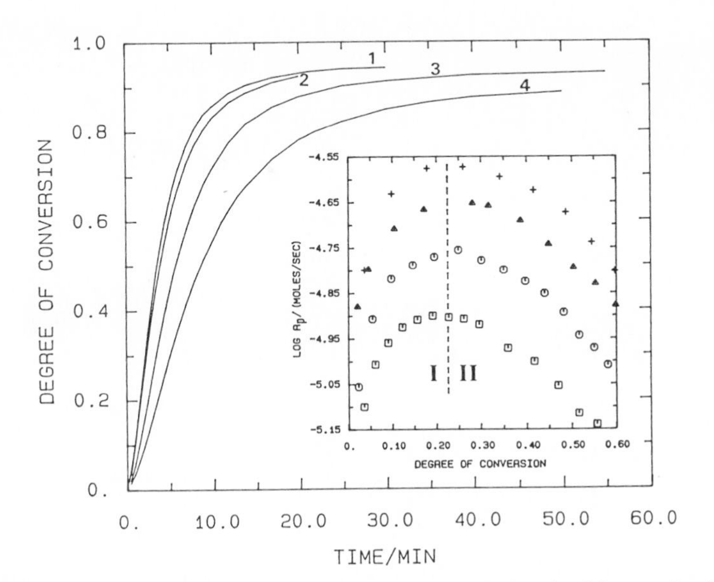

Figure 1. Conversion-time curves and polymerization rate-time curves for: 1. 0.69 (+); 2. 0.48 (▲); 3. 0.27 (⊙); 4. 0.14 (⊡) mM potassium persulfate based on water.

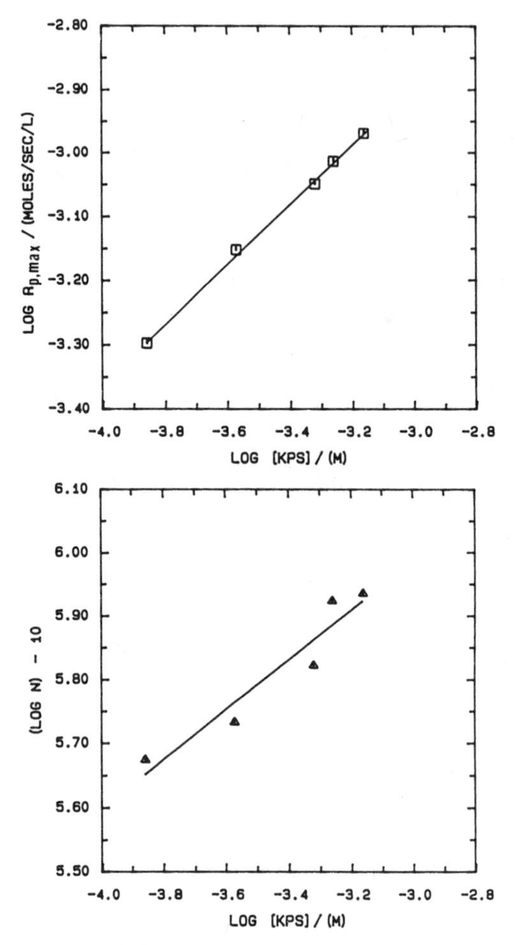

Figure 2. Dependence of maximum rate of polymerization and particle number on potassium persulfate concentration.

according to the 0.47 and 0.40 powers, respectively, of potassium persulfate concentration, which indicated that the number of particles reached a constant value at the maximum polymerization rate, and no more particles were nucleated after Interval I. This dependence was consistent with the 0.40 power predicted by Smith-Ewart case 2. However, the mechanism and kinetics of microemulsion polymerization were different from those of conventional emulsion polymerization. Particle nucleation occurred in the microemulsion droplets, with the fraction becoming particles determined by the initiator concentration. An increase in initiator concentration increased the radical flux to the microemulsion droplets, which in turn increased the fraction of droplets converted to polymer particles. The average number of radicals per particle $\bar{n}$ depended on the partitioning of 1-pentanol between the dispersed and continuous phases, which was difficult to determine. Nevertheless, the values of $\bar{n}$ were calculated to be in the range 0.06-0.15 assuming that all of the 1-pentanol was in the aqueous phase or the particle. These values were much smaller than the 0.5 value of Smith-Ewart case 2.

Figure 3 shows that the latex particle size distributions were broad, with particles ranging from 5 to 40 nm. The droplet sizes of the microemulsions could not be measured by electron microscopy, but earlier measurements by photon correlation spectroscopy of a microemulsion which was identical except for the substitution of brine for the water gave an average droplet size of 3.65 nm (14). The larger particle sizes of the latex particles were attributed to the fact that not all of the microemulsion droplets captured radicals and became polymer particles during Interval I; some served as reservoirs to supply monomer to the polymer particles by diffusion through the continuous phase. The broad particle size distributions were attributed to the long nucleation stage and the absence of a constant-rate interval.

Figure 4 shows that the polymer weight-average molecular weights were 1-2x10^6 and varied with only the -0.07 power of potassium persulfate concentration. The high molecular weights and low dependence on persulfate concentration was attributed to the slow rates of radical entry, which segregated the radicals inside the microemulsion droplets and reduced the probability of bimolecular termination. Similar results were obtained in an earlier study (17) on the effect of polymerization temperature. From the latex particle sizes and polymer molecular weights, each latex particle comprised only 2-3 polymer molecules. Candau et al. (21) obtained polyacrylamide of very high molecular weight (10^7) in the inverse microemulsion polymerization of acrylamide and concluded that each latex particle comprised only one polymer molecule. Broad particle size distributions, high polymer molecular weights, and a low dependence of polymer molecular weight on persulfate concentration were also found for the miniemulsion polymerization of styrene (18).

Effect of 2,2'-Azobis(2-Methyl Butyronitrile) Concentration. Figure 5 shows the variation of conversion and rate of polymerization with time at 70°C for 0.66-2.15 mM 2,2'-azobis(2-methyl butyronitrile) based on water. These curves show the same kinetic features as those of the potassium persulfate system. The polymerization rate increased with increasing 2,2'-azobis(2-methyl butyronitrile) concentration, and only two intervals were found; however, the maximum polymeriza-

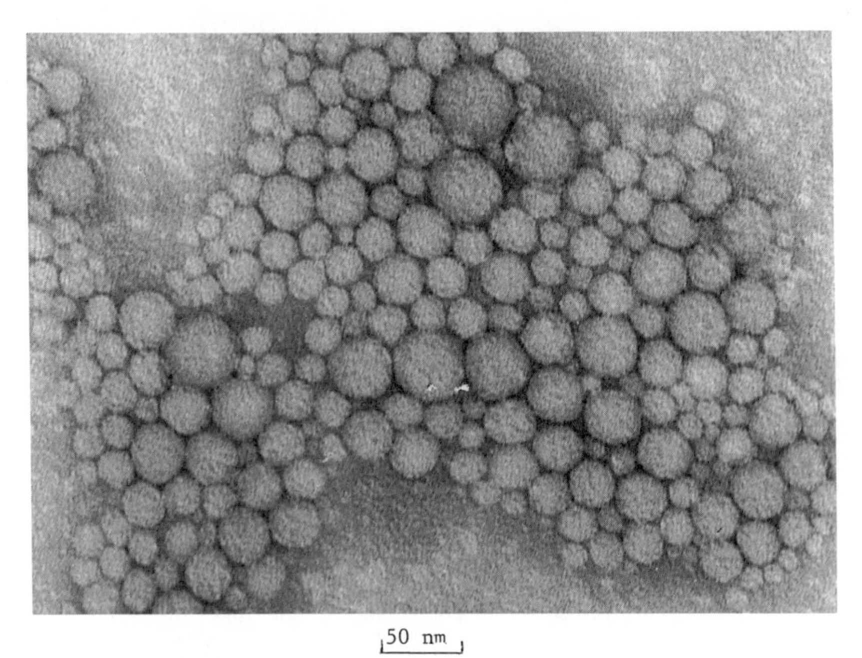

Figure 3. Transmission electron micrograph of polymerized styrene microemulsion with phosphotungstic acid negative-staining.

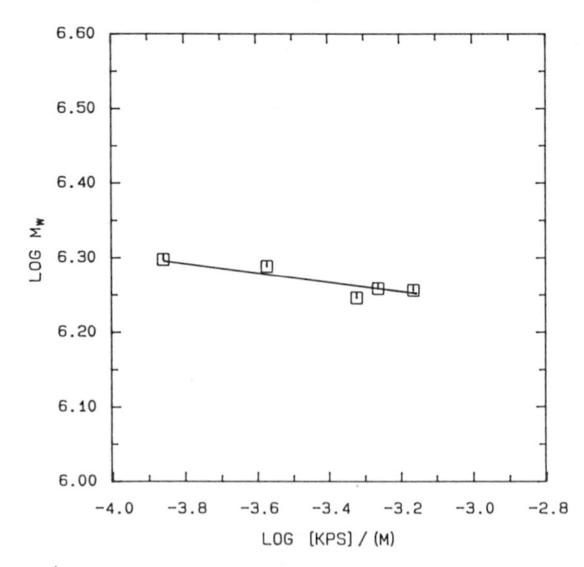

Figure 4. Dependence of weight-average molecular weight on potassium persulfate concentration.

tion rate was attained at 10-15% conversion, as compared to 20-25% for the water-soluble persulfate initiator and 2-15% for conventional persulfate-initiated emulsion polymerization. These differences were attributed to the fact that the 2,2'-azobis(2-methyl butyronitrile) is soluble in 1-pentanol, and the 1-pentanol is slightly soluble in water, which results in the generation of some radicals in the aqueous phase. Thus, two mechanisms were postulated for the nucleation of particles: entry of radicals into the droplets from the aqueous phase and generation of radicals inside the droplets. In this case, the polymerization rate would be determined by the equilibrium partitioning of radicals between the monomer droplets and the aqueous phase. However, two radicals generated in an microemulsion droplet would have a high probability of recombining before initiating polymerization. The recombination would give a slower polymerization rate, and the particle nucleation would depend mainly on the entry of radicals from the aqueous phase. Therefore, the radical generation rates and polymerization rates for potassium persulfate and 2,2'-azobis(2-methyl butyronitrile) initiators were compared. The calculated radical generation rate for 0.27 mM persulfate initiator at 70°C was 7.57×10^{12} radicals/ml/sec (22). For 2,2'-azobis(2-methyl butyronitrile), the calculated radical generation rates were 4.98×10^{13} radicals/ml/sec based on aqueous phase (2.15 mM) and 8.45×10^{14} radicals/ml/sec based on styrene (36.5 mM) (22). Because of the partitioning of this initiator between the monomer and aqueous phases, the actual radical generation rate should lie between these two values. Although the calculated radical generation rate for the 2.15 mM (based on water) 2,2'-azobis(2-methyl butyronitrile) was sevenfold greater than that for the 0.27 mM potassium persulfate, the polymerization rate was slower. Therefore, the higher calculated radical generation rate was probably outweighed by the high probability of recombination in the microemulsion droplets. The fact that the nucleation stage was shorter than for potassium persulfate was attributed to the greater efficiency of radical capture because the 1-pentanol acted as the medium.

Figure 6 shows that the maximum polymerization rate and number of particles varied as the 0.39 and 0.38 powers, respectively, of the 2,2'-azobis(2-methyl butyronitrile) concentration; these values were similar to the 0.47 and 0.40 powers found for persulfate initiator, and higher than the 0.21 and 0.21 powers found for the miniemulsion polymerization of styrene initiated by 2,2'-azobis(2-methyl butyronitrile) (18). Thus, for both the water-soluble potassium persulfate and oil-soluble 2,2'-azobis(2-methyl butyronitrile) initiators, the primary source of radicals was the aqueous phase. With the larger droplet sizes of miniemulsion polymerization, the source of radicals was both the monomer droplet phase and the aqueous phase. The particle sizes and molecular weight distributions were similar to those found for persulfate initiator. Figure 7 shows that the weight-average molecular weight varied with the -0.55 power of 2,2'-azobis(2-methyl butyronitrile) concentration, which is close to the -0.60 power found for the conventional emulsion polymerization of styrene using persulfate initiator. Possible reasons are: the aqueous phase is the primary source of radicals in both cases; the 1-pentanol in the microemulsion polymerization improves the radical capture efficiency by acting as the medium; and the sizes of microemulsion droplets and emulsifier micelles are similar.

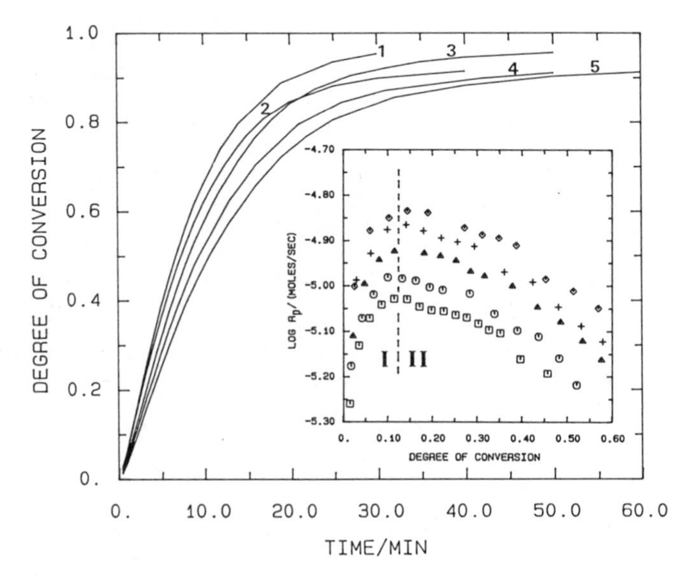

Figure 5. Conversion-time curves and polymerization rate-conversion curves for: 1. 2.15 (◇); 2. 1.68 (+); 3. 1.18 (▲); 4. 0.87 (○); 5. 0.66 (□) mM 2,2'-azobis(2-methyl butyronitrile) based on water.

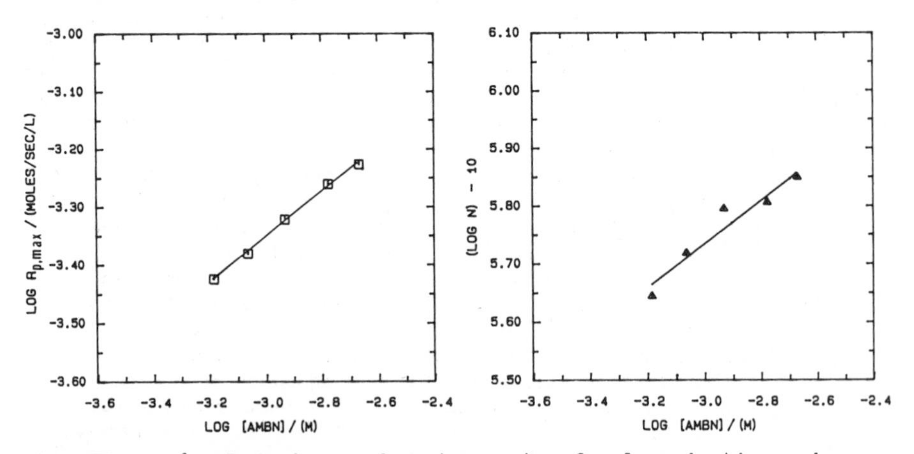

Figure 6. Dependence of maximum rate of polymerization and particle number on 2,2'-azobis(2-methyl butyronitrile) concentration.

Effect of Styrene Concentration. To determine the effect of concentration, the styrene was diluted with toluene while keeping the volume of the oil phase constant, in the expectation that the size of the microemulsion droplets, and thus the competition for radicals, would be similar. The polymerizations were carried out at 70°C using 0.55 mM potassium persulfate based on water. Figure 8 shows that the higher styrene concentrations gave faster polymerization rates; however, Figure 9 shows that the maximum rates of polymerization varied with the 2.33 power of styrene concentration, which cannot be accounted for by dilution alone. Figure 10 shows that the weight-average molecular weight decreased with decreasing monomer concentration, as expected. Table I shows the particle size distribution data for the different styrene concentrations. D_n, D_v, and D_w are the number-average, volume-average, and weight-average diameters determined by electron microscopy, $D_{v(cal)}$ is the volume-average diameter corrected for the volume of toluene, which was assumed to be lost during the electron microscopy, and PDI is the polydispersity index D_w/D_n. The values of $D_{v(cal)}$ and the polydispersity index were similar for the three different styrene concentrations, which indicates that the constant oil phase volume gave the same microemulsion droplet size and hence the same particle nucleation and growth mechanisms.

Table I. Particle Size Distributions in Styrene Microemulsion Polymerization (0.55 mM potassium persulfate; 70°C)

Styrene/ Toluene	D_n(nm)	D_v(nm)	D_w(nm)	$D_{v(cal)}$(nm)	PDI
100/0	20.0	22.3	26.6	22.3	1.333
75/25	18.5	20.6	24.7	22.7	1.338
50/50	16.3	18.4	22.4	23.2	1.370

The foregoing mechanism proposed for the microemulsion polymerization of styrene is based on the assumption that the mimcroemulsions comprised dispersions of small oil droplets in a continuous water medium. However, some workers (23,24) have shown by Fourier-transform pulse-gradient spin-echo NMR self-diffusion methods that 1-pentanol cosurfactant in 2/1 1-pentanol/sodium dodecyl sulfate ratio gives dynamic or bicontinuous lamellar microemulsions. Whether these structures form at the 0.4/1 1-pentanol/sodium dodecyl sulfate ratio used here is not known. If the microemulsions comprised bicontinuous lamellar structures with alternating monomer and water layers, the particle nucleation would occur in the monomer layer according to the diffusion of radicals formed by initiator decomposition, with the growing spherical particles adsorbing surfactant to disrupt the lamellar structure. To distinguish between these two structures, however, requires further study.

Conclusions

Polymerization of styrene oil-in-water microemulsions using potassium persulfate or 2,2'-azobis(2-methyl butyronitrile) initiator gave stable latexes which were bluish and less translucent than the original microemulsions. The mechanism and kinetics of polymerization were

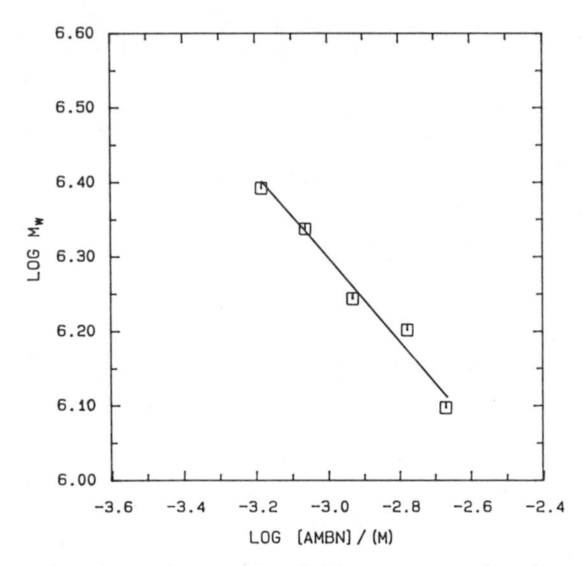

Figure 7. Dependence of weight-average molecular weight on 2,2'-azobis(2-methyl butyronitrile) concentration.

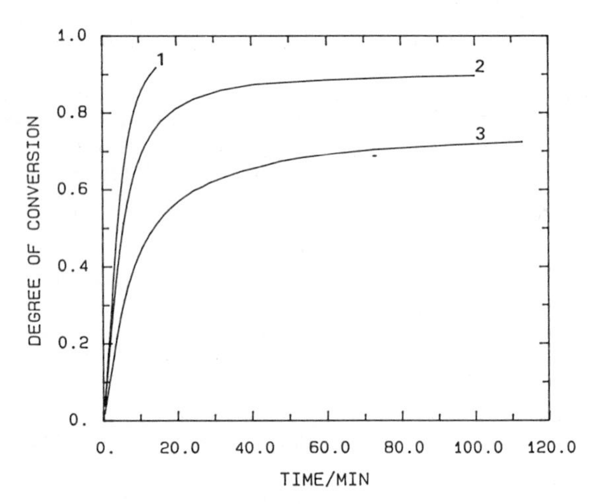

Figure 8. Conversion-time curves for 0.55 mM potassium persulfate with a constant-volume oil phase of: 1. 100/0; 2. 75/25; 3. 50/50 styrene/toluene weight ratio.

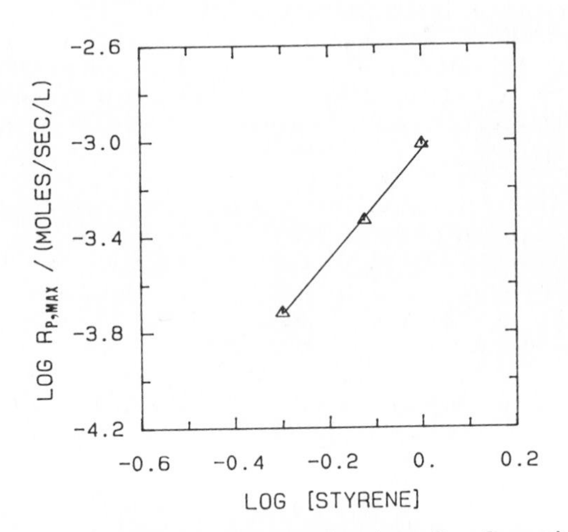

Figure 9. Dependence of maximum rate of polymerization on weight fraction of styrene in the styrene/toluene mixture.

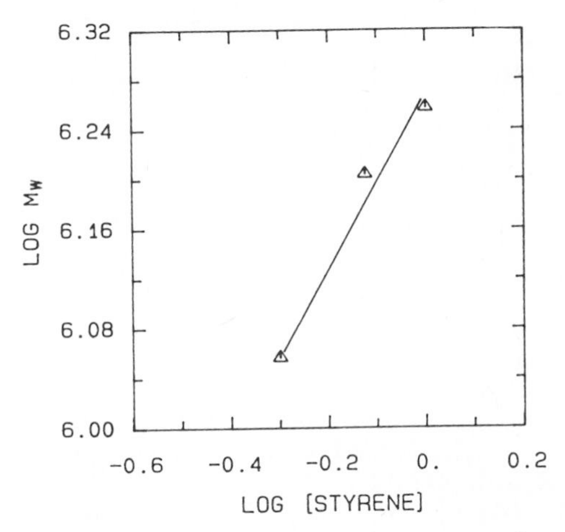

Figure 10. Dependence of weight-average molecular weight on weight fraction of styrene in the styrene/toluene mixture.

different from those of conventional emulsion polymerization or miniemulsion polymerization. The conversion-time curves showed only two intervals, with neither a constant-rate region nor a gel effect. The particle nucleation stage continued to 20-25% conversion for potassium persulfate and 15-20% conversion for 2,2'-azobis(2-methyl butyronitrile); this long nucleation stage was attributed to the slow rate of radical entry into the microemulsion droplets. The small average particle sizes (20-30 nm) and high molecular weights ($1\text{-}2 \times 10^6$) indicated that the particles comprised only 2-3 polymer molecules. The maximum rate of polymerization and number of particles varied with the 0.47 and 0.39, and the 0.40 and 0.38, powers of the potassium persulfate and 2,2'-azobis(2-methyl butyronitrile) concentrations, respectively. The particle size distributions for different styrene concentrations at constant oil phase volume were similar; thus, the droplet sizes and mechanisms of particle nucleation and growth were also similar. A mechanism was proposed for the styrene microemulsion polymerization: the particles were nucleated by capture of radicals from the aqueous phase for both water-soluble and oil-soluble initiators, and the microemulsion droplets which did not capture radicals served as reservoirs to supply monomer to the polymer particles. The possibility of homogeneous nucleation was not ruled out.

Literature Cited

1. Hoar, J. P.; Schulman, J. H. Nature 1943, 152, 102.
2. Improved Oil Recovery by Surfactant and Polymer Flooding; Shah, D. O.; Schechter, R. S., Ed.; Academic: New York, 1977.
3. Surface Phenomena in Enhanced Oil Recovery; Shah, D. O., Ed.; Plenum: New York, 1981.
4. Sunder Ram, A. N.; Shah, D. O. In Emulsions and Emulsion Technology, Part 3; Lissant, K. J., Ed.; Marcel Dekker: New York, 1986; p 139.
5. Microemulsions - Theory and Practice; Prince, L. M., Ed.; Academic: New York, 1977.
6. Micellization, Solubilization, and Microemulsions; Mittal, K. L., Ed.; Plenum: New York, 1977.
7. Solution Chemistry of Surfactants; Mittal, K. L., Ed.; Plenum: New York, 1979; Vol. 2.
8. Microemulsions; Robb, I. D., Ed.; Plenum: New York, 1982.
9. Surfactants in Solution; Mittal, K. L., Ed.; Plenum: New York, 1984; Vol. 3.
10. Macro- and Microemulsions - Theory and Applications; Shah, D. O., Ed.; ACS Symposium Series No. 272; American Chemical Society: Washington, D.C., 1985.
11. Microemulsion Systems; Rosano, H. L.; Clausse, M., Ed.; Marcel Dekker: New York, 1987.
12. Microemulsions: Structure and Dynamics; Friberg, S. E.; Bothorel, P., Ed.; CRC: Boca Raton, 1987.
13. Atik, S. S.; Thomas, J. K. J. Am. Chem. Soc. 1981, 103, 4279.
14. Johnson, P. L.; Gulari, E. J. Polym. Sci., Polym. Chem. Ed. 1984, 22, 3967.
15. Jayakrishnan, A.; Shah, D. O. J. Polym. Sci., Polym. Lett. Ed. 1984, 22, 31.
16. Kuo, P. L.; Turro, N. J.; Tseng, C. M.; El-Aasser, M. S.; Vanderhoff, J. W. Macromolecules 1987, 20, 1216.

17. Guo, J. S.; El-Aasser, M. S.; Vanderhoff, J. W. J. Polym. Sci., Polym. Chem. Ed. in press.
18. Choi, Y. T.; El-Aasser, M. S.; Sudol, E. D.; Vanderhoff, J. W. J. Polym. Sci., Polym. Chem. Ed. 1985, 23, 2973.
19. Delgado, J., Ph.D. Dissertation, Lehigh University, Bethlehem, 1986.
20. Chamberlain, B. J.; Napper, D. H.; Gilbert, R. G. J. Chem. Soc. Faraday Trans. I 1982, 78, 591.
21. Candau, F.; Leong, Y. S.; Fitch, R. M. J. Polym. Sci., Polym. Chem. Ed. 1985, 23, 193.
22. Polymer Handbook, 2nd ed.; Brandrup, J.; Immergut, E. H., Ed.; John Wiley & Sons: New York, 1975.
23. Stilbs, P.; Rapacki, K.; Lindman, B. J. Colloid Interface Sci. 1983, 95, 583.
24. Ceglie, A.; Das, K. P.; Lindman, B. J Colloid Interface Sci. 1987, 115, 115.

RECEIVED August 22, 1988

Chapter 7

Preparation of Microlatex with Functionalized Polyesters as Surfactants

F. Cuirassier[1], Ch. H. Baradji, and G. Riess

Ecole Nationale Supérieure de Chimie de Mulhouse, 3 rue Alfred Werner, 68093 Mulhouse Cédex, France

Carboxy-terminated polyesters, based on isophtalic acid and aliphatic diols were synthesized, in a molecular weight range of M_n 1000-4000. They have the following schematic structures:

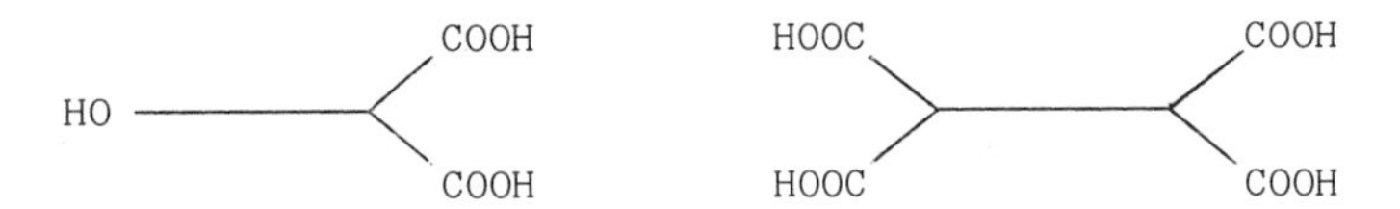

Such telechelic polyesters show typical surfactant behavior after neutralization with various amines (mono- and diethanol amine, etc...). Micellization has been studied by photon correlation spectroscopy. The micelle size, typically in the range of 10-30 nm, and its structure have been examined as functions of the polyester molecular characteristics and its degree of neutralization. A systematic study of the emulsion polymerization of styrene and acrylic monomers in the presence of neutralized polyester has been carried out. High solid content latexes with small particle size (below 100 nm), useful as waterborne coatings, could be obtained by this procedure.

Emulsion polymerization has become an important process for the production of a large number of industrial polymers in the form of polymer colloids or latexes. They are the base of adhesives, paints and especially of waterborne coatings. An interest has been developed in recent years in emulsion polymerization systems in which the classical low molecular weight surfactants are replaced by polymeric surfactants, either hydrophilic-hydrophobic block and graft copolymers (1-4) or functionalized oligomers (5).

[1]Current address: Hoechst–France Research Center, Seine Saint Denis, 64 Avenue Gaston Monmousseau, 93240 Stains, France

0097–6156/89/0384–0100$06.00/0

Waterborne coatings based on latex stabilized with polymeric surfactants may exhibit interesting applicational properties, such as shear stability and adjustable rheological characteristics (6-7).

In addition, the polymeric surfactant, necessary as a stabilizer during the emulsion polymerization, becomes, after drying of the latex, a constituent of the final paint film. This opens therefore the possibility for controlling the mechanical and surface properties of the film. The main advantage of polymeric surfactants is their double role, as stabilizer of the latex and modifier, i.e. as a plasticizer of the final polymer product.

In this contribution, we intend to show the behavior of carboxy terminated polyesters as surfactants in micelle formation and in emulsion polymerization of vinylic and acrylic monomers. The goal is to prepare high solid content latexes of small particle size, i.e., polymer dispersions with an average diameter below 100 nm. Our previous studies in this area have shown such microlatexes, based on acrylic polymers modified by polyesters, are an interesting approach to waterborne coatings leading to high surface gloss paint films (7-9).

This approach involves the following steps:

- The synthesis of carboxy terminated polyesters in the molecular weight range of 1000-4000
- The preparation of the polymeric surfactant by neutralization of the carboxy groups with various amines
- The preparation of latexes by emulsion polymerization of vinylic and acrylic monomers in the presence of these polyesters as polymeric surfactants
- The application of the waterborne coating on a substrate, followed by heat treatment, leading by volatilization of the amine to a water insoluble film in which the polyester can be considered as a modifier of the vinylic or acrylic polymer.

These different aspects will be determined by considering the followings:

- The synthesis and characterization of carboxy terminated polyesters
- The micellar behavior of such functionalized polyesters after neutralization with amines
- Emulsion polymerization in the presence of these polymeric surfactants.

Synthesis and Characterization of Carboxy terminated Polyesters

For this study we have synthezised a series of polyesters of different molecular weights having the following schematic structures:

Such polyesters are prepared by condensation in xylene of isophtalic acid with a slight excess of 1,5-pentanediol. The molecular weights are adjusted by the ratio diol/diacid. The carboxy end groups are introduced by reacting the hydroxy terminated polyesters with suitable amounts (stoicheiometric amount for the tetracid) of trimellitic anhydride.

For the tetracid series the reaction sequence is as follows :

$$HO-R-(O-\underset{\overset{\|}{O}}{C}-C_6H_4-\underset{\overset{\|}{O}}{C}-O-R)_n-OH \quad + \; 2 \; HOOC\text{-}C_6H_3(CO)_2O$$

$R = (CH_2)_5$

$$\downarrow$$

$$HOOC-C_6H_3(COOH)-\overset{O}{\overset{\|}{C}}-O-R-(O-\underset{\overset{\|}{O}}{C}-C_6H_4-\underset{\overset{\|}{O}}{C}-O-R)_n-O-\overset{O}{\overset{\|}{C}}-C_6H_3(COOH)-COOH$$

The characteristics of these polyesters are given in Table I.

Table I. Characteristics of the Carboxy Terminated Polyesters

Sample	Acid index (a)	Hydroxy index (b)	$\bar{M}_n$ (c)	$\bar{M}_n$ (d)	[η] ml/g (e)
MD 60	61	36	1840	1820	9,3
MD 90	89	45	1260	1100	8,2
MD 120	113	69	990	/	7,0
T 60	61	/	3700	3850	13,5
T 90	100	/	2240	/	10,5
T 120	128	/	1750	1700	6,6

(a) (Milliequivalent COOH per g of polyester) x 56108. Determination by KOH titration
(b) (Milliequivalent OH per g of polyester) x 56108. Determination by acetylation
(c) $\bar{M}_n$ calculated according to the acid index
(d) $\bar{M}_n$ determined by vapor pressure osmometry
(e) Intrinsic viscosity in THF at 25°C

Micellization of Functionalized Polyesters

The carboxy terminated polyesters become water dispersable only by neutralization with a base such as amines :

$$PES-COOH + NR_3 \rightleftarrows PES-COO^- \; {}^+NR_3H$$

As for classical surfactants, these neutralized polyesters in an aqueous phase have a tendency to form colloidal dispersions of micellar type. The size of these micelles will be a function of the molecular characteristics of the polyester (molecular weight, number of end groups) and of the neutralization conditions.

The different amines used for neutralization of the polyesters are given in Table II.

Industrially, neutralization of the polyester and the preparation of the colloidal dispersion is carried out by a "phase inversion procedure". It consists in adding, first the given amount of amine to the preheated polyester (80°C), then dropwise the amount of water suitable to adjust the solid content of the final system.

To achieve more homogeneous colloidal systems on a lab scale, the following is a better technique:

- Solubilizing the polyester in a water miscible solvent like THF
- Adding this organic solution to water containing the amine
- Stripping off the organic solvent under reduced pressure
- Adjusting with water the solid content of the colloidal system.

The micellar size of these colloidal dispersions is then determined by photon correlation spectroscopy (Coulter N4). Typical results are given in Figure 1, showing the variation of the particle size as a function of the neutralization degree α .

This figure shows that aggregates are formed below complete neutralization of the polyester carboxy groups. The micelle size reaches a constant value above 100 % neutralization. The amine's slight volatility makes it usually necessary, especially in industrial application, to work at a neutralization degree α of 1.05 to 1.25. It means an excess of 5 to 25 % amine with respect to the theoretical value.

We have systematically examined the micellar size of the different polyesters completely neutralized with the amines listed in Table II. The size of the micelles prepared by the laboratory technique is constant, within experimental errors, for a given completely neutralized polyester with an amine. Stable colloidal systems could be obtained with all samples (Table III) except for sample T 60.

From the micelle size, the surface occupied by the hydrophilic part of one polyester molecule can be calculated, and thus the surface occupied by one carboxy group. This calculation is done assuming the density of the polyester in the micelle to be similar to bulk polyester and all the carboxy groups located on the micelle surface (conformation A and B given in Figure 2).

From Table III it appears that for the samples MD 60, MD 90, T 90 the surface occupied by one carboxy group is the same, i.e., in the range of 0.32 nm^2. This is in agreement with the values of 0.34 to 0.51 nm^2 given by Pal et al. (11) for α , ω dicarboxylic acids.

It should be noted that the dicarboxylic polyesters (MD samples) are in fact a mixture containing theoretically 50 % diacid, 25 % tetracid and 25 % of non-carboxylated chains. If all the carboxy groups are located at the micelle surface, the average surface occupied by one carboxy groups would be 0.32 nm^2 as indicated previously. However if only one chain end of the tetracid is reaching the surface (conformation D in Figure 2) the surface occupied by one carboxy group located at the surface would be of 0.42 nm^2.

We shall consider a_1 = 0.32 nm^2 in the following, a value consistent for two diacid samples (MD 60 and MD 90) and one tetracid (T 90).

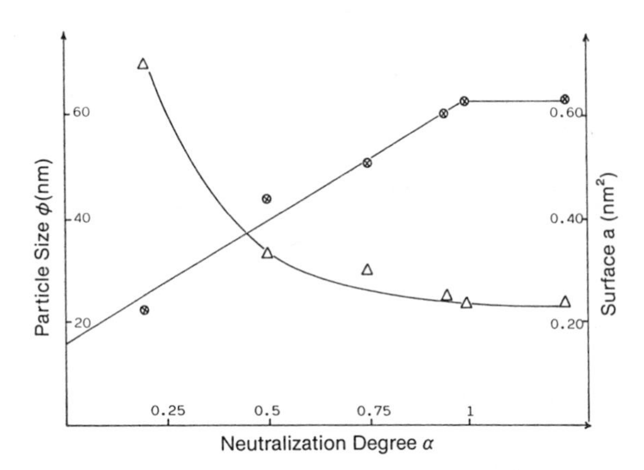

Figure 1. Influence of the neutralization degree α on the particle size (ϕ) of the micelles and on the surface (a) occupied by one polyester chain. System MD 60 - MDEA.
⊗ $a = f(\alpha)$ Δ $\phi = f(\alpha)$

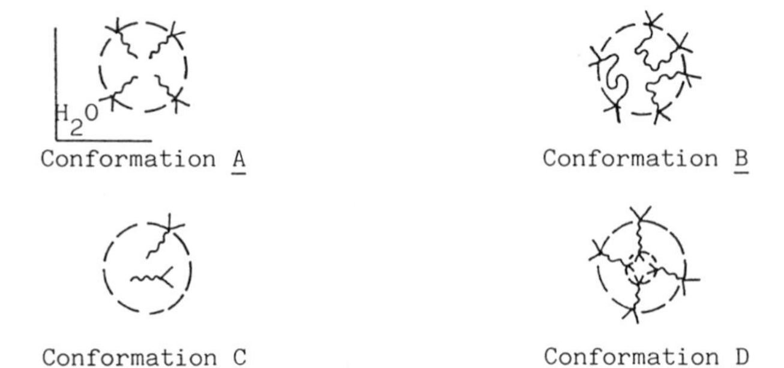

Figure 2. Schematic representation of chain conformation in the micelles for neutralized diacid and tetracid polyesters.

Table II. Amines Used as Neutralization Agents for the Polyesters

Amines	Structure	Code	Solubility* parameter δ $(cal/cc)^{1/2}$
Monoethanolamine (2-amino ethanol)	$H_2N - CH_2 - CH_2 - OH$	MEA	16.5
Diethanolamine Bis (2-hydroxyethyl)amine	$HN -(CH_2 - CH_2 - OH)_2$	DEA	16.5
N,N-dimethylethanol amine N,N-dimethylamino-2 ethanol-1	$(CH_3)_2N - CH_2 - CH_2 - OH$	DMEA	11.5
N-methyldiethanolamine N,N Bis (2 hydroxyethyl) methylamine	$CH_3 - N -(CH_2 - CH_2 - OH)_2$	MDEA	14.5

* Solubility parameter calculated according to SMALL (10).

Table III. Micellar Characteristics : Size, Surface Occupied by Carboxy End Groups and Conformation

	Structure	Average diameter nm	a Surface per molecule nm^2	a_1 Surface per carboxy group nm^2	Conformation*
MD 60 $\overline{M}_n$ = 1840 d = 1.23	diacid	23.7 ± 0.8	0.63 ± 0.02	0.315	A
MD 90 $\overline{M}_n$ = 1260 d = 1.25	diacid	15.9 ± 0.8	0.63 ± 0.03	0.315	A
MD 120 $\overline{M}_n$ = 990 d = 1.25	diacid	14.7 ± 1.1	0.54 ± 0.04	0.27	C 15 % of the chains entrapped
T 90 $\overline{M}_n$ = 2240 d = 1.24	tetracid	14.0 ± 0.5	1.29 ± 0.04	0.32	B
T 120 $\overline{M}_n$ = 1750 d = 1.24	tetracid	15.6 ± 0.4	0.90 ± 0.02	0.225	40 % of chains in conformation B 60 % of chains in conformation D

* see Figure 2

Based on this assumption and conformation C (Figure 2) we can calculate that the sample MD 120 has ≃ 15 % of the chains entrapped in the micelle. This is in agreement with the fact that this diacid sample has the highest hydroxy-to-acid index ratio and thus a slight excess of non-carboxylated chains (see Table I).

The same calculation for the tetracid sample T 120 indicates 60 % of the chains have only one end coming to the micelle surface (conformation D in Fig. 2). The remaining carboxy groups inside the micelle might associate in the form of clusters, as demonstrated for ionomer resins (12).

From Figure 1(the particle size vs neutralization degree α) we can also calculate the surface $\underline{a}$ occupied by one molecule. The variation of $\underline{a}$ vs α given in Fig. 1 shows that for α = 1 one has $\underline{a}$ = 0.63 nm^2, in agreement with the results given in Table III. For α = 0 the value of $\underline{a}$ is 0.16 nm^2. This indicates at least a part of the non-neutralized COOH groups are located on the micelle surface. For classical surfactants, derived from fatty acids, such "mixed micelles" of neutralized and non-neutralized species are a well-established fact (13).

It is of interest to examine the average diameter ϕ and the occupied surface for micelles prepared following the industrial procedure. These values are given in Table IV for the series of diacids neutralized with the various amines. Assuming complete neutralization and a = 0.63 nm^2 , we can calculate the percentage of chains having their carboxy groups on the micelle surface. From Table IV the industrial procedure appears to point to larger size micellar systems, and thus lower "a" values than those obtained by the laboratory technique. They are, however, still stable. The percentage of chains having their carboxy groups on the surface is highest with MD 90 and MD 120, the less viscous polyesters, in combination with MDEA or DMEA, the amines of low volatility and the strongest basic character. This demonstrates that phase inversion occuring at 80°C for the industrial procedure and formation of small particles are controlled by the viscosity of the system and the amine characteristics.

Solubilization of Monomers in Micellar Systems

Solubilization of monomers in the micellar core of surfactants has earlier been well demonstrated (14). It was, therefore, of interest to examine the solubilization of a water-insoluble monomer such as styrene in the core of the polyester micelles. A technique by Funke was followed, in which a micellar polyester solution was titrated with styrene while its optical density recorded. For the system MD 60-styrene, 0.60 ± 0.05 g of styrene is solubilized per gm polyester. Similarly the increase in micellar size, due to swelling by styrene, could be followed by photon correlation spectroscopy. This is shown in Figure 3, corresponding to the plot of:

$$(\phi / \phi_o)^3 = N_o/N \left(1 + d \frac{\text{volume styrene}}{\text{weight polyester}}\right)$$

Table IV. Micellar Characteristics of Carboxy Terminated Polyesters Micelles Prepared by an Industrial Procedure

Amine	Polyester	MD 60	MD 90	MD 120
MEA	φ	68 ± 2	23	19
	a	0.22	0.43	0.42
	%	35	68	66
DEA	φ	43 ± 7	21.5	17.5
	a	0.35	0.47	0.45
	%	55	74	71
MDEA	φ	38	17.5	14
	a	0.40	0.57	0.57
	%	63	91	90
DMEA	φ	29	15	15
	a	0.52	0.66	0.53
	%	82	100	84

φ in nm
a in nm^2

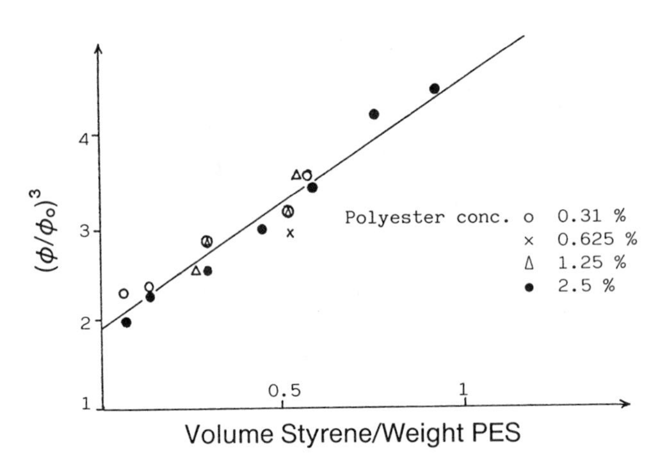

Figure 3. Swelling of micelles with styrene. System MD 60-DEA Plot of $(\phi / \phi_o)^3$ vs volume styrene/weight polyester.

with

ϕ the actual and ϕ_o the initial diameter of the micelle
N the actual number and N_o the initial number of micelles per unit volume
d density of the polyester

Figure 3 corresponds to the system MD 60-DEA-styrene, where neutralization was achieved by the phase inversion technique (industrial procedure). It can be seen that swelling of the micelles with monomer leads to a linear relationship of $(\phi / \phi_o)^3$ vs monomer concentration. It is also interesting to note that the slope, which is proportional to the density of the polyester, leads to a value of d = 1.19, in good agreement with 1.23 determined for bulk polyester. The extrapolation to zero concentration of styrene indicates, from the value of $N_o/N = 2$, that solubilization of monomer in such a micellar system involves an additional aggregation of the particles during the initial stage of swelling. This phenomena is probably to be correlated to the fact that styrene monomer, a good solvent for the polyester and even of the neutralized polyester, solubilizes a part of this polymeric surfactant in the core of the micelle. In contrast, it was found that swelling with BMA, generally leads to values of $N_o/N < 1$ indicating a disaggregation phenomena of the micelles prepared by the industrial procedure.

Emulsion Polymerization in the Presence of functionalized Polyesters - Preparation of Latex

Having shown that carboxy terminated polyesters present typical surfactant characteristics after neutralization with amines, it was of interest to examine their application possibilities in emulsion polymerization. Such polymerizations for styrene and for butyl-methacrylate (BMA) have been carried out either by a batch technique or by a semi- continuous procedure. The reaction conditions were the following :

15 % polyester neutralized with various amines
34.75 % monomer (styrene or BMA)
50 % water
0.25 % $K_2S_2O_8$ Polymerization temperature : 80°C

Since the percentages are given in weight %, such a receipe should lead to a latex with a solid content of 50 %.

The reaction time is 16 hours for the batch procedure. In the semi-continuous procedure a seed latex is first prepared with 1/10 of the monomer amount during 30 minutes at 80°C. The remaining monomer is then added continuously during 2.5 hours.The polymerization is completed by heating the system for an additional 1 hour.

Under these reaction conditions, the monomer conversion reaches at least 95 %.

The obtained latex is then characterized by its coagulum content and its particle size determined by photon correlation spectroscopy or electron microscopy. The results are given in Table V and VI.

Table V. Emulsion Polymerization of Styrene in Batch in the Presence of Polyester Surfactants

Polyester / Amine	Polyester diacid			Polyester tetracid	
	MD 120	MD 90	MD 60	T 90	T 120
DMEA	92*	(-)	167	(-)	(-)
MDEA	(-)	64	81	(-)	(-)
DEA	(-)	80	130	(-)	(-)

Particle Size in nm Solid Content 50 %
(-) coagulation of the latex
* the corresponding latex prepared by the semi-continuous process gives a particle size of 77 nm by photoncorrelation spectroscopy, the number and weight average diameters determined by electron microscopy are respectively $\bar{d}_n$ = 73 nm and $\bar{d}_w$ = 79 nm.

Table VI. Emulsion Polymerization of BMA in the Presence of Polyester Surfactants

Polyester / Amine	Polyester diacid			Polyester tetracid	
	MD 120	MD 90	MD 6 0	T 90	T 120
DMEA	-	-	245	-	-
	(-)	140	(214)		
MDEA	64	150	300	-	-
	(68)	(79)	(125)		
DEA	-	166	157	330	-
	(-)	(73)	(134)		
MEA	-		160	276	-
	(-)		(174)		

Particle Size in nm Solid Content 50 %
- coagulation of the latex
values in () are the latex diameter for the semi-continuous process, the others being the latex diameter for the batch process.

These results indicate that:

- Diacid polyesters are more efficient than tetracid polyesters, as for styrene and BMA emulsion polymerization
- Stable, high solid content latexes can be prepared with the MD series of polyesters, especially with MD 60, independently of the neutralization amine
- The semi-continuous polymerization procedure leads to a smaller particle size for BMA than the batch process
- For a same polymerization process (batch process) smaller particles are generally obtained with styrene compared to BMA. It may be attributed to polystyrene being less compatible than polybutylmethacrylate with the polyester surfactant
- The particle size of the latex diminushes with decreasing polyester molecular weight
- Microlatex, i.e. particles having a diameter of less than 100nm, can be obtained, especially with polyesters MD 90 and MD 120.

It is well known in emulsion polymerization practice that the efficiency of a surfactant depends on its hydrophilic-lipophilic balance (HLB), an indication of its relative solubility in water compared to that in monomer or polymer (15). Based on this fact, we have represented in Figure 4 the particle size for the BMA latex as a function of the hydrophilic character of the amine and the polyester. The hydrophilic character of the amine is mainly indicated by its solubility parameter (δ), whereas that of the polyester is essentially a function of its molecular weight, and thus its carboxy content.

Figure 4 shows an optimum balance exists when the polyester is neutralized with MDEA. For this system, Table VII gives the change of the specific surface for the latex and the corresponding surface occupied by one polyester molecule.

Table VII. Specific Surface of Polybutylmethacrylate Latex - Surface Occupied by One Polyester Molecule as a Function of the Molecular Weight of Polyester - System MD 60, MD 90, MD 120 Neutralized with MDEA

Polyester	Particle size of latex nm	Specific surface* nm^2 $.10^{-21}$	Surface occupied by one polyester molecule nm^2
MD 120	68	4.15	0.45
MD 90	79	3.57	0.50
MD 60	125	2.26	0.46

* Specific surface given for 100 g of latex with a solid content of 50 %

Within the experimental error limits, the surface occupied by a polyester molecule is constant and in the range of 0.45-0.50 nm^2. Assuming, as previously shown, that a neutralized end group has a surface of 0.63 nm^2, about 75 % of the polyester chains are therefore located on the particle surface. As expected, the specific surface of the latex decreases with increasing polyester surfactant molecular weight.

In a more fundamental approach, i.e. by using systems of lower solid content, we examined the batch polymerization of styrene and BMA, to establish a correlation between the polyester content and the final latex particle size.

The polymerization conditions were as follows:

- 10 weight % monomer with respect to the aqueous phase
- Polyester MD 60 neutralized with DMEA (industrial procedure)
- Polyester concentration in water varied between 0.05 and 1.25 weight %
- $K_2S_2O_8$ at a concentration of 1.3 . 10^{-3} mole/l with respect to monomer
- Reaction temperature 80°C.

Results given in Table VIII show, as expected, decreasing latex particle size with increasing polyester surfactant concentration (Figure 5).

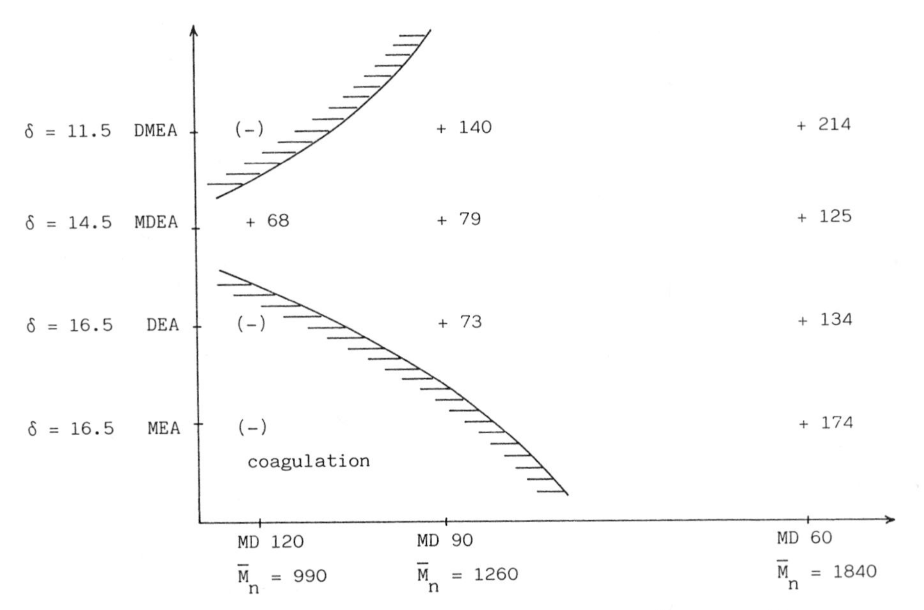

Figure 4. Emulsion polymerization of BMA (semi-continuous procedure). Particle size of the latexes for different polyesters neutralized with various amines.
\+ stable latexes (-) coagulation during polymerization
δ given in $(cal/cm^3)^{1/2}$

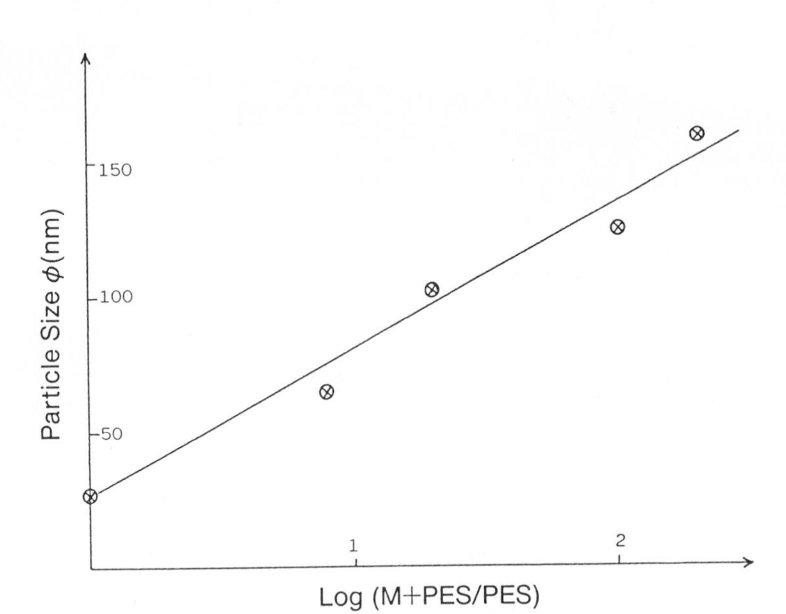

Figure 5. Emulsion polymerization in batch of BMA. Particle size of the latexes vs log (M+PES / PES) with M the weight amount of monomer and PES the weight amount of polyester.

Table VIII. Emulsion Polymerization in Batch of Styrene and of BMA in the Presence of Polyester Surfactants

% Polyester	Number * of initial micelles $N_o.10^{-14}$	Latex BMA		Latex styrene	
		ϕ nm	$N.10^{-14}$	ϕ nm	$N.10^{-14}$
0.05	0.32	161	0.39	/	/
0.1	0.64	133	0.71	/	/
0.5	3.2	103	1.6	76.5	3.8
1.25	8	69	5.4	71.5	5.0

ϕ = average diameter determined by photoncorrelation spectroscopy
N = number of particles per cm^3 of latex
* values calculated for 1 cm^3 of polyester solution, by assuming an average micelle diameter of 29 nm

Furthermore in contrast to classical low molecular weight surfactants, polymeric surfactants produce a number of latex particles of the same order of magnitude as the number of initial micelles. This can be attributed to the lower diffusion rate of these polymeric surfactant molecules during the emulsion polymerization process (2, Baah, F., University of Haute Alsace, unpublished).

Conclusion

Carboxy terminated polyesters, of diacid and tetracid type, have been synthesized and characterized by chain-end titration, by vapour pressure osmometry and by viscometry.

Neutralizing these polyesters with different hydroxyamines made it possible to obtain stable micellar systems. The size of these micelles was shown to essentially be a function of the structure and the polyester molecular weight. The smallest micelles are obtained with diacid polyester of lower molecular weight. A model for the chain conformation of polyester in a micelle was proposed, taking into account the area occupied by one end group located at the micelle surface. Furthermore, such micelles were demonstrated to be able to solubilize non-neutralized polyester chains, as well as monomers like styrene or butylmethacrylate (BMA).

The behavior of neutralized polyesters having typical surfactant characteristics, were examined in emulsion polymerization of styrene and BMA. The special features of these polymeric surfactants were shown, especially the fact that the particle size of the latex can be regulated by adjusting the concentration and the polyester molecular weight.

It was also shown that stable latexes of high solid content, and small particle size could be practically obtained by this emulsion polymerization technique. Such microlatexes based on acrylic polymers modified by polyesters are an interesting approach to waterborne coatings leading to high gloss paint films (9).

Acknowledgments

The authors wish to express their gratitude to "AKZO COATINGS" Montataire (France) for financial support of this work.

Literature Cited

1. Riess, G.; Hurtrez, G.; Bahadur, P. In Encyclopedia of Polymer Science Engineering; 2nd ed.; Wiley: New York, 1985; Vol. 2, p 324
2. Rogez, D. Ph.D. Thesis, University of Haute Alsace, May 1987.
3. Thyebault, H. Ph.D. Thesis, University of Haute Alsace, September 1987.
4. Riess, G.; Thyebault, H. French Patent 86/08556, June 1986.
5. Yu, Y-Ch.; Funke, W. Angew. Makromol. Chem. 1982, 103, 187.
6. Laible, R. Proc. 11th Int. Conf. in Organic Coatings Science and Technology; Patsis, A., Ed., 1985; p 12.
7. Baradji, Ch.H. Ph.D. Thesis, University of Haute Alsace, May 1984.
8. Cuirassier, F. Ph.D. Thesis, University of Haute Alsace, December 1984.
9. Baradji, Ch.H.; Cuirassier, F.; Riess, G.; Bouvier, D.; Toth, A.; Trescol, J.J. to AKZO Netherland Patent 83 04021, November 1983.
10. Small, P.A. J. Appl. Chem. 1953, 3, 71.
11. Pal, R.P.; Chatterjee, A.K.; Chattoraj, D.K. J. Colloid Interface Sci. 1975, 52 46.
12. Lundberg, R.D. In Encyclopedia of Polymer Science Engineering; 2nd ed.; Wiley: New York, 1985; p 393
13. Skoulios, A. Ann. Phys. 1978, 3, 421.
14. Harkins, W.O. J. Polymer Sci. 1950, 5, 217.
15. Blackley, D.C. In Emulsion Polymerization: Theory and Practice; Wiley: New York, 1975.

RECEIVED September 12, 1988

Chapter 8

Liquid Crystalline Phases and Emulsifying Properties of Block Copolymer Hydrophobic Aliphatic and Hydrophilic Peptidic Chains

Bernard Gallot[1] and Hussein Haj Hassan

Centre de Biophysique Moléculaire, Centre National de la Recherche Scientifique, 1A Avenue de la Recherche Scientifique, 45071 Orléans Cédex 2, France

Amphiphilic lipopeptides with a hydrophobic paraffinic chain containing from 12 to 18 carbon atoms and a hydrophilic peptidic chain exhibit lyotropic mesophases and good emulsifying properties. The X-ray diffraction study of the mesophases and of dry lipopeptides showed the existence of three types of mesomorphic structures : lamellar, cylindrical hexagonal and body-centred cubic. Two types of polymorphism were also identified : one as a function of the length of the peptidic chain and the other as a function of the water content of the mesophases. The emulsifying properties of the lipopeptides in numerous pairs of immiscible liquids such as water/ hydrocarbons and water/base products of the cosmetic industry showed that small amounts of lipopeptides easily give three types of emulsions : simple emulsions, miniemulsions and microemulsions.

Many surfactants have been used to formulate microemulsions (1). They were of three types : anionic surfactants such as petroleum sulfonates, sodium octyl benzene sulfonate, sodium dodecyl sulfate, alkaline soaps ; cationic surfactants such as dodecyl ammonium and hexadecyl ammonium chlorides or bromides ; and nonionic surfactants such as polyoxyethylene glycols. Furthermore, many exhibit liquid-crystalline properties (2) and in some cases the structure of the mesophases has been established (3). Nevertheless, nearly nothing is known about their compatibility with blood and tissues, and, from our own experience, some exhibit a high lytic power for red cells (4).

[1]Current address: Laboratoire des Matériaux Organiques, Centre National de la Recherche Scientifique, Boîte Postale 24, 69390 Vernaison, France

0097–6156/89/0384–0116$06.00/0

In order to obtain surfactants able to emulsify base products for cosmetic industry without presenting adverse side effects to the skin and tissues, it was necessary to synthesize new surfactants. We have chosen as new surfactants the amphiphilic lipopeptides (5-6).

Amphiphilic lipopeptides $C_n(AA)_p$ are formed by a hydrophobic lipidic chain C_n containing from 12 to 18 carbon atoms linked through an amide bond to a hydrophilic peptidic chain $(AA)_p$ with a number average degree of polymerization p between 1 and 90. The repeating unit of the peptidic chain is an amino-acid residue and the general formula of the lipopeptides $Cn(AA)_p$ is :

$$H-(CH_2)_n-NH-(CO-\underset{R}{CH}-\underset{R'}{N})_p-H$$

with R' = H (except for sarcosine where R' = CH_3) and R is the side chain of the amino-acid. The following are the various amino-acid residues : sarcosine (Sar) with R = H ; lysine bromhydrate (K) with R = $(CH_2)_4-NH_2,HBr$; sodium salt of glutamic acid (E) with R = $(CH_2)_2-COONa$; hydroxyethylglutamine (Eet) with R = $(CH_2)_2-CO-NH-(CH_2)_2OH$ and hydroxypropylglutamine (Epro) with R = $(CH_2)_2-CO-NH-(CH_2)_3OH$.

Lipopeptides present three main advantages. The two parts of the lipopeptide molecules (lipidic and peptidic chains) are present in many biological molecules and macromolecules and one can expect a good compatibility with biological fluids and tissues (4) and an absence of toxicity of their degradation products. The HLB of lipopeptides can be easily adjusted by varying the degree of polymerization p of the peptidic chains and the nature of the amino-acid side chains R. The incompatibility between the hydrophobic paraffinic chains and the hydrophilic peptidic chains leads to a phase separation at the molecular scale and to the existence of mesophases (7).

In this paper, we will describe the lyotropic liquid crystalline and the emulsifying properties of lipopeptides with peptidic chains.

EXPERIMENTAL METHODS

Preparation of Mesophases. Lipopeptides are dissolved in a small excess of water and, when total homogeneity is achieved, the desired concentration is obtained by slow evaporation at room temperature. Then the sample is left at room temperature in tight cells to reach equilibrium.

X-ray Diffraction Studies. They are performed under vacuum with a Guinier type focussing camera equipped with a bent quartz monochromator giving a linear collimation of the $CuK_{\alpha 1}$ (λ = 1.54 Å) radiation (8).

Preparation of Emulsions. The mixture oil-lipopeptide is heated to 70°C under agitation for complete homogeneization. It is then cooled to 45°C and water is added. The agitation is maintained throughout the preparation and until the system is cooled to room temperature (9).

Preparation of Miniemulsions. They are prepared by two methods :

- In the first method, the ionic lipopeptide and cetyl alcohol are mixed with water for an hour at 63°C (pre-emulsification stage). The oil is then added at 63°C and the agitation is continued for an additional hour (9,10).

- In the second method, the mixture oil/lipopeptide/cetyl alcohol is homogeneized at 70°C for several minutes ; then it is cooled at 50°C and water is added, agitation is carried on until complete homogeneization. The system is then cooled to room temperature under agitation (9).

Preparation of Microemulsions. Oil, lipopeptide and water are mixed in the same way as in emulsion preparation ; the mixture is then titrated, at room temperature, with the cosurfactant until transparency is obtained (9).

Stability of Emulsions and Miniemulsions. The stability of emulsions and miniemulsions is determined by following, as a function of time, the variation of the emulsified volume at fixed temperatures between - 10°C and + 50°C.

RESULTS AND DISCUSSION

Liquid-Crystalline Properties

Structure and Polymorphism of Lipopeptides. Amphiphilic lipopeptides $C_n(AA)_p$ exhibit mesophases in aqueous solution for water concentrations smaller than about 60 %. The structure of the mesophases and of the dry lipopeptides obtained by evaporation of the mesophase water at a slow rate was determined by X-ray diffraction. Lipopeptides X-ray diagrams obtained are similar to those exhibited by classical amphiphiles (11). They have allowed us to establish the existence of three types of liquid-crystalline structures : lamellar, hexagonal and cubic.

The lamellar structure consists of plane, parallel equidistant sheets ; each sheet of thickness d results from the superposition of two layers : one of thickness d_A contains the hydrophilic peptidic chains and the water, while the other layer of thickness d_B contains the hydrophobic paraffinic chains.

The hexagonal structure consists of long and parallel cylinders of diameter $2R_H$, filled with the hydrophobic paraffinic chains of the lipopeptides and assembled in a hexagonal array of parameter D, while the space between the cylinders is occupied by the hydrophilic peptidic chains and the water.

The body-centred cubic structure consists of spheres of diameter $2R_C$ filled with the hydrophobic paraffinic chains of lipopeptides and assembled on a body centred cubic lattice of side a, while the space between the spheres is occupied by the hydrophilic peptidic chains and the water.

The lattice parameters d, D and a are directly obtained from X-ray patterns, while the other parameters : d_A, d_B, 2R and S (average surface occupied by a chain at the interface between the hydrophilic and hydrophobic domains) are calculated using formulae based on simple geometrical considerations (11,12).

The type of structure adopted by the lipopeptides is determined by the ratio of the volumes of the hydrophilic domains (containing the peptidic chains and the water) and the hydrophobic domains (containing the paraffinic chains). Therefore the lipopeptides exhibit two types of polymorphism : one as a function of the length of the peptidic chains and the other as a function of the water content of the mesophases. When the degree of polymerization p of the peptidic chains increases dry lipopeptides (obtained by evaporation of the water) exhibit successively lamellar, hexagonal and cubic structures in the case of liposarcosine (12) and lamellar and hexagonal structure in the case of lipolysine and lipo(glutamic acid) (13). Furthermore the addition of water to lipopeptides is able to transform the lamellar structure into a hexagonal one (12-13) or a hexagonal structure into a body-centred cubic one (12).

Factors Governing the Geometrical Parameters of the Mesophase. The factors governing the geometrical parameters of the liquid-crystalline structures are the water content of the mesophases, the length of the paraffinic chains and the length of the peptidic chains.

Influence of the Water Concentration. When the water concentration increases :

- The lattice parameter : d for the lamellar structure, D for the hexagonal structure and a for the cubic structure increases.
- The characteristic parameter of the hydrophobic domains : d_B for the lamellar structure, $2R_H$ for the hexagonal structure and $2R_C$ for the cubic structure decreases.
- The characteristic parameter of the hydrophilic domains : d_A for the lamellar structure, $D-2R_H$ for the hexagonal structure and $a-2R_C$ for the cubic structure increases.
- The average surface S available for a molecule at the interface increases for the 3 types of structure.

The figures 1 and 2 illustrate such a behaviour in the case of the lamellar and hexagonal structures of $C_{18}K_2$, and of the body-centred cubic structure of $C_{17}Sar_{50}$ respectively.

Influence of the Length of the Paraffinic Chains. Sets of lipopeptides $C_n(AA)_p$, with the same degree of polymerization p for the peptidic chains, but with a number of carbon atoms n of their paraffinic chains equal to 12, 14, 16, 17 and 18 have been studied. When the number n of carbon atoms increases, d, D, d_B and $2R_H$ increase, while the characteristic parameter of the hydrophilic peptidic domains and S both remain constant (9,12).

Influence of the Length of the Peptidic Chains. Three sets of lipopeptides $C_{17}Sar_p$, $C_{18}K_p$ and $C_{18}E_p$ with a constant length of the paraffinic chains and different values of the degree of polymerization p of the peptidic chains have been studied. For the 3 types of structures (lamellar, hexagonal and body-centred cubic) when p increases the lattice parameter, the characteristic parameter of the hydrophilic peptidic chains and the specific surface all increase. The characteristic parameter of the hydrophobic paraffinic chains decreases (12,13).

Emulsifying Properties

The emulsifying properties of lipopeptides were tested in many oil/water systems. The oil was aromatic such as toluene and styrene, paraffinic such as decane and dodecane, or a base product of the cosmetic industry. Lipopeptides give 3 types of emulsions : macro-

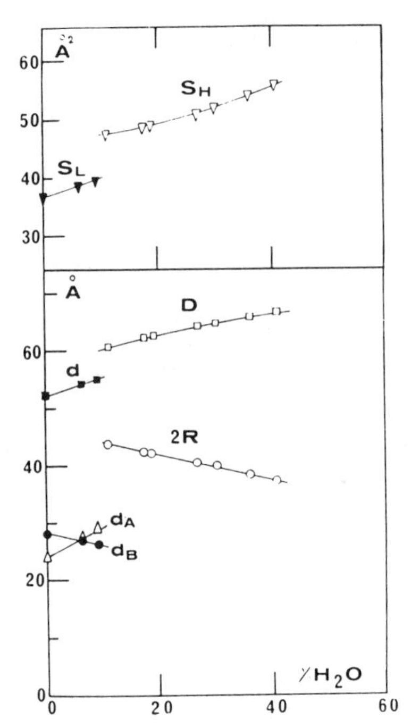

Figure 1. Variation of the parameters of the lamellar and hexagonal structures of lipopeptide $C_{18}K_2$ versus water concentration.

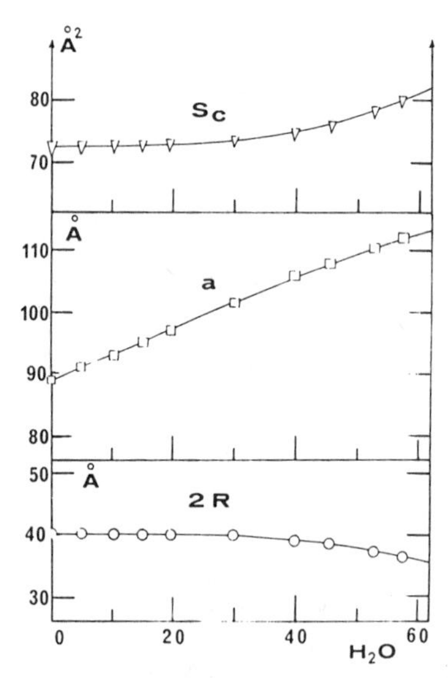

Figure 2. Variation of the parameters of the body-centered cubic structure of lipopeptide $C_{17}Sar_{50}$ versus water concentration.

emulsions with droplet diameters higher than 1000 nm, miniemulsions with droplet diameters between 100 and 400 nm, and microemulsions with droplet diameters between 10 and 100 nm. We will sum up the main results obtained with the 3 types of emulsions.

Type of Emulsions. The emulsions vary from a fluid milklike to a thick cream, depending upon the nature of the oil, the ratio oil/water, the nature and the concentration of the lipopeptides.

All the emulsions are of the oil in water (O/W) type as shown by the dilution method, the selective dyes method and the conductivity method (9). Such a result is in agreement with the HLB values (between 8 and 15) of the lipopeptides (9).

Stability of Emulsions. Stabilities varying from 2 months to more than 24 months were found.

The main factors governing the stability of the emulsions are : the length of the paraffinic chains, the nature, the degree of polymerization and the end group of the peptidic chain and the nature of the oil.

Influence of the Length of the Paraffinic Chains. The comparative study of lipopeptides with the same peptidic chains but with paraffinic chains containing from 12 to 18 carbon atoms has shown that emulsions obtained with lipopeptides with a paraffinic chain containing 12 or 14 carbon atoms are stable for less than 4 days. Emulsions obtained using lipopeptides with paraffinic chains containing 16 or 18 carbon atoms are stable for longer than 3 months. Nevertheless, the domain of stability of the emulsions is slightly higher for the 18 carbon atoms paraffinic chains than for the 16.

C_{12} and C_{14} lipopeptides are more soluble in water than their C_{16} and C_{18} counterparts and can be more easily carried into the aqueous phase destroying the hydrophilic-hydrophobic equilibrium between the lipopeptides and the water and oil phases.

Influence of the Peptidic Chain End Group. The influence of the nature of the peptidic chain end group has been studied for 4 types of lipopeptides : liposarcosine, lipolysinebromhydrate, lipoglutamic acid sodium salt and lipohydroxyethylglutamine ; similar results were obtained.

Figure 3 illustrates the results obtained in the emulsification of the O/W isopropyl myristate/water system by a liposarcosine with a degree of polymerization of 2 and a 18 carbon atoms paraffinic chain. The domain of stability of the emulsions decreases from the liposarcosine chlorhydrate (dotted line), to the liposarcosine (full line) and to the liposarcosine whose terminal amine function has been acetylated (points).

When the polarity of the end group of the peptidic chain decreases, the emulsifying power of the lipopeptide decreases.

Influence of the Nature of the Amino-acid Side Chain. The influence of the nature of the amino-acid side chain on the emulsifying properties of lipopeptides is illustrated in the figure 4 for the system O/W isopropyl myristate/water, and for 4 lipopeptides with a paraffinic chain containing 18 carbon atoms and with a degree of polymerization of 2 for the peptidic chains. The emulsions stability region decreases from lipohydroxyethylglutamine (full line) to lipolysine (dotted line), to lipoglutamic acid (points) and to hydroxypropylglutamine (crosses). This behaviour is related to the

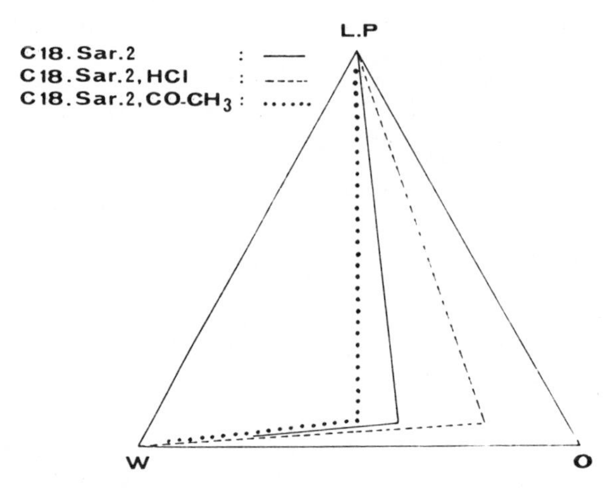

Figure 3. Influence of the nature of the end group of the peptidic chains on the domain of stability of emulsions.

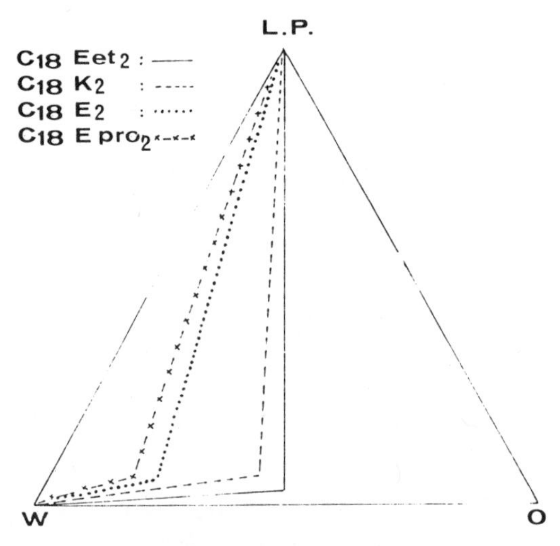

Figure 4. Influence of the nature of the amino-acid side chain on the domain of stability of emulsions.

number of hydrophilic sites per amino-acid residue and to the hydrophilicity of the amino-acid residue.

For less polar oils such as dodecane similar results were obtained. The hydrophilicity of the peptidic chain plays an important part in the emulsifying properties of lipopeptides for common oils.

Influence of the Degree of Polymerization of the Peptidic Chain. The influence of the degree of polymerization p of the peptidic chains of lipopeptides has been studied for 3 types of lipopeptides : liposarcosine, lipolysine bromhydrate and lipoglutamic acid sodium salt with paraffinic chains containing 18 carbon atoms. The degree of polymerization p of the peptidic chains was between 1 and 10. Lipopeptides with $p \geqslant 5$ give emulsions stable for less than 3 days. Lipopeptides with p = 1, 2 or 3 give stable emulsions. Figure 5 illustrates the results obtained in the case of the system O/W isopropyl myristate/water and of lipolysine bromhydrates. One can see that the domain of stability of the emulsions decreases when p increases from 1 to 3. For lipoglutamic acid sodium salts similar results were obtained. For liposarcosines, in contrast, the domain of stability of emulsions is nearly the same for p = 1,2 and 3.

Influence of the Oil Nature. The influence of the oil's nature on the domain of stability of emulsions O/W has been studied for 3 types of lipopeptides with a paraffinic chain containing 18 carbon atoms : lipolysine bromhydrate, liposarcosine chlorhydrate and liposarcosine. The results obtained were similar for the 3 types of lipopeptides. The figure 6 illustrates the result obtained in the case of liposarcosine with a degree of polymerization of 2. The domain of stability of the O/W emulsions decreases from isopropyl myristate and butyl stearate (full line) to Mygliol (doted line), to Cosbiol (cross), to dodecane (points) and to styrene. Polar oils are easier to emulsify than non polar and aromatic ones.

Miniemulsions

Lipopeptides emulsify with difficulty aromatic oils such as styrene or toluene. Furthermore they are not able to emulsify some oils such as vaseline, ricin, wheat germ and silicon oils. To emulsify such oils we have used a binary emulsifying system consisting of a mixture of a fatty alcohol (cetyl alcohol) and a ionic lipopeptide (liposarcosine chlorhydrate, lipolysine bromhydrate or lipoglutamic acid sodium salt). With concentrations of lipopeptide and cetyl alcohol of 1 to 3 % we have obtained miniemulsions similar to those obtained by El Aasser and al. with sodium lauryl sulfate and cetyl alcohol (10).

Type of Miniemulsions. All the miniemulsions were found to be of the O/W type.

Stability of the Miniemulsions. The main factors governing the stability of the miniemulsions are : the length of the paraffinic chains, the nature of the peptidic chains, the degree of polymerization of the peptidic chains, the mixed emulsifier concentration, the molar ratio lipopeptide/cetyl alcohol, the nature of the oil and the method of preparation of the miniemulsions.

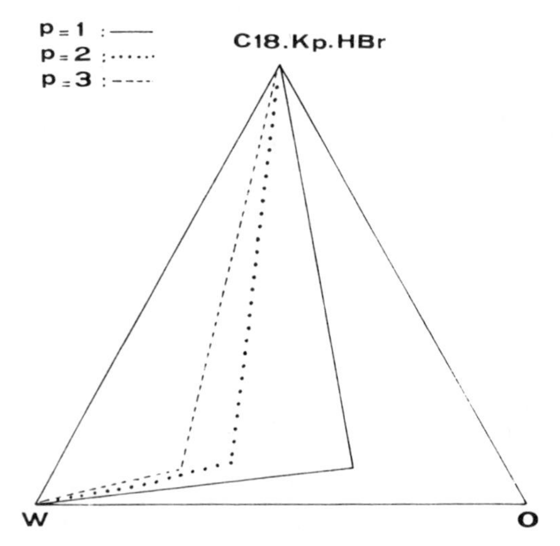

Figure 5. Influence of the degree of polymerization of the peptidic chains on the domain of stability ef emulsions.

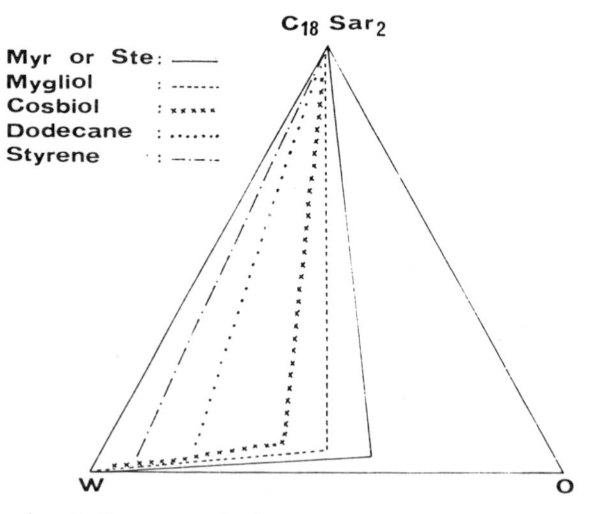

Figure 6. Influence of the nature of the emulsified oil on the domain of stability of emulsions.

Influence of the Length of the Paraffinic Chains. Lipopeptides with paraffinic chains containing 18 or 16 carbon atoms give with oils very difficult to emulsify (styrene, ricin oil, vaseline oil, silicon oil..) miniemulsions stable for more than 2 months. In contrast, lipopeptides with paraffinic chains containing 14 or 12 carbon atoms give miniemulsions stable for less than 20 days.

Influence of the Nature of the Peptidic Chain. The stability of the miniemulsions decreases when liposarcosine is replaced by lipolysine and lipoglutamic acid. This result is illustrated in Table I for two different oils (styrene and wheat germ oil) for a mass ratio O/O+W of 0.2. Although the amounts of lipopeptide and cetyl alcohol used to obtain miniemulsions have been increased from liposarcosine to lipoglutamic acid and to lipolysine, the stability of the miniemulsions decreases from more than 60 days for liposarcosine, to 12 days for lipolysine and 10 days for lipoglutamic acid.

Table I. Influence of the Nature of the Lipopeptide.

	Styrene			Wheat Germ Oil		
	LP %	$C_{16}OH$ %	Stab.	LP %	$C_{16}OH$ %	Stab.
$C_{18}Sar_2$,HCl	2	1	60	2.5	1.5	60
C_{18}(K,HBr)$_2$	4	1.3	12	4.5	2.0	12
C_{18}(E,Na)$_2$	3	1.2	10	4.0	1.8	10

LP : lipopeptide ; $C_{16}OH$: cetyl alcohol ; Stab.: stability in days.

Influence of the Degree of Polymerization. The amounts of lipopeptide and cetyl alcohol necessary to stabilize miniemulsions increases with the degree of polymerization p of the peptidic chains of lipopeptides. The Table II illustrates this behaviour for the systems styrene/lipolysine bromhydrate and wheat germ oil/lipoglutamic acid sodium salt.

Table II. Influence of the Degree of Polymerization of the Peptidic Chains.

O/O+W = 0.2		p = 2	p = 3
Styrene	C_{18}(K,HBr)$_p$	4.5 %	7.0 %
	$C_{16}OH$	2.0 %	3.0 %
Wheat germ oil	C_{12}(E,Na)$_p$	4.0 %	6.0 %
	$C_{16}OH$	1.5 %	2.5 %

Influence of the Mixed Emulsifier Concentration, The stability of the miniemulsions increases with the concentration of the mixed emulsifier. Table III illustrates these results for the emulsification of two different oils : styrene and silicon oil by liposarcosine chlorhydrate with a degree of polymerization of 2.

Table III. Influence of the Concentration in Weight of Mixed Emulsifier on the Stability of Miniemulsions. LP = $C_{18}Sar_2$,HCl

$\frac{O}{O+W}$ = 0.2	$\frac{LP}{O+W}$ %	$\frac{C_{16}OH}{O+W}$ %	Stability in days
Styrene	2.0 4.0	1.0 2.0	100 > 180
Silicon oil	2.5 5.0	1.5 3.0	70 > 120

Influence of the Molar Ratio Lipopeptide/Cetyl Alcohol. As already shown by different authors in the case of classical emulsifiers such as sodium lauryl sulfate (10,14,15), the mixed emulsifier system lipopeptide/cetyl alcohol gives stable miniemulsions for molar ratios LP/C_{16}OH between 2/1 and 1/3.

Influence of the Ratio Oil/Water. The mass ratio O/O+W cannot exceed 60 % for aromatic oils and 50 % for cosmetic oils. The amounts of lipopeptide and cetyl alcohol necessary to obtain miniemulsions vary only slightly with the amount of oil in the system oil/water.

Influence of the Oil Nature. Aromatic oils (toluene and styrene) are easier to miniemulsify than cosmetic oils. They require smaller amounts of lipopeptide and cetyl alcohol to give stable miniemulsions.

Influence of the Method of Emulsification. Miniemulsions prepared by the second method are more stable than those prepared with the preemulsification method.

The Table IV gives 2 examples of the influence of the method of emulsification on the stability of miniemulsions prepared with the same amounts of lipopeptide (LP) and cetyl alcohol (C_{16}OH). When the oil is styrene the stability of the miniemulsion increases from 60 to 240 days. When the oil is vaseline oil, the stability of the miniemulsion increases from 20 to 70 days.

To understand the difference of stability of the miniemulsions prepared by the two methods we have studied the miniemulsions by freeze fracture and electron microscopy (9) and measured the size of the particles. For all the systems studied, the dimensions of the particles are smaller for the miniemulsions prepared by the second method ; for instance in the case of styrene (Table IV) the diameter Ø of the particles is 310 nm against 840 nm. Such a difference in the particle size explains the difference of stability of the miniemulsions prepared by the two methods.

We have also found by freeze fracture and electron microscopy that the size of the particles increases when the amount of emulsified oil increases, and decreases when the concentration of the mixed emulsifier increases. As an example, for the system $C_{18}Sar_2$, HCl/cetyl alcohol/styrene/water, the average diameter Ø of the particles increases from 700 nm to 900 nm when the mass ratio O/(O+W) increases from 0.2 to 0.5, but decreases from 900 nm to 750 nm when the surfactant concentration increases from 2 % to 4 %.

Table IV. Influence of the Method of Preparation of Miniemulsion. LP = $C_{18}Sar_2$,HCl.

O/O+W = 0.2		Method 1	Method 2
Styrene	LP	2 %	2 %
	$C_{16}OH$	1 %	1 %
	Stability	60 days	240 days
	Ø	840 nm	310 nm
Vaseline oil	LP	2.5 %	2.5 %
	$C_{16}OH$	1.5 %	1.5 %
	Stability	20 days	70 days

Microemulsions

Ionic and nonionic lipopeptides give microemulsions when an aliphatic alcohol or amine with less than 7 carbon atoms is used as cosurfactant. Nevertheless the best cosurfactants are butanol, propanol, butylamine and propylamine.

The enlargement of the microemulsion region is influenced by the following factors :

- Increase in the hydrophilicity and HLB of the lipopeptide
- Decrease of the paraffinic chain length
- Increase in the polarity of the peptidic chains end group.

An example is the case of liposarcosine/n-butylamine/isopropylmyristate/water system. Microemulsions region increased from $C_{18}Sar_2$ to $C_{18}Sar_{10}$ as the hydrophilicity of the peptidic chain increased, from $C_{18}Sar_2$ to $C_{18}Sar_2$,HCl as the polarity of the peptidic chain increased, and from $C_{18}Sar_2$,HCl to $C_{12}Sar_2$,HCl as the paraffinic chain length decreased.

Literature Cited

1. Langevin, D. Mol. Cryst. Liq. Cryst. 1986, 138, 259.
2. Ekwall, P. Advances in Liquid Crystals 1975, 1, 1.
3. Hendricks, Y.; Charvolin, J. J. Phys. 1981, 42, 1427.
4. Gallot, B.; Haj Hassan, H., unpublished results.
5. Gallot, B.; Douy, A. French Patent 82 159 76, 1982 ; Chem. Abstr. 1984, 171762h.
6. Gallot, B.; Douy, A. U.S. Patent 4 600 526, 1986.
7. Gallot, B. In Liquid Crystalline Order in Polymers; Blumstein, A., Ed.; Academic: New York, 1978; Chapter 6, 191.
8. Douy, A.; Mayer, R.; Rossi, J.; Gallot, B. Mol. Cryst. Liq. Cryst. 1969, 7, 103.
9. Haj Hassan, H. Ph.D. Thesis, Orléans University, Orléans, France, 1987.
10. El-Aasser, M.S.; Lack, C.D.; Choi, Y.T.; Min, T.I.; Vanderhoff, J.W.; Fawkes, F.M. Colloids and Surfaces 1984, 12, 79.

11. Luzzati, V.; Mustacchi, H.; Skoulios, A.; Husson, F.; Acta Crystallog. 1960, 13, 660.
12. Douy, A.; Gallot, B. Makromol. Chem. 1986, 187, 465.
13. Gallot, B.; Douy, A.; Haj Hassan, H. Mol. Cryst. Liq. Cryst. 1987, 153, 347.
14. Grimm, W.L.; Min, T.I.; El-Aasser, M.S.; Vanderhoff, J.W. J. Colloid Interface Sci. 1983, 94, 531.
15. Brouwer, W.M.; El-Aasser, M.S.; Vanderoff, J.W. Colloid and Surfaces 1986, 21, 69.

RECEIVED August 10, 1988

LIQUID CRYSTALS

Chapter 9

Mesophase Formation in Solutions of Semirigid Polymers

Poly(γ-benzyl-L-glutamate) Liquid Crystals

D. B. DuPré and R. Parthasarathy[1]

Department of Chemistry, University of Louisville, Louisville, KY 40292

Experimental volume fractions for the appearance of uniformly anisotropic solutions of poly-γ-benzyl-glutamate are discussed in terms of rigid rod theories (Flory) and a more recent extension of the virial approach to wormlike chains of restricted deflection (Khokhlov-Semenov-Odijk). The inherent flexibility of this α-helical polymer is quantified by the persistence length which is solvent dependent. Both theoretical approaches predict lower volume fractions for mesophase separation than are generally observed. Theories are in better correspondence with experiment when the comparison is partitioned for molecular contour lengths above and below the measured persistence length. Virial and lattice models approach one another in the high molecular weight limit when the axial ratio of the macromolecule in the lattice formulation is taken as the persistence length over the diameter of the semiflexible chain.

Molecular elongation has long been recognized to be the dominant influence in the formation of orientationally ordered fluids (liquid crystals). Theories of mesophase stability have been developed that treat constituent molecules as hard or soft rods of length to breadth ratio, $x = L/d$. When this aspect ratio becomes large as in the case of extended chain polymers, flexibility along the main axis of the molecule will undoubtedly be introduced at some point and redound on mesophase formation. Mechanisms of chain flexibility include kinks in repeat units at special sites (broken rods) and loss of persistence due to cumulative effects of slight deviations of monomers from parallelism along the chain contour. The consequence of chain flexibility on liquid crystal formation may be

[1]Current address: Department of Chemistry, Temple University, Philadelphia, PA 19122

0097–6156/89/0384–0130$06.00/0

examined through the minimum volume fraction of polymer necessary for the appearance of an anisotropic phase in the case of lyotropic liquid crystals or the nematic-isotropic transition temperature in the case of thermotropic liquid crystals. The inherent flexibility of a polymer may be quantified through the persistence length (1,2), q, which is amenable to measurement by numerous physical methods.

The influence of chain flexibility on lyotropic mesophase formation in cellulose and poly-n-hexylisocyanate polymers has been examined in papers by Ciferri, Conio, Krigbaum *et al.* (3-10). Cellulose and its derivatives are polymers that might best conform to the broken rod model, whereas poly-n-hexylisocyanate is a persistent polymer (q ~ 50-300 Å) with worm-like character. In this article we will present a similar analysis from our studies of the rod-like polymer, poly-γ-benzyl-L-glutamate, PBG, (q ~ 700-1500 Å). Data have been collected in our laboratory on phase equilibria and persistence lengths of this polymer over a wide range of molecular weights and solvents (11-18; and Parthasarathy, R.; Houpt, D.J.; DuPré, D.B., *Liq. Cryst.*, in press). Measurements of elastic properties (18-24; and Parthasarathy, R.; Houpt, D.J.; DuPré, D.B., *Liq. Cryst.*, in press) of lyotropic liquid crystals of this polymer have revealed that the macromolecule, though extraordinarily rigid in solvents that support the α-helical conformation, is best described geometrically as a long and slender, semi-rigid rod.

Persistence Length Measurements

Persistence lengths of polymers in solution may be determined by a variety of physical methods including light scattering, dielectric constant and translational diffusion measurements, flow birefringence and viscometry. Values obtained from these studies for a given polymer are not often in good agreement even in the same solvent (25,26). The discrepancies are most likely due to the different and invariably inadequate models employed for the geometric shape and dynamics of the macromolecule. The simplifying assumption of rigid rod behavior is apparently inadequate at low molecular weights (26). For this discussion, therefore, we will rely on persistence length data obtained in a previous study (Parthasarathy, R.; Houpt, D.J.; DuPré, D.B., *Liq. Cryst.*, in press) using a viscometric method based on the wormlike chain model of polymer dynamics. Intrinsic viscosity, [η], measurements as a function of polymer molecular weight were performed for PBG solutions in a variety of solvents that have not been implicated in substantial macromolecular association. (A small amount of trifluoroacetic acid is added to dioxane to prevent such association.) The hydrodynamic theory of Yamakawa and Fujii (27), with numerical corrections as reported by Conio *et al.* (7), allows the determination of the persistence length from logarithmic plots of [η] versus molecular weight. Values so obtained in five solvents studied are listed in the first column of Table I. The Yamakawa-Fujii method requires an estimate of the diameter, d, of the wormlike chain. We have taken d to be 18 Å for PBG in all of the solvents except m-cresol where a better fit of data to theory

Table I. Volume Fractions at Mesophase Separation for Solutions of Polybenzylglutamate as a Function of Solvent and Molecular Weight

Solvent		Mol. wt.	N_p	ϕ_B expt	Odijk		Flory		
					ϕ_i	ϕ_a	v_p^*	v_p^*, with $x = q/d$	v_p^*, with $x = 2q/d$
Nitrobenzene		85K	0.49	0.25	0.179	0.183	0.232		
		110K	0.63	0.215	0.156	0.158	0.182		
q = 1200 Å	L<q	150K	0.83	--	0.136	0.140	0.139		
d = 18 Å	L>q	190K	1.08	--	0.123	0.127	0.108		
		210K	1.20	--	0.118	0.123	0.0976		
		260K	1.48	0.17	0.111	0.116	0.0792	0.116	0.059
m-Cresol		85K	0.65	--	0.170	0.174	0.196		
	L<q	110K	0.84	0.277	0.151	0.155	0.153		
	L>q	150K	1.11	0.26	0.135	0.140	0.116		
q = 900 Å		190K	1.44	0.25	0.124	0.136	0.0902		
d = 15 Å		210K	1.60	--	0.120	0.127	0.0817		
		260K	1.98	0.23	0.114	0.122	0.0662	0.129	0.0656
Cyclohexanone	L<q	85K	0.83	0.34	0.234	0.239	0.232		
	L>q	110K	1.08	--	0.211	0.218	0.182		
		150K	1.43	0.28	0.192	0.201	0.139		
q = 700 Å		190K	1.86	0.26	0.178	0.190	0.108		
d = 18 Å		210K	2.05	--	0.174	0.187	0.0976		
		260K	2.54	0.18	0.167	0.181	0.0792	0.195	0.100

Table I. continued

Solvent		Mol. wt.	N_p	ϕ_B expt	Odijk ϕ_i	Odijk ϕ_a	Flory v_p^*	Flory v_p^*, with $x = q/d$	Flory v_p^*, with $x = 2q/d$
Pyridine		85K	0.39	--	0.164	0.169	0.232		
		110K	0.50	0.312	0.140	0.144	0.182		
q = 1500 Å		150K	0.67	--	0.121	0.123	0.139		
d = 18 Å		190K	0.87	--	0.107	0.110	0.108		
	L<q	210K	0.96	0.206	0.103	0.106	0.0976		
	L>q	260K	1.19	0.184	0.950	0.0987	0.0792	0.0937	0.0474
Dioxane 6% TFA		85K	0.53	--	0.186	0.190	0.232		
		110K	0.68	0.291	0.163	0.166	0.182		
	L<q	150K	0.91	0.29	0.143	0.147	0.139		
q = 1100 Å	L>q	190K	1.18	--	0.130	0.135	0.108		
d = 18 Å		210K	1.31	0.247	0.125	0.131	0.0976		
		260K	1.62	0.219	0.118	0.124	0.0792	0.127	0.0644

was obtained with a diameter of 15 Å. [X-ray measurements place the diameter of the α-helix of PBG between 15-25 Å (25).] While all our values for q are obtained from the same, consistent method, it should be noted that the Yamawaka-Fujii fit produces q values that are often lower than those obtained by light scattering (12). The estimated error in q, as determined by this procedure, is ±100 Å.

Polymer Composition at Phase Separation

The molecular weight dependence of the critical concentration for the establishment of uniformly anisotropic solutions of PBG is shown in Table I for various solvents that we have examined. Volume fractions (ϕ) of polymer quoted in this compilation correspond to the B-point in the nomenclature of Robinson (28-29). The B-point differs from the A-point, a lower concentration where the anisotropic phase just begins to form and is in equilibrium with isotropic polymer solution.

There are two major theoretical approaches to the understanding of orientational ordering of extended chain polymers: one is a model wherein rod-shaped particles are confined to a lattice under the restriction that segments may not overlap with one another; the other utilizes a virial expansion that accounts for mutual orientational correlations and interactions of pairs, triplets, etc. of elongated particles at varied concentrations.

The lattice model was introduced by Flory (30-31) who calculated the free energy of the system from principles of statistical mechanics. Flory found that a collection of rods becomes metastable with respect to an ordered phase at a critical volume fraction, v_p^*, of rods that is dependent on the axial ratio, $x = L/d$, where L is the length of an individual rod and d is its diameter. An approximate expression for this dependence is given by (30):

$$v_p^* \approx (8/x)(1-2x) \qquad (1)$$

which holds within 2% for values of $x \geq 10$. On diagrams of phase equilibria versus x, this metastable v_p^* is bracketed by rod compositions in coexisting isotropic and anisotropic phases of v_p and v_p' (Flory's notation), corresponding to Robinson's ϕ_A and ϕ_B points, respectively. The order parameter, S, at the transition to the ordered phase is predicted to be ~ 0.95 (31).

Values of v_p^* calculated on the basis of $x = L/d$ are presented in Table I. For PBG in the α-helical state, the contour length of the macromolecule may be calculated on the basis of a 1.5 Å unit translation along the extended axis per amino acid residue (25). The diamter of the PBG α-helix has been estimated variously to be between 15-25 Å as noted above (25). Values of d used in the determination of persistence lengths are listed in the solvent categories of Table I. Values of v_p^* calculated from Equation 1 using $x = L/d$ are invariably too low for the solvents of this study. Better agreement with the lattice theory has been reported for critical volume fractions of PBG in dimethylformamide and m-cresol (32). Experimental volume fractions for liquid crystal formation of PBG in dioxane are lower, however, than those calculated from Flory's theory (33). PBG is known to undergo extensive intermolecular association in DMF with moisture content and dioxane, making the true aspect ratio of the statistical object in solution these solvents uncertain. If the effective macromolecular length increases due to association, the volume fraction for incipient liquid crystal formation would be expected to be lower.

A modification of the lattice model to account for semi-rigid polymers replaces the contour length, L, in the axial ratio with the Kuhn segmental length (34,35): *i.e.*, x is replaced with $x_K = 2q/d$ in Equation 1. This notion treats the flexible polymer as a freely jointed chain with rigid segments of length 2q. This model is expected to apply only for high molecular weights where at least several Kuhn segments would be established between chain ends in each molecule. Another alternate interpretation of the lattice results for semi-flexible polymers would be to replace the contour length of the polymer with the persistence length itself, at least for molecular weights where the contour length of polymer is greater than, but not too much larger than q (*i.e.*, where $L \gtrsim q$). This treatment is tantamount to supposing that for chains just a little longer than q, it is only the persistent part of the chain that matters in liquid crystal formation. The contour lengths of PBG at the highest molecular weights available in our studies do not exceed 1.5 to 2.6 times our measured values of q. Values of v_p^* so calculated for PBG are also shown in Table I. It is found that v_p^* volume fractions calculated on the basis of x = q/d are closer to the experimental values at high molecular weight than those calculated on the basis of $x = x_K$. Data for PBG in cyclohexanone provides the best correspondence when x = q/d. In both interpretations of chain flexibility, the calculated volume fractions are, however, generally too low in the lattice model [*cf.*, Table II].

The free energy of a collection of long rods may also be calculated in a virial expansion in the density which takes into account overlap of pairs, triplets, quartets, *etc.*, of rods in ever closer proximity as the volume fraction of rods increases. This approach, due to Onsager (36), relies on the reasonableness of truncation of the series at the second, or perhaps third, virial coefficient. For hard rods, the contribution of the third virial coefficient goes as ~10 d/L and hence is small for rods with axial ratios x > 10 (37). The development is quantitatively reliable for long extended chain polymers (x > 100) which undergo anisotropic phase separation at volume fractions in the range ~0.1 to 0.3. (The Onsager theory would be exact for infinitely long rods.) In this approach, the volume fractions of polymer in the coexistence region in isotropic and anisotropic phases, ϕ_i and ϕ_a, respectively, were found to be (36):

$$\phi_i = 3.34/x \quad (2)$$

$$\phi_a = 4.49/x \quad (3)$$

The order parameter at this phase transition (which is first order) is S = 0.8 (36) for large axial ratios.

Polymer flexibility has been introduced into the virial consideration in extensions of Khokhlov and Semenov (38,39) and,

TABLE II. Ratio of Theoretical to Experimental Volume Fractions at Mesophase Separation in the High Molecular Weight Limit

Solvent	Odijk		Flory		
	ϕ_i	ϕ_a	v_p^* with $x = L/d$	v_p^* with $x = q/d$	v_p^* with $x = 2q/d$
Nitrobenzene	0.65	0.68	0.47	0.68	0.35
m-Cresol	0.50	0.53	0.29	0.56	0.29
Cyclohexanone	0.93	1.01	0.44	1.08	0.56
Pyridine	0.52	0.54	0.43	0.51	0.26
Dioxane/6% TFA	0.54	0.57	0.36	0.58	0.29

more recently, of Odijk (40). In these theories, contortions of worm-like chains are subject to a mean ordering field established by neighboring chains at concentrations where chains might physically overlap as they rotate or undulate about their centers of mass. For chains, or chain segments, longer than a characteristic length, the ordering field constrains deflections to be within a cone of solid angle, $\sqrt{\langle\theta^2\rangle}$. The deflection length, λ, is related to $\langle\theta^2\rangle$ and the persistence length of the polymer through:

$$\lambda = q \langle\theta^2\rangle/2 < q \tag{4}$$

The effect of restricted undulation on the worm-like character of the polymer is to make orientational ordering along the entire expanse of the macromolecule more difficult than it would be if the chain where perfectly straight and rigid. On entering a nematic condition, segments of a persistent chain greater than λ in length must be aligned not only at the beginning and the end, but also at all contour points in between. This is an additional restriction on orientational ordering of elongated, semiflexible particles, not incorporated into the rigid rod model or the extension to freely jointed assemblies of rigid rodlets of the Kuhn length. With this additional entropy cost, it follows that a higher concentration of segments will be necessary to achieve liquid crystallinity in a collection of wormy chains.

Khokhlov-Semenov and Odijk have developed expressions for the concentration for coexisting isotropic and anisotropic phases of flexible macromolecules. For very long chains where $L/q \rightarrow \infty$, these concentrations (given as volume fractions of polymer here) are:

$$\phi_i = 5.4/(q/d) \tag{5}$$

$$\phi_a = 6.2/(q/d) \tag{6}$$

with a calculated order parameter of S = 0.61 in the anisotropic phase at this point. The persistence length replaces the contour length of the polymer in the axial ratio of equations of the form of Equations 2 and 3. Generally, one must also go to higher concentrations of flexible polymer chains to achieve liquid crystal phase separation, obtaining less inherent order in the anisotropic component.

For polymers of finite length (40),

$$\phi_i = (d/q)c_i$$

$$c_i = \frac{3.34 + 5.97\,N_p + 1.585\,N_p^2}{N_p(1 + 0.293\,N_p)} \tag{7}$$

$$\phi_a = (d/q)c_a$$

$$c_a = \frac{4.486 + 11.24\,N_p + 17.54\,N_p^2}{N_p(1 + 2.83\,N_p)} \tag{8}$$

where N_p is the number of persistence lengths within the span of a chain contour ($N_p = L/q$ in the notation of this paper). Equations 7 and 8 are plotted in Figure 1 as a function of the more familiar ratio, $x = L/d$ for a selected value of the persistence length and molecular diameter suitable for discussion of PBG liquid crystals. The plot is also compared to the metastable volume fraction v_p^* predicted by the Flory theory for corresponding axial ratios. Higher volume fractions of polymer at phase separation are found in the Khokhlov-Semenov-Odijk theory for contour lengths greater than q. The virial results fall below that of the lattice model of completely rigid rods, however, when $L < q$. [This result is generally true for all q and d values.]

The theoretical effect of a change in contour length (molecular weight) for a given persistent polymer chain is seen in Figures 2 and 3, where the transition width [difference in critical volume fractions, $(\phi_a-\phi_i)/\phi_i$] and order parameter are plotted as a function of N_p. Both of these parameters exhibit dramatic behavior as soon as any flexibility is introduced in a rigid chain (i.e., when $q < \infty$; $N_p > 0$). The width of the coexistence region drops rapidly to a minimum at a contour length of about 0.6 q; the order parameter plunges and reaches a minimum for L about 0.5 q. A considerably more complicated macromolecular length dependence, scaled to the persistence length, is thus embodied in the theoretical approach of Khokhlov-Semenov and Odijk.

Experimental results obtained for PBG solutions are compared with the Odijk calculations in Table I (and Figure 1 for the solvent nitrobenzene). The extended Onsager approach with account for chain flexibility is in better accord with experiment for molecular weights where $L > q$. The lattice model is found to be superior for lengths, $L < q$. Limiting (high molecular weight) values of v_p^* calculated on the basis of $x = q/d$ conform well, however, to those calculated from Odijk's Equation 8. Both approaches generally yield values that are too low in relation to our experimental results, except for the solvent cyclohexanone (cf. Table II).

On the basis of this data and guidance from elements of theories discussed herein, experiment is indicating an important influence of chain flexibility on phase separation in elongated polymers. The influence is not as dramatic, however, as the extended Onsager approach would suggest. Rigid rod behavior appears to hold better over an extended molecular weight region where the macromolecular contour length is below the inherent, but solvent dependent, persistence length of the polymer. Predicted volume fractions rise above $L \gtrsim q$ in the wormlike chain model, yielding values closer to experiment in this region. The lattice model figured on the basis of the persistence length as a single rigid entity of these moderately long macromolecules is in better correspondence with experiment at high molecular weights and is in better conformity with limiting values based on semiflexible Onsager modifications current in the literature.

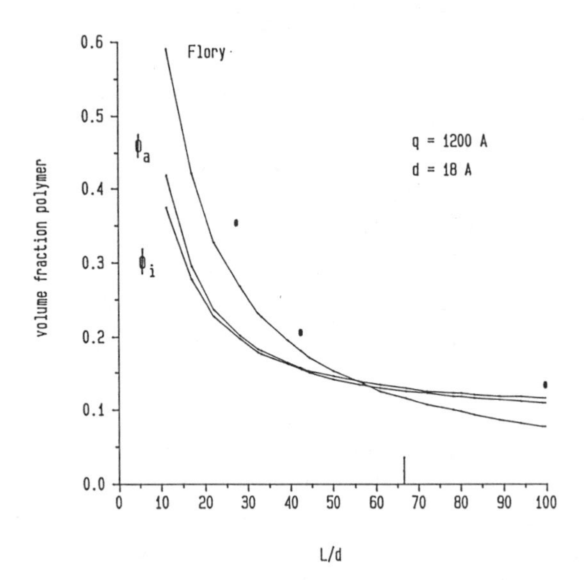

FIGURE 1. Volume fractions of polymer at mesophase separation as function of axial ratio of rod-like or semirigid chain. ϕ_i and ϕ_a, the volume fractions of polymer in coexisting isotropic and anisotropic phases, respectively, are calculated on the basis of Odijk's theory with q = 1200 Å and d = 18 Å. The metastable volume fraction, v_p^*, calculated on the basis of x = L/d from Flory's lattice model is also shown for comparison. Filled circles are experimental data points for PBG in nitrobenzene. The vertical bar on the abscissa is the axial ratio at which L = q for this polymer in this solvent.

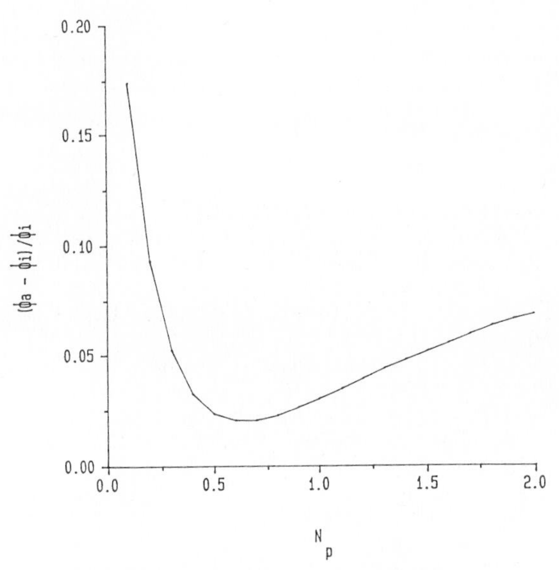

FIGURE 2. Width of biphasic zone expressed as $(\phi_a - \phi_i)/\phi_i$ at mesophase separation calculated on the basis of the Khokhlov-Semenov-Odijk theory as a function of the number of persistence lengths, N_p, in the polymer chain.

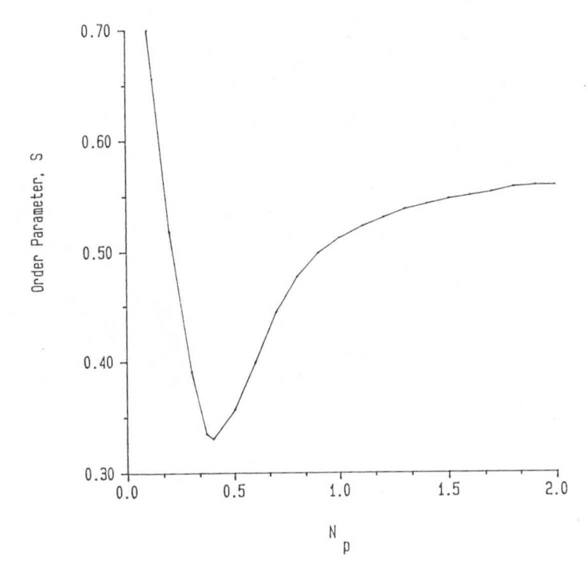

FIGURE 3. Order parameter, S, at mesophase separation from Khokhlov-Semenov-Odijk theory as a function of N_p.

Acknowledgments

This work was supported in part by the National Science Foundation under grant DMR-8412051. Persistence length measurements were performed by Ms. Deborah J. Houpt.

Literature Cited

1. Porod, G. Monatsh. Chem. 1949, 80, 251.
2. Kratky, O.; Porod, G. Recl. Trav. Chim. 1949, 68, 1106.
3. Conio, G.; Bianchi, E.; Ciferri, A.; Tealdi, A.; Aden, M.A. Macromolecules 1983, 16, 1264.
4. Aden, M.A.; Bianchi, E.; Ciferri, A.; Conio, G. Macromolecules 1984, 17, 2010.
5. Bianchi, E.; Ciferri, A.; Conio, G.; Cosani, A.; Terbojevich, M. Macromolecules 1985, 18, 646.
6. Bianchi, E.; Ciferri, A.; Conio, G.; Lanzavecchia, L.; Terbojevich, M. Macromolecules 1986, 19, 630.
7. Conio, G.; Bianchi, E.; Ciferri, A.; Krigbaum, W.R. Macromolecules 1984, 17, 856.
8. Krigbaum, W.R.; Hakemi, H.; Ciferri, A.; Conio, G. Macromolecules 1985, 18, 973.
9. Bianchi, E.; Ciferri, A.; Conio, G.; Krigbaum, W.R. Polymer 1987, 28, 813.
10. Ciferri, A.; Marsano, E. Gazz. Chim. Ital. 1987, 117, 567.
11. Duke, R.W.; DuPré, D.B. J. Chem. Phys. 1974, 60, 2759.
12. DuPré, D.B.; Duke, R.W. J. Chem. Phys. 1975, 63, 143.
13. DuPré, D.B.; Duke, R.W.; Samulski, E.T. J. Chem. Phys. 1977, 66, 2748.
14. Patel, D.L.; DuPré, D.B. Mol. Cryst. Liq. Cryst. 1979, 53, 323.
15. DuPré, D.B. Polym. Sci. Eng. 1981, 21, 717.
16. Subramanian, R.; Wittebort, R.J.; DuPré, D.B. J. Chem. Phys. 1982, 77, 4694.
17. DuPré, D.B. J. Appl. Polym. Sci.: Appl. Polym. Sci. Symp. 1985, 41, 68.
18. Subramanian, R.; DuPré, D.B. J. Chem. Phys. 1984, 81, 4626.
19. DuPré, D.B. In Polymer Liquid Crystals; Ciferri, A.; Krigbaum, W.R.; Meyer, R.B., Eds.; Academic: New York, 1982; Chapter 7.
20. Fernandes, J.R.; DuPré, D.B. Mol. Cryst. Liq. Cryst. (Letters) 1981, 72, 67.
21. Fernandes, J.R.; DuPré, D.B. In Liquid Crystals and Ordered Fluids, Vol. 4; Griffin, A.C.; Johnson, J.F., Eds.; Plenum: New York, 1984; pp 393-399.
22. Fernandes, J.R.; DuPré, D.B. Polymer Preprints 1983, 24(2), 298.
23. DuPré, D.B.; Fernandes, J.R. J. Appl. Polym. Sci.: Appl. Polym. Sci. Symp. 1985, 41, 221.
24. DuPré, D.B.; Fernandes, J.R. In Polymeric Liquid Crystals; Blumstein, A., Ed.; Plenum: New York, 1985; pp 415-422.
25. Block, H. Poly(γ-Benzyl-L-Glutamate) and Other Glutamic Acid Containing Polymers; Gordon and Breach: New York, 1983.
26. Vitovskaya, M.G.; Tsvetkov, V.N. Eur. Polym. J. 1976, 12, 251.

27. Yamakawa, H.; Fujii, M. Macromolecules 1973, 6, 407; 1974, 7, 128.
28. Robinson, C. Trans. Faraday Soc. 1956, 52, 571.
29. Robinson, C. Mol. Cryst. 1966, 1, 467.
30. Flory, P.J. Proc. Royal Soc. London 1956, A234, 73.
31. Flory, P.J.; Ronca, G. Mol. Cryst. Liq. Cryst. 1979, 54, 289.
32. Flory, P.J. Adv. Polym. Sci. 1984, 59, 1.
33. DuPré, D.B.; Samulski, E.T. In Textbook on Liquid Crystals; Saeva, F., Ed.; Marcel Dekker: New York, 1979; Chapter 5.
34. Flory, P.J. Macromolecules 1978, 11, 1141.
35. Matheson, R.R.; Flory, P.J. Macromolecules 1981, 14, 954.
36. Onsager, L. Ann. NY Acad. Sci. 1949, 51, 627.
37. Straley, J.P. Mol. Cryst. Liq. Cryst. 1973, 22, 333.
38. Khokhlov, A.R.; Semenov, A.N. Physica 1982, 112A, 605.
39. Khokhlov, A.R.; Semenov, A.N. Physica 1981, 108A, 546.
40. Odijk, T. Macromolecules 1986, 19, 2313.

RECEIVED August 10, 1988

Chapter 10

Cellulosic Mesophases

S. Ambrosino and Pierre Sixou

Laboratoire de Physique de la Matière Condensée, U.A. 190 Parc Valrose, 06034 Nice Cédex, France

The field of Liquid Crystal Polymers is in rapid expansion. Numerous theoretical and experimental studies have been done to understand the relationship between their chemical structure and physical properties. Cellulosic polymers form a special type of liquid crystal polymer. They exhibit "cholesteric" mesophase properties either on temperature change (thermotropics) or in solution (lyotropics). The influence of several parameters--such as molecular weight, degree of polymerization and substitution, dilution in a mixture of solvents, etc.--on mesophase properties is discussed. They are variations of optical and rheological properties, and of the cholesteric pitch. The second part of the investigation involves mixtures of mesomorphic cellulosic polymers of different degrees of substitution. The influence of degree of substitution on the compatibility of these systems is studied. The case of ternary mixtures--for example, two polymers (mesomorphic or not) in a common solvent--has also been discussed. Experimental results of the study of the different textures, parameters influencing compatibility, etc. of the hydroxypropylcellulose/ethyl cellulose/acetic acid mixture are given.

In the past few years, there have been important developments in the field of polymer mesophases. One of the essential problems is to understand the relationship between the molecular structure and the properties of mesophases and their solids.

Several theoretical works predict the onset of mesophases (1-5) and take into account the molecular structure of polymers. There are two kinds of structures that are often cited: the rigid or semi-rigid linear polymers and the connected ones or those containing mesogenic side chains (6).

0097-6156/89/0384-0142$06.00/0

In both cases, pretransitional properties (7), the type of liquid crystalline phase and order parameters (8-9), elastic (10) and rheological properties (11-12) have been the object of numerous studies or review articles. Some applications in very different fields (high modulus fibers, self-reinforced polymeric materials, electronic display devices, non-linear optics, etc.) have already been realized or are being pursued.

The cellulosic polymers (13) form a type of main chain liquid crystalline polymers of special interest for the following reasons:

- Both thermotropic and lyotropic behavior in a large variety of solvents are found.
- Transition temperatures often lie near room temperature, which allows easy observation.
- The properties of mesophases can easily be changed by often minor modifications of the chemical structure (nature of substituent, degree of substitution, etc.).
- The cholesteric mesophases have interesting optical properties.
- Commercial products can be found in a large range of molecular weights (degree of polymerization extending from a few unities to one thousand or more).

These polymers are thus often considered as good "model systems." In what follows, we shall present experimental data concerning those mesophases and especially their mixtures.

Formation and Properties of Cellulosic Mesophases

The cellulosic derivative chains are relatively rigid. If one approximates the chain by a worm-like chain model, the obtained persistence lengths (q) are of an order of magnitude of several to more than thirty monomers (i.e., more than a hundred and fifty angstroms). The persistence length of the same chain can be modified by an order of magnitude of two, depending on the solvent employed. For a given solvent, a chain in the nature of the substituant can also modify the rigidity of the chain strongly (three times) (14).

In addition to the rigidity, steric effects and flexibility of side groups seems to influence the formation and properties of cellulosic mesophases: they allow (or not) the existence of a mesophase before crystallization, influence the temperature for onset of a mesophase, and contribute to the value of the cholesteric pitch.

Thermotropic Phases.

Figure 1 is a representation of the temperature of the isotropic-cholesteric transition in the case of hydroxypropylcellulose (HPC) and fully acetylated acetoxypropylcellulose (APC) (15). The range of molecular weights is rather large: in the case of HPC, it extends from low masses (2000) to very high ones (1,000,000). Two main features appear:

- A saturation at high molecular weight: comparison with

theory (16) is difficult, but the slope change should occur near a chain length of the order of q (in monomer units).
The presence of hydroxyl groups raises the temperature of isotropic behavior. The persistence lengths of the two compounds are comparable (slightly higher for HPC). The higher APC transition temperature to the isotropic phase is probably due to steric effects of its different substituted part.

The thermotropic phases obtained with these polymers are cholesteric. It has been observed that the pitch decreases with increasing molecular weight indicating a saturation for high molecular weight. The forces governing the cholesteric twist and those determining the stability of the mesophase are of different origin. Yet, it is apparent that they are altered in the same manner with increasing chain length.

For partially substituted HPC, we observed that the temperature at which the mesophase appears is modified by the degree of substitution in HPC. However, this temperature is always between that of non- and fully substituted HPC. The question is whether the cholesteric pitch changes in the same way. We have, therefore, studied the effect of degree of acetylation on the pitch, and an unpredicted effect was observed (17). Hydroxypropylcellulose, having a degree of polymerization about 350, exhibits reflection colors between 140°C and 160°C. Partial acetylation of HPC leads to a dramatic decrease in the temperature at which cholesteric reflection colors are seen (90-110°C). Complete acetylation of hydroxyl groups does not lead to further decrease in the temperature range at which colors are exhibited, but rather, shifts this range to higher temperatures. The optical properties of a partially acetylated sample are not intermediary between a fully and a non-acetylated one. The same observation can be made in the lyotropic case. This observation could mean that the asymmetric part in the interaction potential is not affected in the same way as the symmetrical part when the acetylation degree is modified.

Lyotropic Phases. Lyotropic cellulosic mesophases can be observed in a large variety of solvents with derivatives that can be thermotropic (ethylcellulose, hydroxypropylcellulose, acetoxypropylcellulose, etc.) or not (cellulose acetate). Figure 2 shows an example of a phase diagram of a mixture consisting of a thermotropic polymer--ethylcellulose (EC)--dissolved in an ordinary organic solvent--acetic acid. The anisotropic phase is separated from the isotropic by a two-phase region. The size of the two-phase gap depends on parameters such as temperature, molecular weight, polydispersity, etc. In figure 2, the extent of the two-phase region is not represented, but the average values are indicated.

At constant temperature, when the polymer concentration is increased, anisotropic droplets appear and coalesce. Layers of the cholesteric structure can be observed inside the droplets,

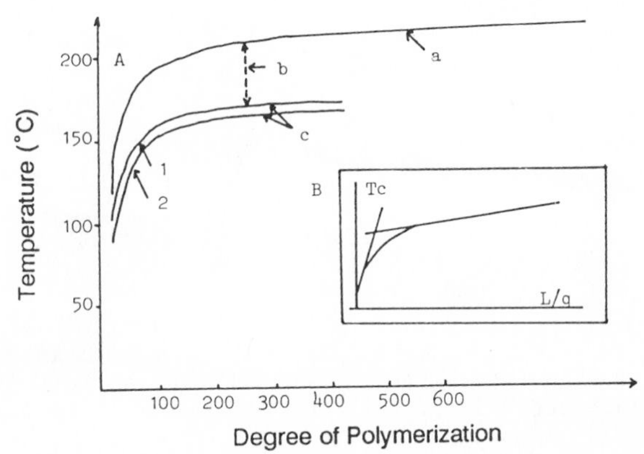

Figure 1: Transition temperature as a function of polymerization degree:
A/ a-Hydroxypropylcellulose (HPC)
b-Acetoxypropylcellulose (APC) partially substituted
c-Acetoxypropylcellulose fully substituted
1-On heating
2-On cooling
B/ Theoretical prediction
L: Chain length
q: Persistence length

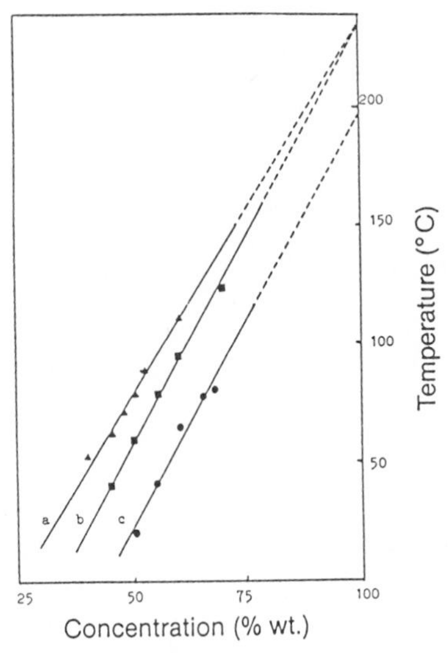

Figure 2: Phase diagram EC/AA
a-Ethoxyl content 48%; viscosity (5% solution in 80/20 toluene/ethanol) : 22cp
b-Ethoxyl content 48%; viscosity (5% solution in 80/20 toluene/ethanol) : 10cp
c-Ethoxyl content 46%; viscosity (5% solution in 80/20 toluene/ethanol) : 100cp

and the state of organization is quite similar to that described by Livolant and Bouligand (18-20) in the case of other cholesteric compounds. When the size of the droplets increases, typical defects of cholesterics (dislocations and disclinations) appear inside the spherulites. These defects can be observed as well when the polymer concentration increases--that is, when the pure cholesteric phase is reached. The effects of temperature depend on the initial polymer concentration when constant. At low concentrations the spherulite diameter decreases while the pitch increases with temperature (21). At very high concentrations, an imperfect planar orientation exists. If the temperature is increased to the isotropic phase and then is slowly decreased, a square mesh texture is observed in the polarizing microscope (22). At high enlargement, the pitch can be seen inside the polygons of the mesh. A schematic representation of these textures has been given by Bouligand (19). The polygonal field texture can be studied as a function of concentration or of salt addition to the mixture (23). An increase in polymer concentration leads to a decrease of the pitch and an increase of the mesh size, possibly due to an increase of the elastic constant. An illustration of this first effect is shown in figure 3. Salt addition also produces a decrease of the pitch and of the mesh size. Typical textures of cellulosic mesophases are shown in figure 4.

The lyotropic mesophase prepared from a polymer dissolution in a mixture of solvents is an interesting case. At a given polymer concentration, the organization of macromolecules is governed by the surface of the droplets and intermolecular interactions. It can be varied by changes in the relative proportions of the solvents.

Textures and rheological properties are strongly influenced by the relative proportions of the two solvents. To emphasize this effect, we studied the case of a mixture consisting of a polymer dissolved in a binary critical liquid mixture (24). Water and isobutyric acid (IBA) have been chosen as solvents because they form a mesophase with HPC and they have a critical temperature, approximately 26°C for an IBA proportion of approximately 0.4. The phase diagram shows different regions: one- or two-phase, mesomorphic, or not. The two-phase region can be segregated. The mesomorphic phase can form a gel or show cholesteric colors (25). The most important information given by this phase diagram is that the behavior of phase separation of IBA/H_2O/HPC system is governed by the extreme preference of HPC for IBA as compared with water. This results in the formation of the ordered phase in dilute HPC solutions as the IBA content in the solvent is reduced. Mesophases can be obtained for low polymer concentrations of a few percent.

Both with thermotropics and lyotropics, a large number of properties can be studied. For example, NMR can be used to obtain the order parameter of the cholesteric phase (26). Rheooptic studies (27) (for example, the variation of the refractive index under shear) describe the way cholesteric phase reorganizes to nematic after orientation by a strong shear.

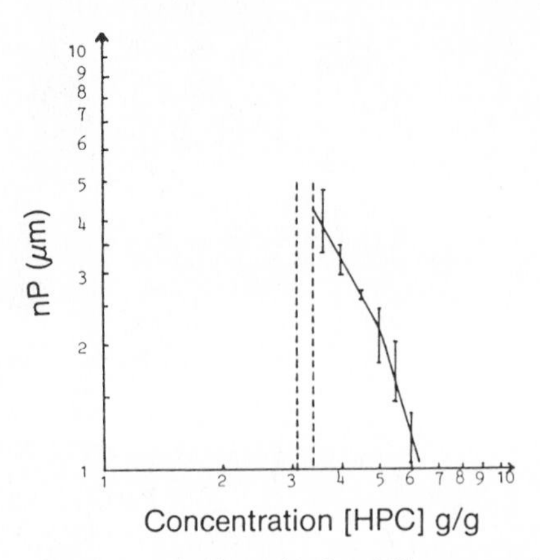

Figure 3: Variation of the pitch with concentration for HPC/ acrylic acid

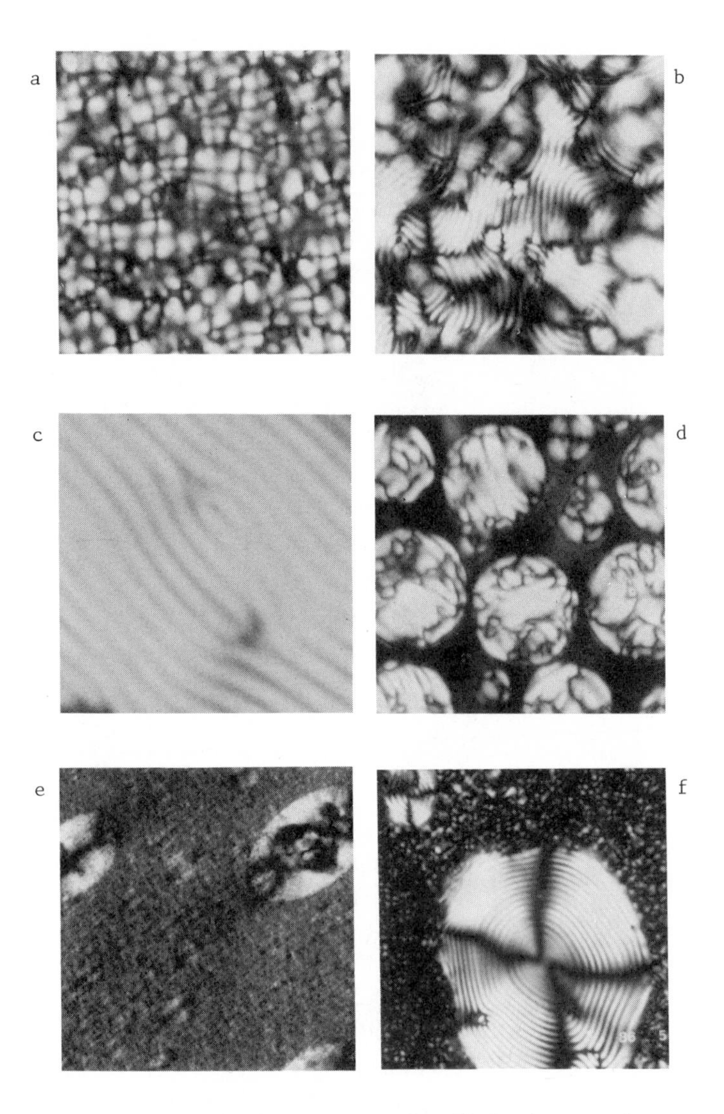

Figure 4: Typical textures of cellulosic mesophases
a) Polygonal field
b) Details of the pitch
c) Defects
d) Spherulites
e) Spherulites in a few field
f) Organization of the layers in a spherulite

Rheological properties have also been studied. It has been observed for some concentrations that the first difference of normal stress becomes negative for a certain range of shear stress. This effect had already been observed in the case of HPC (28). Figure 5 illustrates the case of another cellulosic derivative: ethylcellulose.

Mixture of Two Mesomorphic Cellulosic Polymers

The development of blends and polymer mixtures is a rapidly expanding field. Indeed, a mixture is often a simple way to obtain an ensemble of properties so that a high-performance material corresponding to a given application can be constructed with a cost only slightly higher than that of the constituents. Non-mesomorphic polymer mixtures have been extensively studied (29-31), particularly their thermodynamic stability. Incompatibility is the rule in most cases, especially for high molecular masses of the constituents, unless specific interactions exist between the different constituents.

Phase diagrams can be calculated from the free energy of mixing. An expression has been derived (16) which includes several terms: the entropy of mixing, the isotropic interactions, the entropy of chain conformation and the contribution of anisotropic interactions. The first two terms are characteristic of non-mesomorphic polymer mixtures, while the last represent additional contributions related to chain rigidity and liquid crystal properties. The details of the method and numerical examples (32) have been published elsewhere. The extension to the case of ternary mixtures--for example, two polymers (mesomorphic or not) in a common solvent--can also be described. Figure 6 shows very schematically typical phase diagrams. In this figure, the case of a mixture of two non-mesomorphic polymers and a mesomorphic polymer in a non-mesomorphic polymer have been added for comparison.

An interesting experimental example is given by HPC and EC mixtures in a common solvent: acetic acid. The textures of these two polymers are well identified as polygonal fields, but the size of the mesh is very different. Therefore, it is possible to recognize droplets of a rich phase of EC in a background of a rich phase of HPC (and inversely). Figure 7 shows the phase diagram at 25°C. An important point deals with two-phase regions that can be either isotropic-isotropic or anisotropic-isotropic or anisotropic-anisotropic. Phase diagrams obtained by changing the relative proportions of the two polymers and the temperature are shown in figure 8, a, b, c, for a given concentration of solvent (30%, 40%, or 50%). There is incompatibility of the two polymers, and an increase in temperature does not allow to observe a critical point of phase separation. The solvent evaporates and the polymers degrade at high temperature.

This incompatibility appears to be the rule for cellulosic polymers studied here. HPC can be acetylated to give APC and

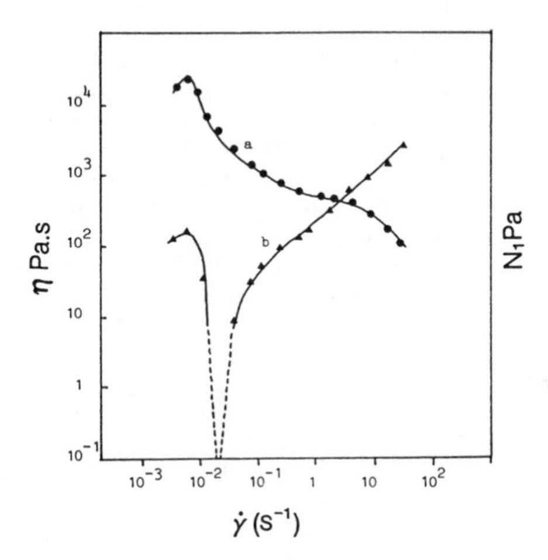

Figure 5: Viscosity (η) and first normal stress difference N_1 as a function of shear rate ($\dot{\gamma}$)
a-Variation of η with $\dot{\gamma}$
b-Variation of N_1 with $\dot{\gamma}$
for EC 40% in acetic acid

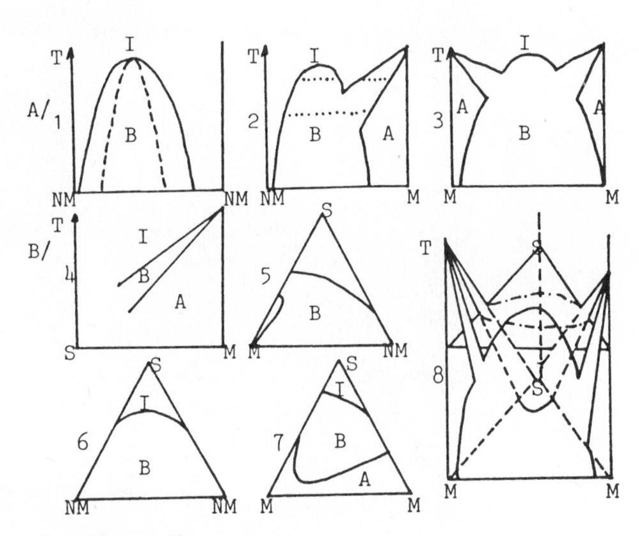

Figure 6: Phase diagrams
A/Pure polymers:
1-Two "non mesomorphic" (NM)
2-One "mesomorphic" (M)/one "non mesomorphic"
3-Two "mesomorphic"
B/Lyotropics:
4-One "mesomorphic"/solvent
5-One "mesomorphic"/one "non mesomorphic"/solvent
6-Two "non mesomorphic"/solvent
7-Two "mesomorphic"/solvent
8-Temperature effect
I: Isotropic phase
b: Biphasic phase
a: Anisotropic phase
The X axis represents the concentration of the mixture of the two polymers (mesomorphic or non mesomorphic).
The Y axis represents the temperature.

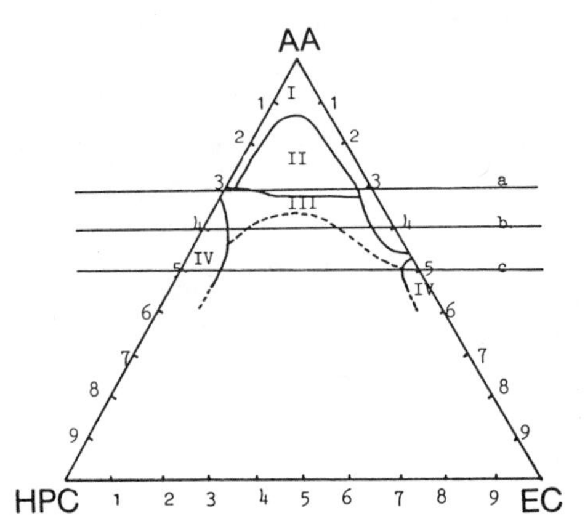

Figure 7: Ternary phase diagram HPC/EC/AA at room temperature (EC: ethoxyl content 48% viscosity (5% solution) : 10 cp

I-isotropic phase

II-isotropic/isotropic phases

III-anisotropic/isotropic phases or anisotropic/ anisotropic phases

IV-anisotropic phase

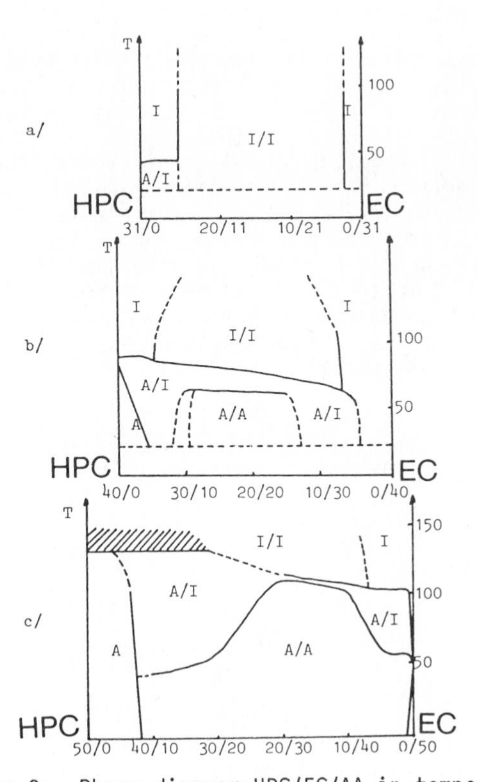

Figure 8: Phase diagram HPC/EC/AA in temperature
EC: ethoxyl content 48%
viscosity (5% solution) : 10 cp
a-HPC/EC=31%(AA)
b-HPC/EC=40%(AA)
c-HPC/EC=50%(AA)

the question is: Are these two compounds immiscible whatever the degree of acetylation? The experimental results show that HPC and APC (with a degree of polymerization of 175) are incompatible. However, partially substituted APC (67%) and completely substituted HPC seem to be compatible. In general, an increase in the molecular weight leads to phase separation. Here, shortening the polymer chains appears to lead to phase separation, possibly due to a more rapid diffusion.

The influence of the method of preparation was also studied. Preparation of the mixture with or without solvent (methanol), leads to different textures--thus showing the greater affinity of the polymer containing the larger percentage of hydroxypropyl groups for methanol.

In conclusion, although cellulosic mesophases begin to be well characterized, the need remains to fully understand the mechanisms behind the formation of these mesophases. In so far as mesophases are "labeled," due to their birefringence and to their special cholesteric textures, the study of their mixtures can be of great interest for the understanding of phase separation phenomena as well as the dynamics of nucleation and growth. The birefringence can also be an advantage in the study of surface of interfacial phenomena.

Literature Cited

1. Wang, X. J.; Warner, M. J. Phys. A 1986, 19, 2215-2227.
2. Warner, M.; Gunn, J.M.F.; Baumgartner, A.B. J. Phys. A 1985, 18, 3007-3026.
3. Ten Bosch, A.; Maissa, P.; Sixou, P. J. Phys. Lett. 1983, 44, 105-111.
4. Ten Bosch, A.; Maissa, P.; Sixou, P. Phys. Letters 1983, 9, 298-300.
5. Ten Bosch, A.; Maissa, P. In Fluctuations and Stochastic Phenomena in Condensed Matter; Garido, L., Ed.; Lecture Note in Physics No. 268; Springer-Verlag: Berlin, 1988; pp 333-350.
6. Wang, X.J.; Warner, M. J. Phys. A 1987, 20, 713-721.
7. Gilli, J.M.; Maret, G.; Maissa, P.; Ten Bosch, A.; Sixou, P.; Blumstein, A. J. Phys. Lett. 1985, 46, 329-334.
8. Ten Bosch, A.; Maissa, P.; Sixou, P. J. Chem. Phys. 1983, 79, 3462-3466.
9. Ten Bosch, A.; Maissa P.; Sixou, P. In Polymeric Liquid Crystals; Blumstein, A., Ed.; Plenum: New York, 1985, pp 377-387.
10. Ten Bosch, A.; Sixou, P. J. Chem. Phys. 1987, 86 (11), 6556-6559.
11. Doi, M.; Edwards, S.F. The Theory of Polymer Dynamics; International Series of Monographs on Physics 73; Oxford Science, 1986; Chapter 10, pp 350-380.
12. Maissa, P.; Ten Bosch, A.; Sixou, P. J. Polym. Sci. Polym. Lett. 1983, 21, 757-765.
13. Gilbert, R.D.; Patton, P.A. Progr. Polym. Sci. 1983, 9, 115-131.

14. Sixou, P.; Ten Bosch, A. In Cellulose Structure Modification and Hydrolysis; Young, R.A.; Powell, R.M., Eds; John Wiley: New York, 1986; Chapter 12, p 205.
15. Lavins, G.V.; Sixou, P. Mol. Cryst. Liq. Cryst. 1988, in press.
16. Ten Bosch, A.; Pinton, J.F.; Maissa, P.; Sixou, P. J. Phys. A. Math. Gen. 1987, 20, 4531-4537.
17. Laivins, G.V.; Sixou, P.; Gray, D.G. J. Polm. Sci. 1986, 24, 2779-2792.
18. Livolant, F. J. Phys. (France) 1987, 48, 1051-1066.
19. Bouligand, Y. J. Phys. (France) 1972, 33, 525-547.
20. Bouligand, Y. J. Phys. (France) 1972, 33, 715-736.
21. Fried, F.; Sixou, P. to be published.
22. Fried, F.; Sixou, P. J. Polym. Sci. 1984, 22, 239-247.
23. Seurin, M.J.; Sixou, P. Eur. Polym. J. 1987, 23 (1), 77-87.
24. Laivins, G.V.; Sixou, P. J. Polym. Sci. 1988, 26, 113-125.
25. Laivins, G.V.; Sixou, P. J. Polym. Sci. B. Polym. Phys. 1988, 26, 113-125.
26. Dayan, S.; Fried, F.; Gilli, J.M.; Sixou, P. J. Appl. Polym. Sci. Appl. Polym. Symp. 1983, 37, 193-210.
27. Gilli, J.M.; Laivins, G.V.; Sixou, P. Liq. Cryst. 1986, 1 (5), 491-497.
28. Fried, F.; Sixou, P. Mol. Cryst. Liq. Cryst. 1988, in press.
29. Walsh, D.J.; Higgins, J.S.; Maconnachie, A. In Polymers Blends and Mixtures; Nato Asi Series; E Applied Sciences N°89; Martinus Nijhoff: Dordrech, 1985.
30. Paul, D.R.; Newman, S. In Polymer Blends; Academic: New York, 1978; Vol. 1, Chapters 1-4, p 7.
31. Olabisi, O.; Robeson, L.M.; Shaw, M.T. In Polymer-Polymer Miscibility; Academic: New York, 1979.
32. Maissa, P.; Ten Bosch, A.; Sixou, P. Presented at the International Conference on Liquid Crystal Polymers, Bordeaux, July 1987; paper 3/15.

RECEIVED August 26, 1988

Chapter 11

Lyotropic Mesophases of Cellulose in the Ammonia–Ammonium Thiocyanate Solvent System

Effects of System Composition on Phase Types

Kap-Seung Yang, Michael H. Theil, and John A. Cuculo

Fiber and Polymer Science Program and Department of Textile Chemistry, North Carolina State University, Raleigh, NC 27695–8302

Cellulose can be dissolved in an NH_3/NH_4SCN solvent over a solvent composition range 70-80% (w/w) NH_4SCN. Beyond certain minimum cellulose concentrations, depending on solvent composition, liquid crystalline phases will form. Cellulose having a 210 degree of polymerization will form a mesophase at a concentration of 3.5% (w/v) at 25°C in a solvent containing 70% NH_4SCN. The minimum cellulose concentration for mesophase formation was highest when the solvent contained 75.5% NH_4SCN. Under most conditions, these mesophases were the twisted cholesterics that typically develop from chiral mesogens. Their helicoidal pitch was higher in solvents richer in NH_4SCN and at lower cellulose concentrations. Nematic phases, which can be construed as untwisted cholesteric phases, were easily prepared from solutions in which the cellulose concentration range was 8-16% and the solvent was 24.5% NH_3 and 75.5% NH_4SCN (w/w). Evidence suggests that the nematic phase forms when interchain cellulose interactions are suppressed. Fibers extruded from nematic solutions were more highly oriented, more fibrillar in texture and appreciably stiffer than those from cholesteric solutions. The former had moduli comparable to that of Fortisan, a strong regenerated cellulose fiber.

Mesophases containing high molecular weight components have been known for many years. The earliest studied involved polyelectrolytes such as tobacco mosaic virus (*1*). Approximately 30 years ago mesophases were observed

0097–6156/89/0384–0156$08.00/0

involving the nonionic helical polymer poly-γ-benzyl-L-glutamate (PBLG) (*2,3*). Shortly thereafter, Flory (*4,5*) developed a lattice theory that described mesophase formation in a system of stiff chains and low molecular weight solvent.

Stimulated by the formation of ultrahigh-strength/high-modulus fibers from nematic solutions of rigid macromolecules (*6-8*), a great deal of attention has been focused on the structure and properties of macromolecular mesophases. More recently such anisotropic solutions have also been obtained from solutions of semi-rigid polymers such as cellulose (*9-11*) and cellulose derivatives (*12*). Concurrently there has been much activity in the area of new solvents for cellulose (*10,13,14,15*). Hudson et al. reported that ammonia (NH_3)/ammonium thiocyanate (NH_4SCN) mixtures to be an excellent solvents for cellulose. They demonstrated that this family of solvents has several significant practical advantages including conveniently low vapor pressures at room temperature (*16*). Cellulose in NH_3/NH_4SCN does not appear to degrade, and essentially no other chemical reaction between cellulose and the solvent occurs (*17*). Isotropic cellulose solutions have been wet spun to produce fibers with properties close to those of conventional rayon (Liu, C. K.; Cuculo, J. A. submitted to *J. Polym. Sci., Polym. Chem. Ed.*). Recently a cholesteric mesophase including.cellulose was formed in a solvent consisting of 27% NH_3 and 73% of NH_4SCN by weight (*18*). A cholesteric structure consists of a set of quasi-nematic layers (*19*) whose individual directors are turned through a fixed angle from one layer to the next. The layers which are turned through 2π are equivalent, and the distance between these particular layers is defined as the pitch of a helicoidal cholesteric structure.

The chirality of the constituent molecules of a thermotropic mesophase determines its cholesteric sense. When equal moles of mirror image isomers are mixed (a racemic mixture), the twisting power falls to zero and the cholesteric structure changes to a compensated nematic one (*20*). More surprising is that the cholesteric nature of a lyotropic mesophase of PBLG depends on the nature of the solvent. For example, PBLG mesophases in chloroform and in dioxane are right handed cholesterics while in methylene chloride and in ethylene chloride, they are left handed. In a mixture of dioxane and methylene chloride (20% dioxane, v/v), the cholesteric structure changed to a compensated nematic structure (*21*). Samulski and Samulski (*22*) have discussed the mechanism of formation of the compensated nematic phase by considering the magnitude of the dielectric constants of PBLG and solvent molecules. Toriumi et al.(*23,24*) also reported that the conditions for formation of the compensated nematic phase was markedly influenced by temperature and by the hydrogen bonds formed between the carbonyl ester group in PBLG and m-cresol. Similarly, the current work describes transformations of lyotropic mesophases as effected by varying the ratio of NH_3/NH_4SCN, cellulose concentration, degree of polymerization (*DP*) of cellulose, storage time, and temperature.

Experimental

Materials. Origin and characteristics of the cellulose samples used are listed in Table I. The low molecular weight cellulose sample A, was prepared from American Enka rayon by hydrolysis in 4 M HCl for 15 minutes and then washed with an eluotropic series of solvents. Sheets of samples C and D were shredded in a Wiley Mill (40 mesh) before use. Sample B, with degree of polymerization (*DP*) of 210, was used in this study except where noted. Ammonium thiocyanate (Witco Chemical) was dissolved in condensed anhydrous ammonia (Air Products). Unless otherwise noted, all chemicals were ACS reagent grade.

The solvent compositions chosen for study were in the range of 70 to 80% (w/w) NH_4SCN (0.33 to 0.47 mole fraction) reported by Cuculo and Hudson (*15*) to be good solvents for cellulose. Prior to use, all cellulose samples and the NH_4SCN were dried under vacuum for 4 hours at 80°C except where noted. Solvents relatively richer in NH_4SCN were prepared by adding NH_4SCN to a stock solution.

Table I. Characteristics of Cellulose Samples Investigated

Sample	Supplier	*DP*	*DP* determination method
A	American Enka (hydrolyzed)	35	*Gel permeation chromatography
B	Whatman Chemical Ltd. (cellulose powder, CC41, microgranular grade)	210	*ASTM method (Designation D1795-62) (*25*)
C	Avtex Fiber	450	*TAPPI method 230 os-76
D	ITT Rayonier (Cellunier Q, sulfite pulp)	765	*Sedimentation coefficient of a nitrated sample

*Determined by the supplier

Solution Preparation. A known amount of cellulose was combined with NH_3/NH_4SCN solvent in a tightly capped centrifuge tube and was evenly dispersed by use of a Vortex Genie agitator. The tube was then placed in dry-ice for 24 hours and subsequently warmed in hot water (ca. 50°C) to bring about flow. The solution was then inspected under a polarizing microscope. Upon complete dissolution of the cellulose, the sample was routinely frozen and brought to and maintained at 25.0±0.1°C. This ensured a similar thermal history for each sample and established a precise reference time frame. All solution concentrations are reported as percents (g of cellulose per 100mL of NH_3/NH_4SCN).

Microscopical Examination. Solutions were aged at 25°C. A portion of each cellulose solution was carefully placed between a microscope slide and a cover slip and then examined between the crossed polarizers of an Olympus microscope, Model BHSP. Cholesteric pitch sizes were measured from photomicrographs showing the characteristic fingerprint pattern (*3*).

He-Ne Laser Beam Diffraction. Samples were placed in a rectangular quartz cell (path length 2 mm). The ringed diffraction pattern of a transmitted He-Ne laser beam (Metrologic, λ=632.8 nm) recorded on Polaroid film also served to identify the cholesteric structures. Cholesteric pitch sizes were calculated by applying the sample to film distance and diameter of the diffraction ring to the Bragg equation. They were fairly close to those found from photomicrographs.

Optical Rotatory Power Measurement. Cellulose solutions were placed between microscope cover slips separated by a Parafilm spacer (thickness 0.127 mm). Optical rotatory measurements of the solutions were made with a Perkin-Elmer 241 polarimeter using a sodium light source (598 nm) and a mercury light source (356,436,546, and 578 nm). The specific rotation at a given wavelength of light $[\theta]$ was calculated from the observed rotation θ by

$$[\theta] = \frac{\theta}{l\,C} \tag{1}$$

where l = cell length in dm and C = concentration of sample solution in g/mL.

Centrifugation. The volume fraction of the anisotropic phase Φ was determined by centrifugation at 15,000 rpm (RCF=26,890) and 25.0±0.5°C in a du Pont Sorval Model RC-5 centrifuge. Separation of phases was indicated by the appearance of a visible phase boundary. The period of centrifugation to assure complete separation of the phases was determined visually. Prior to centrifugation, the centrifuge tube was marked to indicate the total volume of the cellulose solution before centrifugation V_t. After centrifugation, the supernatant (isotropic phase) was decanted. The tube with the anisotropic phase still present was then quickly filled to the mark with fresh solvent. The solvent was then removed and its volume which corresponded to the volume of the isotropic phase V_{iso} was measured. The anisotropic volume V_{aniso}, V_{iso}, and Φ are interrelated by

$$V_{aniso} = V_t - V_{iso} \tag{2a}$$

$$\Phi = \frac{V_{aniso}}{V_t} \tag{2b}$$

Minimum Cellulose Concentration for Mesophase Formation. The approximate value of the minimum cellulose concentration for mesophase formation (MCCM) was visually determined by preliminary experiments. For the precise determination of the MCCM, a series of cellulose solutions was prepared in concentration increments 0.5g/100mL in the neighborhood of the incipience of mesophase formation. The MCCM were determined by averaging two cellulose concentrations, the highest concentration showing the isotropic phase and the lowest concentration showing the anisotropic phase. The presence of the mesophase was not always immediately discernible near the MCCM. However, it could be identified upon high speed centrifugation. Biphasic systems, consisting of coexisting isotropic and anisotropic phases, formed in the solutions prepared from solvent containing 70.0-74.0% NH_4SCN were separable by centrifugation. However, mesophase solutions from solvent containing over 75.0% NH_4SCN were not. In these cases, polarizing microscopical observations were substituted for centrifugation. The MCCM were determined by the averaging procedure just described.

Fiber Formation. Solutions were extruded from a wet spinning apparatus (Bradford Univ. Research Ltd., equipped with a six hole spinneret, hole diameter 0.23 mm) by wet or air-gap spinning. Methanol was the coagulant. Alternatively, fibers were extruded from a syringe into methanol through a 2.5 cm air-gap.

Mechanical Properties of Fiber. Tenacities and initial moduli were obtained for fiber samples using an Instron tensile tester Model TT-B. The extension rate was 2.54

cm/min and the gauge length was 2.54 cm. Measurements were made under standard conditions, 25°C and 65% relative humidity.

Scanning Electron Microscopy (SEM). An ISI-40 SEM was used to take photomicrographs of cellulose fibers fractured in liquid nitrogen and coated with a gold-palladium alloy.

Results and Discussion

Microscopical Observation. The cellulose/NH_3/NH_4SCN mesophase solutions displayed several types of patterns. The patterns are variously indicative of cholesteric, nematic, conjugated, and aggregated anisotropic phases. Due to the twisted arrangement of their molecular layers, cholesteric mesophases normally show striated patterns under a polarizing microscope. In this study, characteristic cholesteric phase patterns included ones resembling fingerprints (Figures 1a and b), and banded spherulites (Figure 4). The nematic phase showed high birefringence originating from its uniaxial molecular orientation. It may also display Schlieren (Figure 2a), thread-like (Figure 2b), and uniformly dispersed highly birefringent (Figure 1c) patterns. The term, conjugated patterns (Figures 3a,b and c), is used to represent combinations of different types of patterns. The aggregated anisotropic phase pattern (Figure 5) was formed in the solution prepared from hydrolyzed cellulose of *DP* 35. The occurrence of specific pattern depends upon solvent composition, cellulose concentration, storage time, cellulose *DP*, prior chemical treatment such as hydrolysis, and storage temperature.

General Behavior of Solutions. Several interesting observations were made when varying the ratio of NH_3/NH_4SCN in the cellulose/NH_3/NH_4SCN system at polymer concentrations of 12, 14, and 16g/100mL. With increasing NH_4SCN concentration in the solvent, the following progression of changes in appearance were noted: the pitch size, as indicated by fingerprint pattern spacings, increased (from Figure 1a to 1b), it then transformed to a uniformly dispersed highly birefringent solution (Figure 1c), which finally yielded to a conjugated combination of spherulitic patterns and uniformly dispersed high birefringence (Figure 1d).

A fingerprint pattern was observed in each mesophasic solution in which the NH_4SCN content in the solvent was less than approximately 75.5% (w/w). The required time for formation of a fingerprint pattern depended on solvent composition, cellulose concentration, *DP* of cellulose and storage temperature. It decreased with decreasing NH_4SCN concentration and increasing cellulose

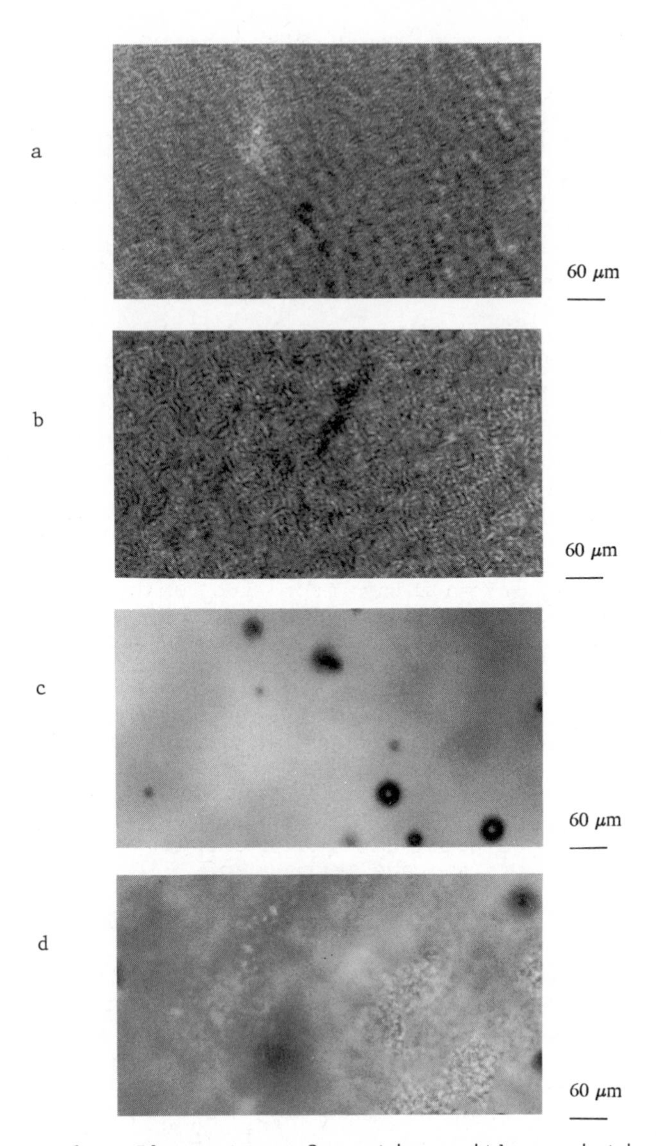

Figure 1. Phase transformation with variation of solvent composition: 27.0/73.0 (**a**), 26.0/74.0 (**b**), 24.5/75.5 (**c**), 23.0/77.0 (**d**); cellulose concentration, 12g/100mL; DP, 210; storage time at 25°C, 10 days; **c** and **d** with 1° red plate; ↗↙, direction of the highest refractive index of the 1° red plate.

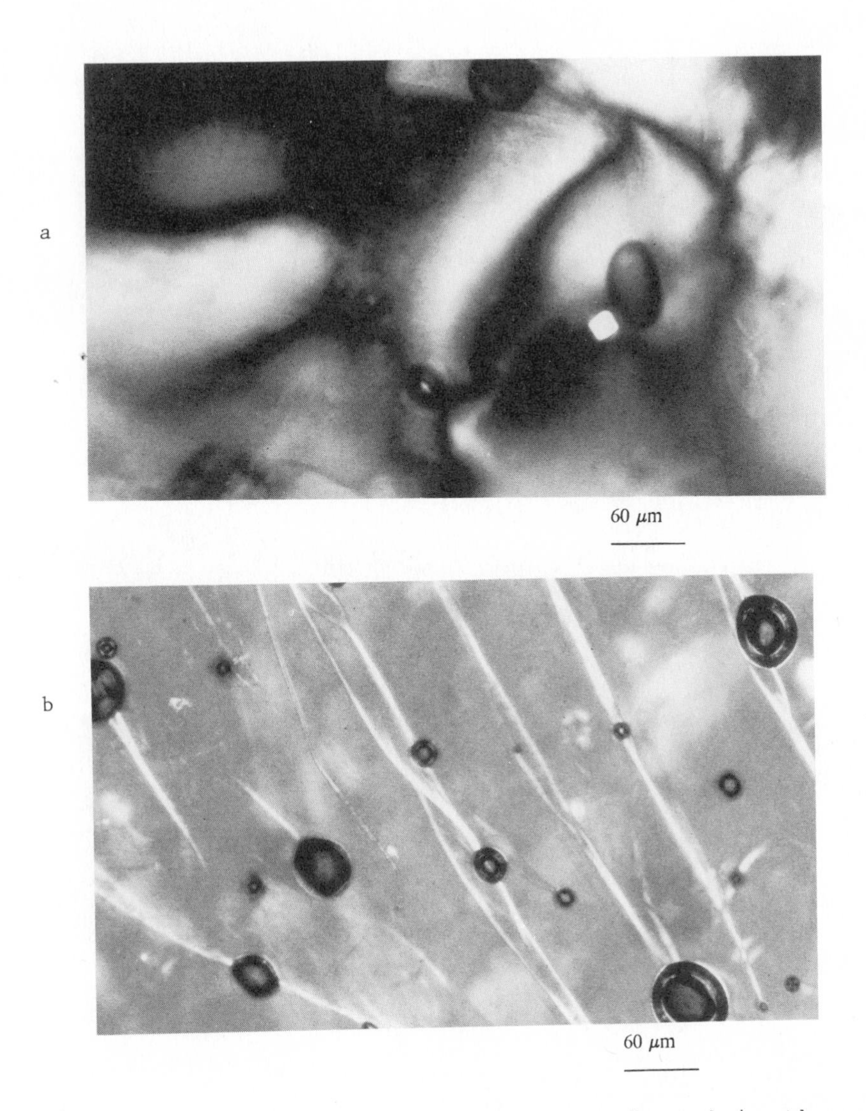

Figure 2. Nematic phase patterns formed in the cellulose solutions prepared from solvent composition of 24.5/75.5: cellulose concentration, 14g/100mL; DP, 450 (**a** and **b**); immediately after dissolution.

a

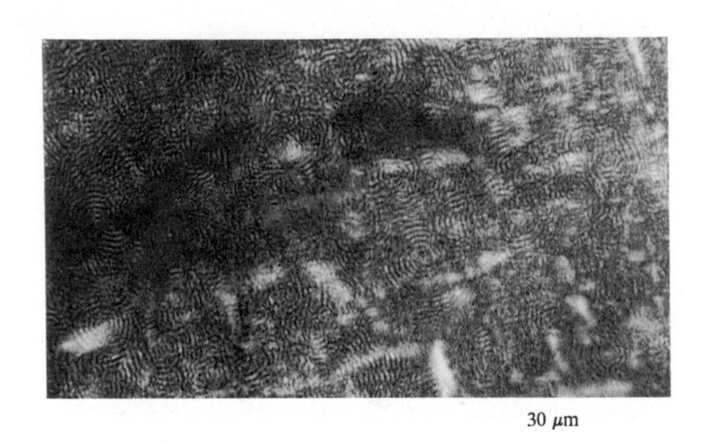

Figure 3a. Conjugated pattern: solvent composition, 27.0/73.0; cellulose concentration, 12g/100mL; DP, 210; storage time at 25°C, 14 days.

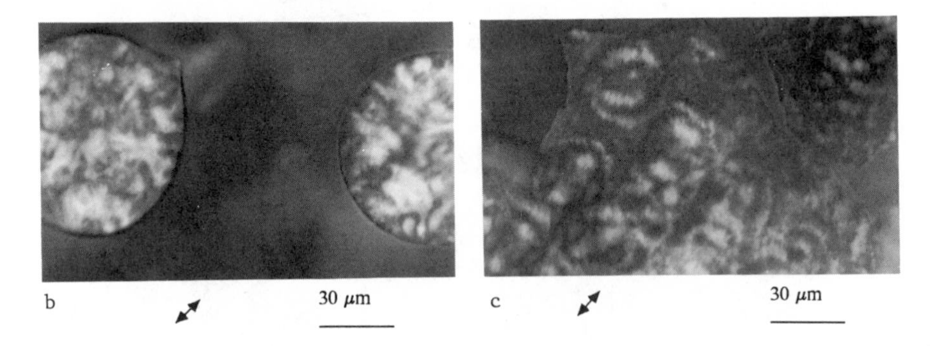

Figure 3b&c. Conjugated patterns formed at relatively high cellulose concentration: solvent composition, 24.5/75.5; cellulose concentration, 18g/100mL (**b**), 20g/100mL (**c**); DP, 210; storage time at 25°C, 10 days, with 1° red plate; ⤢, direction of the highest refractive index of the 1° red plate.

60 μm

Figure 4. Spherulitic pattern: solvent composition, 24.0/76.0; cellulose concentration, 14g/100mL; DP, 210; storage time at 25°C, 20 days.

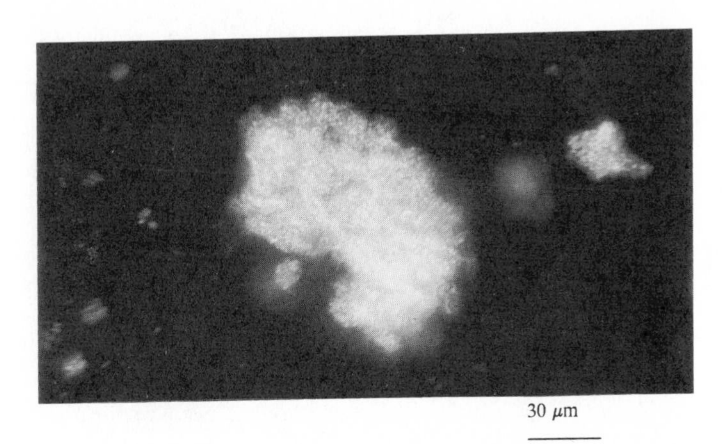

30 μm

Figure 5. Aggregated anisotropic phase pattern: solvent composition, 23.6/75.0/1.4, $NH_3/NH_4SCN/H_2O$; cellulose concentration, 20g/100mL, immediately after dissolution.

concentration. Supporting evidence that the anisotropic solutions with fingerprint patterns incorporated cholesteric mesophases included their high negative specific rotations when compared with the low specific rotation of +35° for cellobiose in the 30.0/70.0 solvent. Further, a linear relationship was found between specific rotation and the reciprocal of the square of the wave length of the incident light. Such a relationship is interpretable by de Vries' theory (*26*) to show that these mesophases were cholesteric. Finally, the ring diffraction pattern generated by the incident He-Ne laser beam also provided evidence of a cholesteric phase (*3,27,28*).

Nematic phase birefringence, which was observed in the cellulose solutions made from the 24.5/75.5 solvent, appeared immediately following dissolution and persisted throughout a two week period. The uniformly dispersed birefringent patterns were the most prevalent ones in the nematic solutions at all DPs studied, but Schlieren and thread-like patterns indicating nematic phases were readily observed in solutions of *DP* 450 cellulose.

Conjugated patterns formed in solutions of moderate cellulose concentration, 12g/100mL-16g/100mL, in solvents containing less than 75.5% NH_4SCN (Figure 3a), or in fairly concentrated cellulose solutions, 18 and 20g/100mL, prepared from solvents containing approximately 75.5% NH_4SCN (Figures 3b and c).

The aggregated phase pattern was evident within a few minutes after dissolution for solutions prepared from hydrolyzed cellulose of *DP* 35 in 23.6/75/1.4 $NH_3/NH_4SCN/H_2O$ (see Figure 5).

Spherulitic phase patterns (Figure 4) appeared in solutions prepared from solvents containing greater than approximately 75.5% NH_4SCN or in solutions prepared from solvent containing approximately 75.5% NH_4SCN if they were stored over two months. The density of spherulite content increased with increasing NH_4SCN concentration and cellulose concentration. Anisotropic cellulose solutions containing the spherulites showed a high negative specific rotation. On that basis, the cellulose solutions with the spherulitic patterns were established as cholesteric.

Varying the solvent composition also affects the MCCM and the anisotropic phase volume fraction. The MCCM showed a maximum at 75.5% NH_4SCN over the range of 0.70 to 0.79 weight fraction NH_4SCN in NH_3/NH_4SCN (Figure 6).

The turbidity of the cellulose solutions increased with increasing deviation in either direction from the 24.5/75.5 solvent composition (Figure 7) and correlated directly with the observed magnitude of $[\theta]$. Increasing storage time increased the turbidity of the solutions prepared from solvent containing either more or less than

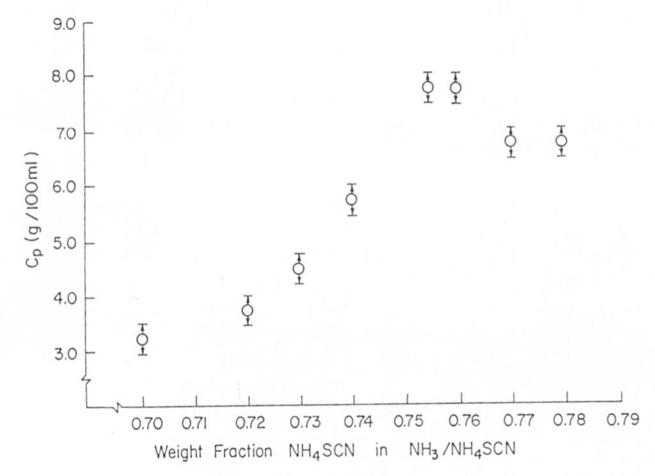

Figure 6. Minimum cellulose concentration for mesophase formation as determined by solvent composition: DP, 210; storage time at 25°C, 30 days.

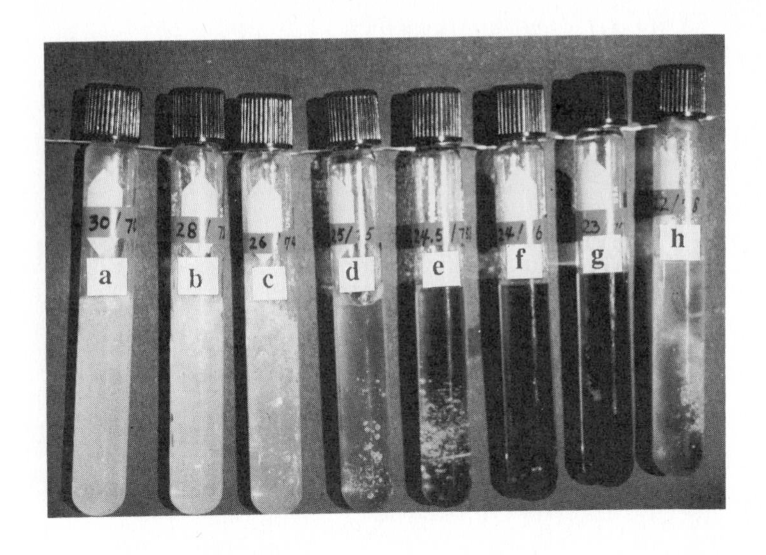

Figure 7. Turbidity dependence of cellulose solution on solvent composition: 30/70 (**a**), 28/72(**b**), 26/74(**c**), 25/75(**d**), 24.5/75.5(**e**), 24/76(**f**), 23/77(**g**), 22/78 (**h**); cellulose concentration, 12g/100mL; storage time at 25°C, 20 days.

75.5% NH_4SCN. There was no noticeable change in turbidity with time for solutions prepared in the 75.5% NH_4SCN solvent.

In the course of transferring cellulose solution prepared from 75.5% NH_4SCN from one vessel to another via syringe, the formation of a fine fibrillar texture was observed (Figure 8).

It was possible to induce thermoreversible changes between the nematic and the cholesteric state in a cellulose solution. The spherulitic pattern existing at solvent composition of 22.0/78.0 and at 25°C was transformed into a uniformly dispersed highly birefringent state by chilling to -78°C for 10 minutes and then thawing that frozen solution. This appearance, indicating uniaxial molecular orientation, became spherulitic again when the sample was kept at 25°C for 10 minutes. In comparison, cellulose solutions prepared from solvent having lower NH_4SCN content, 70-71%, and which displayed small pitch fingerprint patterns, persisted without any change in pattern when similarly treated. In another experiment, lowering the system temperature from 25°C to 10°C appeared to increase the solvent composition range over which the system is nematic.

<u>Cholesteric Pitch</u>. The pitch size P determined from the fingerprint pattern in the solvent composition range 70.0-75.0% NH_4SCN increased with increasing NH_4SCN concentration (Figure 9) and decreased with increasing cellulose concentration C_p (Figure 10). Experimental results demonstrated that cellulose solutions prepared from solvent richer in NH_4SCN [up to 75.5% (w/w)] and lower in cellulose concentration showed the larger P (see Figures 9 and 10). This increase can be explained as the untwisting of the cholesteric phase in an approach to nematic character.and may be interpreted as evidence for impending formation of the nematic phase.

Robinson (*3*), in his pioneering study of polypeptide mesophase solutions, observed that P varied inversely as the square of the polymer volume fraction as shown in

$$1/P = v_2{}^x \tag{3}$$

in which case $x = 2$. In other studies, x was found to be between 1 and 2 (*29*) for a single stranded polypeptide. Subsequent work by Robinson (*21*) and others (*29,30*) have shown that the magnitude of the pitch, the relationship between pitch and polypeptide concentration, and even the sense of the helicoidal arrangement for a given polypeptide may be a function of solvent, temperature and molecular weight.

Figure 8. Fibrillar texture in the solution: solvent composition, 25.0/75.0; cellulose concentration, 16g/100mL; DP, 210; storage time at 25°C, 14 days.

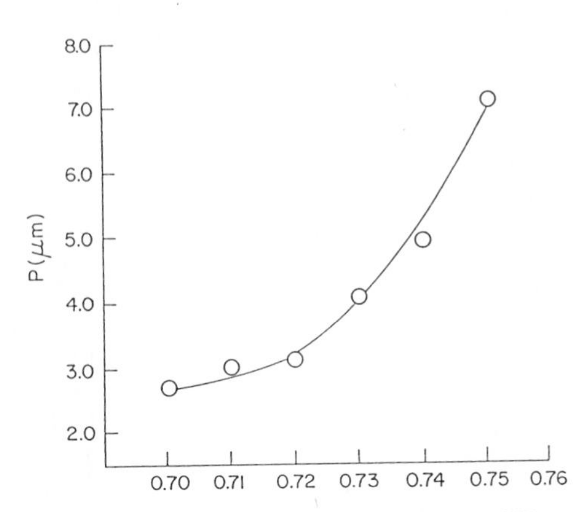

Figure 9. Pitch size dependence on solvent composition: cellulose concentration, 10g/100mL; DP, 210; storage time at 25°C, 17 days.

Cellulose based mesophases have displayed a somewhat steeper inverse dependence of pitch to concentration than that reported by Toriumi (*29*). A value of $x = 3$ has been reported for hydroxypropyl cellulose in water (*31*), for acetoxypropyl cellulose in acetone (*32*) and for hydroxypropyl cellulose in acetic acid and in methanol (*28*). Our data for cellulose in NH_3/NH_4SCN show a reasonably straight line when $P^{-1/3}$ was plotted vs. cellulose concentration (Figure 11), which is in accord with earlier reports (*28,29,31,32*).

The slope in Figure 11 increased with decreasing NH_4SCN concentration when solvent compositions 29.3/70.7 and 28.0/72.0 for samples stored 30 days are compared, and neither of these plots pass through the origin. Thus, in our study Equation 3 may be rewritten as

$$1/P = AC_p^3 + B \tag{4}$$

where A and B are constants. The steeper slope associated with the former solvent composition indicates a more sensitive dependence of pitch size on cellulose concentration. From this observation, it may be conjectured that cellulose-cellulose interactions occur more easily in the solution prepared from a solvent poorer in NH_4SCN.

Again, in Figure 11, the solid lines represent 14 day storage time and the dotted lines represent 30 day storage. The occurrence of the slope change for the solvent composition 29.3/70.7 may be caused by a dependence on cellulose concentration and/or by a dependence on storage time. Measurements of P on solutions containing water (23.6/75/1.4) are also included in this figure for comparison. In the range of relatively high cellulose concentration, the data do not fit the straight line well. The deviation from the linearity is not clearly understood but might arise from experimental problems.

Optical Rotatory Power. De Vries (*26*) developed a theory of the optical properties of planar cholesterics based on a model consisting of a large number of helicoidally arranged birefringent layers. According to this theory the wave length λ_0 of normally reflected light in air is given by

$$\lambda_0 = \overline{n}\, P \tag{5}$$

where, $\overline{n}$ is the average refractive index of the mesophase.

The rotatory power of the cholesteric mesophase at a wave length larger or smaller than that of the reflected

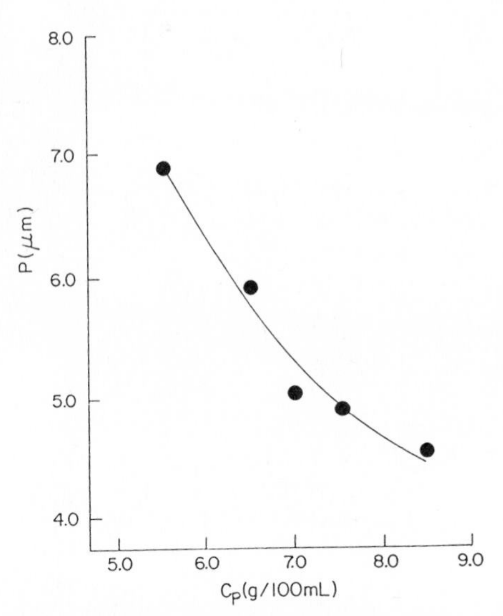

Figure 10. Pitch size dependence on cellulose concentration: solvent composition, 30/70; DP, 210; storage time at 25°C, 30 days.

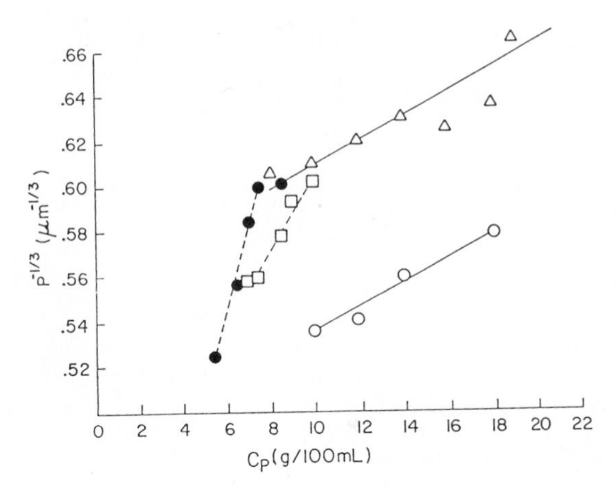

Figure 11. $P^{-1/3}$ vs.cellulose concentration: 29.3/70.7, 30 day storage (●); 29.3/70.7, 14 day storage (△); 28.0/72.0, 30 day storage (□); 23.6/75.0/1.4, 14 day storage (o).

light is given by

$$[\theta] = -\frac{\pi(\Delta n)^2 P}{4\lambda^2\{1-(\lambda/\lambda_0)^2\}} \quad (6)$$

where λ is the wavelength of incident light and Δn is the specific layer birefringence. When λ_0 is much larger than λ, the $(\lambda/\lambda_0)^2$ term will be negligible. For a cholesteric mesophase solution showing large pitch size (larger than 2μm in this study), Equation 6 may be approximated as

$$[\theta] = -\frac{\pi(\Delta n)^2 P}{4\lambda^2} \quad (7)$$

$[\theta]$, without consideration of P of a sample, would be a qualitative means of identifying the cholesteric solutions. Microscopical and laser light diffraction data show that P increased with decreasing cellulose concentration (see Figure 10) and with increasing NH_4SCN concentration (see Figure 9). However, negative specific rotation increased with increasing cellulose concentration (see Figure 12) and with decreasing NH_4SCN concentration (see Figure 13). Therefore, an increase in the magnitude of specific rotation comes from an increase in Δn rather than from P. At a solvent composition exhibiting a fingerprint pattern (28.0/72.0) the specific rotation increased with cellulose concentration at all wavelengths measured (Figure 12).

The specific rotation was found to depend on the solvent composition, cellulose concentration and storage time. The dependence on storage time was more significant with increasing departure from the solvent composition 24.5/75.5. In fact, the solution prepared from the 24.5/75.5 solvent did not show significant change in specific rotation, even up to 213 hours at 25°C. The solution prepared from solvent either richer or poorer than 75.5% NH_4SCN showed an increase in negative specific rotation with increasing difference from that solvent composition (Figure 13). Consider solutions with solvent compositions showing the uniformly dispersed highly birefringent (24.5/75.5), fingerprint (29.5/70.5), and spherulitic (23.5/76.5) patterns. It is interesting to plot their respective specific rotation behavior versus time as in Figure 14. Several features stand out: the constancy of specific rotation for solutions made from the 24.5/75.5 solvent, and tendencies for the specific rotation to increase to a greater extent and more rapidly with increasing departure from that solvent composition. The cellulose solutions' low specific rotation and rotatory stability with time in

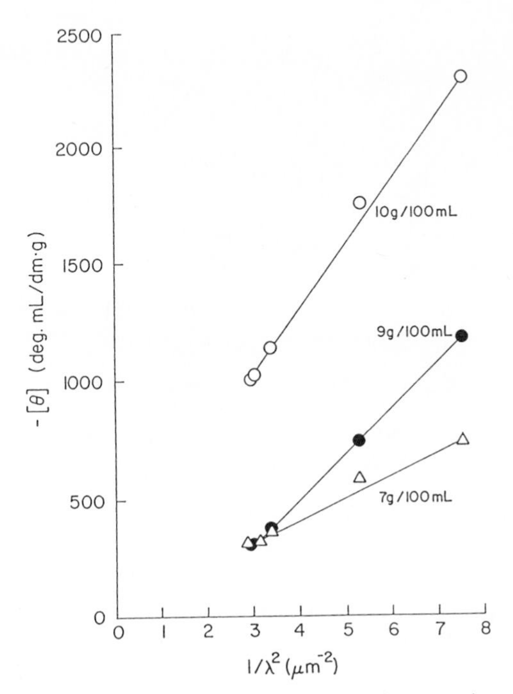

Figure 12. Optical rotatory dispersion dependence on cellulose concentration: solvent composition, 28.0/72.0; DP, 210; storage time at 25°C, 30 days.

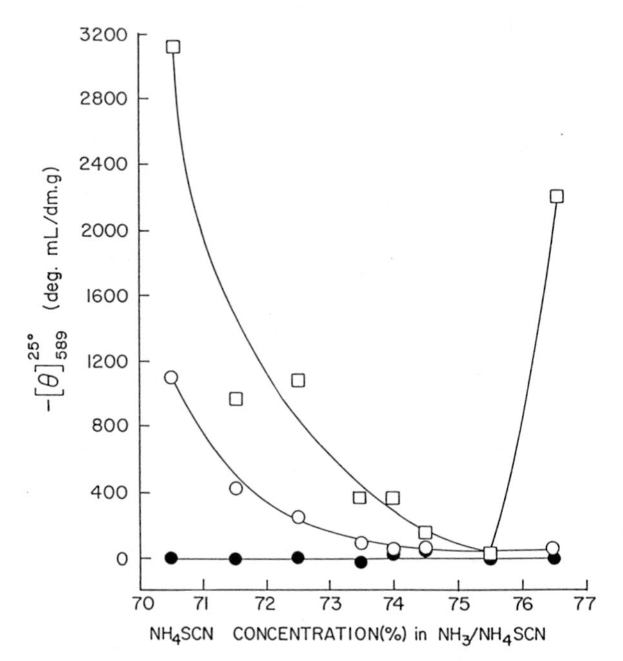

Figure 13. Specific rotation dependence at 25°C on solvent composition and on storage time: cellulose concentration, 12g/100ml; DP, 210; 1 hour (●), 24 hours (o), 213 hours (□).

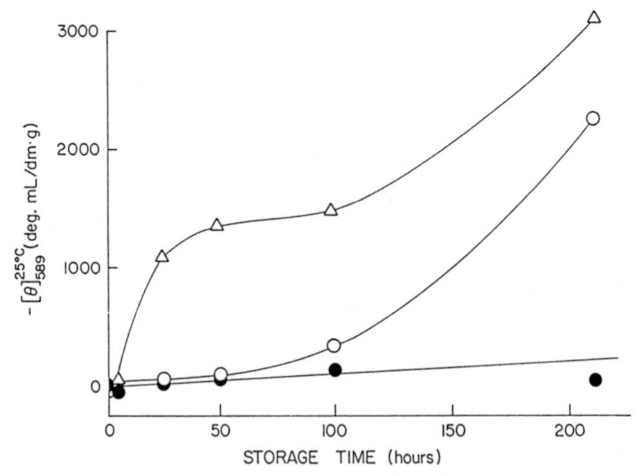

Figure 14. Specific rotation dependence on storage time at 25°C and at three specific solvent compositions: 29.5/70.5 (Δ), 24.5/75.5 (●), 23.5/76.5(o); Cellulose concentration, 12g/10mL; DP, 210.

solvents containing 75.5% NH_4SCN, indicates that the stated solvent composition is a good one at which cellulose-cellulose interactions are minimized.

The cellulose solution containing water (25.0/73.0/2.0) showed a more rapid increase in negative specific rotation with storage time compared with the solution prepared from solvent containing no water (25.5/74.5) (Figure 15).

Plotting [θ] against $1/\lambda^2$ gives a linear relationship as shown in Figure 12. The values for Δn were calculated from the slope and the cholesteric pitch and are shown in Table II. Δn (corrected for cellulose concentration) increased with increasing cellulose concentration and with decreasing NH_4SCN content.

Table II. Specific Layer Birefringence* of Cellulose/NH_3/NH_4SCN†

C_p (g/100mL)	5.5	6.5	7.0	7.5	8.0	8.5	9.0	10.0
NH_3/NH_4SCN								
29.3/70.7	18.6	29.6	32.7	36.4	-	39.0	-	-
28.0/72.0	-	-	20.2	-	22.6	-	30.1	40.7

*$\Delta n \times 10^4$
†Samples aged for 30 days

Some insight into the significance of layer birefringence may be obtained by consideration of the three variables, Δn, specific anisotropic phase volume fraction (anisotropic phase volume fraction/cellulose concentration, Φ/C_p), and C_p. The data shown in Table III may be interpreted to explain mesophase formation

Table III. $\Delta n \times 10^4$ and Φ/C_p Dependence on Cellulose Concentration*

C_p (g/100mL)	5.5	6.5	7.0	7.5
Φ_{aniso}	0.40	0.71	0.81	0.83
Φ_{aniso}/C_p	7.27	10.96	11.57	11.02
$\Delta n \times 10^4$	18.61	29.60	32.69	36.42

*29.3/70.7 Solvent, Samples aged 30 days

behavior with increasing cellulose concentration. An isotropic phase was obtained below the MCCM. A mesophase was obtained with further increase in the cellulose concentration. At first, this increase appears to contribute to both an increase in anisotropic phase volume fraction and also to an ordering of the twisted layers comprising the cholesteric phase. The eventual decrease in the Φ/C_p seems to indicate that continued increases in cellulose concentration contribute predominantly to ordering that results in an increase in specific layer birefringence.

Polymer-Solvent Interaction. Terbojevich et al.(*33*) discussed the effect of the salt concentration, cellulose *DP* and prior hydrolysis on molecular aggregation in the N,N-dimethylacetamide (DMAc)/lithium chloride (LiCl)/cellulose system and concluded.that LiCl promotes destabilization of cellulose aggregates which leads to the formation of a pseudo-complex involving DMAc. Further, they noted that stable aggregation was common with the samples that had been subjected to acid hydrolysis. They also found that the critical concentration for mesophase formation increased with increasing LiCl concentration (*34*).

In the current study, the aggregated anisotropic phase occurred in solutions prepared from acid hydrolyzed cellulose of *DP* 35. The higher minimum cellulose concentration for mesophase formation was observed in cellulose solutions richer in NH_4SCN (see Figure 6). In these aspects, the cellulose/NH_3/NH_4SCN system resembles the DMAC/LiCl/cellulose system.

In his classic paper, Flory predicted the phase behavior in solutions of rod-like particles (*5*). The resulting phase diagram related the solvent-solute interaction parameter χ_1 (*35*) to the volume fraction, v_2, for polymer rods with an axial ratio of 100. A positive χ_1 makes a positive or excess free energy contribution to mixing. Good solvents are characterized by small χ_1 values. Two of Flory's major predictions are that the minimum polymer concentration required for mesophase formation will increase as χ_1 decreases, sharply at first, then more gradually, and at certain χ_1 values two different anisotropic phases coexist. Our microscopical observations of conjugated phases may reflect the validity of the latter prediction.

The former prediction is now discussed. In the cellulose/NH_3/NH_4SCN system, the observation that the minimum cellulose concentration for mesophase formation increased with increasing NH_4SCN in the solvent up to approximately 75.5%, indicates that this specific solvent composition may be regarded as a good solvent composition.

Cholesteric mesophases form only from molecules having chiral centers. Since the helicoidal structures of cholesterics are supramolecular, the chiral centers in these systems must, in some way, interact with each other. It is reasonable then to expect that one way of obtaining an untwisted cholesteric, i.e., a nematic phase, would be by reducing interactions between molecules.

Our findings do seem to indicate that, when polymer-polymer interactions are repressed, as in a good solvent, the formation of a nematic phase is made possible, and, conversely, conditions that encourage polymer-polymer interaction increase the twist of the cholesteric phase. Formation of the nematic phase is observed within a certain range of cellulose concentration, 8-16g/100mL. These nematic patterns gave way to conjugated patterns consisting of both fingerprint plus uniformly dispersed highly birefringent regions, as cellulose concentration increased through 18g/100mL in the same 24.5/75.5 solvent(see Figures 3b and c). Further, when water, a non-solvent, is added to the cellulose solution, it greatly increased the specific rotation observed at otherwise comparable conditions (Figure 15). It is reasonable to expect that the water enhanced polymer-polymer interactions, thus encouraging the formation of the cholesteric phase. The MCCM is highest for the solvent composition that favors nematic phase formation. In each case, conditions that seem to be favoring interactions among polymer chains favor cholesteric phase formation. Poor solvents, i.e.,compositions yielding solutions with high χ_1, would promote enhanced cellulose-cellulose interactions even at low polymer concentrations. This tendency to interact could cause association of the cellulose chains to form parallel arrays (*33*). Their chiral centers could thus interact and lead to progressive angular displacements between successive layers, in other words a cholesteric phase.

Fiber Formation. Preliminary spinning experiments were performed to compare properties of fibers spun from the cholesteric and the nematic solutions. The solvent compositions chosen for preparation of the spinning dopes were 25.0/75.0 and 29.3/70.7 at which respectively, nematic and cholesteric phases formed. A relatively low viscosity was required to sustain good spinnability. Thus, a less viscous partly nematic solution obtained by increasing the solution temperature up to 40°C was used rather than a wholly nematic solution. A solution containing 14g/100mL of cellulose (*DP*=765) was spun with a laboratory wet spinner through 2.5cm air-gap into a methanol bath. In fact, at the spinning temperature of *ca*. 40°C, polarizing microscopy revealed that the spinning dopes were biphasic; they were respectively, nematic/isotropic and cholesteric/isotropic. These are the first fibers produced, and the experimental techniques are crude. A subsequent paper will report the results of a methodical study of fiber formation from the NH_3/NH_4SCN solvent system. Yet, this preliminary work has revealed several important differences between the fibers produced from the two systems. The tenacity and modulus of the fiber obtained from the 25.0/75.0 solvent were twice those of the fiber from the 29.3/70.7 solvent. The surface of the fiber from the former solvent composition was the smoother (Figure 16a) of the two (Figure 16b). SEM of fracture surfaces revealed that the fiber resulting from the solvent composition 25.0/75.0 (Figure 17a) appeared to have the more fibrillar microstructure (Figure 17b). When compared under a polarizing microscope, the fibers prepared from the two solvent compositions showed distinct differences (Figure 18). Chains in the fibers from solvent composition 25.0/75.0 are the more longitudinally oriented ones. This is illustrated by light extinction of the fiber from the former solvent when it is rotated by 45° from Figure 18a to 18b, and the lack of this property in the fiber from the latter solvent. (See Table IV.)

Table IV. The Effects of Solvent on Fiber Properties

NH_3/NH_4SCN	*DP*	C_p g/100mL	Denier	Tenacity g/d	Modulus g/d	Birefringence rank
29.3/70.7	210	16	98	0.56	22.4	low
25.0/75.0	210	16	110	0.98	44.0	high

The nematic and cholesteric phase types attendant to their respective solvent compositions, could affect the tenacity, the modulus and the structure of fibers derived from them. The smooth and highly oriented fiber resulting from the 25.0/75.0 solvent was formed from the nematic phase with its highly unidirectional structure. The distinct fibrillar texture (see Figure 8) observed under a polarizing microscope when a cellulose solution in the 25.0/75.0 solvent was extruded from a syringe supports the idea that the advantageously ordered fiber structure originated from the nematic phase. In general, the fiber extrusion experiments support our conclusions about the identities of the phase types formed in different solvent compositions.

The properties of the fibers produced are compared in Table V with those of Fortisan, a regenerated cellulose fiber that is widely considered to be stiff and strong. The fiber spun from the nematic phase has an initial modulus (149-167 g/d) very close to the lower range of Fortisan. It is recognized that fiber tensile strength or breaking strength depends greatly on processing conditions and perfection of fiber structure itself. Modulus, on the other hand, measured at a small

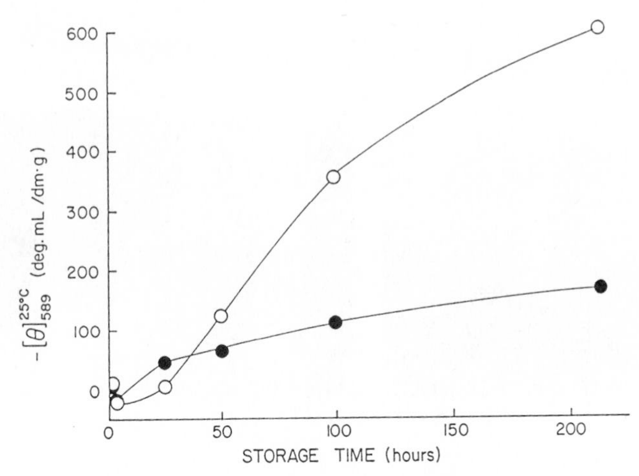

Figure 15. Specific rotation dependence at 25°C on the water content and on storage time: 25.5/74.5 (●), 23.0/75.0/2.0 (o), cellulose concentration, 12g/100mL; DP, 210.

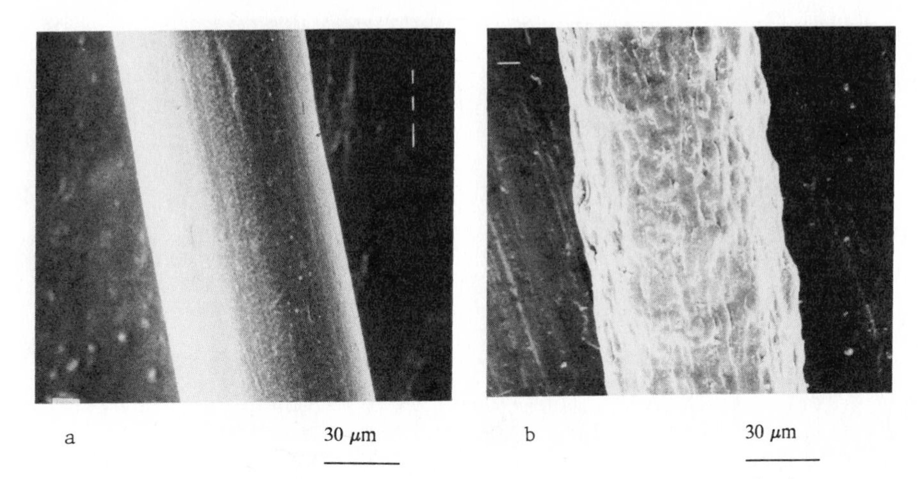

Figure 16. Fiber surface: solvent composition, 25.0/75.0 (**a**); 29.3/70.7 (**b**); cellulose concentration, 16g/100mL; DP, 210.

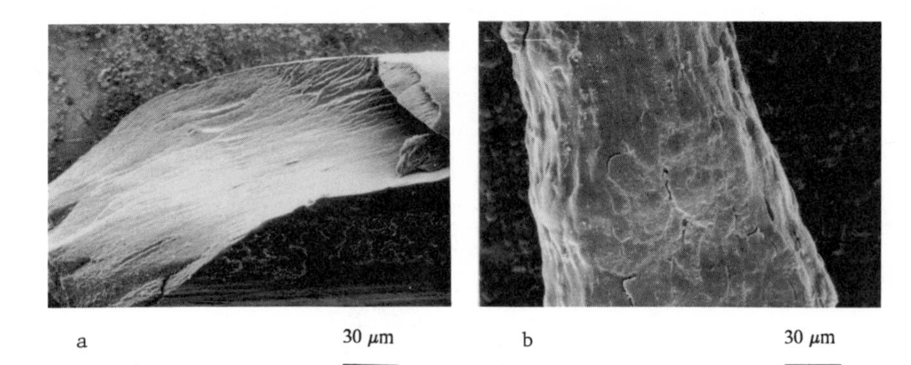

Figure 17. Fiber fracture surface: solvent composition, 25.0/75.0 (**a**), 29.3/70.7 (**b**); cellulose concentration, 18g/100mL (**a**), 16g/100mL (**b**); DP, 210.

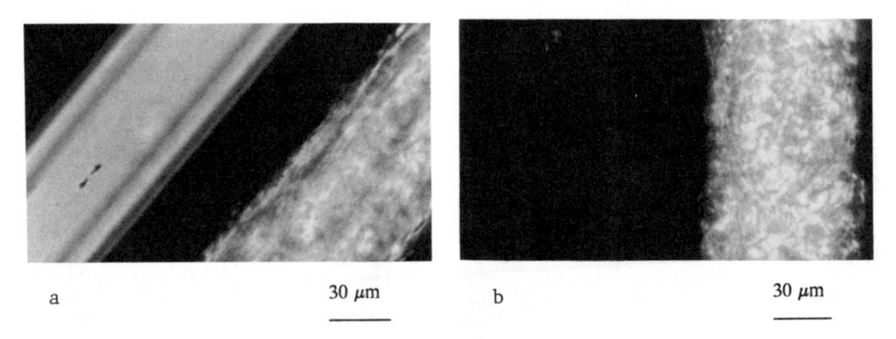

Figure 18. Polarizing microscopical comparison between the two fibers from the two different solvent compositions: 25.0/75.0 (left), 29.3/70.7(right); cellulose concentration, 16g/100mL; DP, 210; **b**, rotated 45° from **a**.

deformation, is not strongly related to perfection of the fiber structures. As a result, moduli of even crudely prepared fibers may be better estimated than breaking strength in the early stages of development of a potentially strong fiber.

Table V. Comparison between Cellulose Fibers Spun from Nematic Phase with Fortisan

Characteristics	Fiber Sources: NH_3/NH_4SCN(24.5/75.5)	Fortisan(*36*)
DP	765	
Spinning method	dj-ws*	
Denier, d/f	8.6	
Tenacity		
dry, g/d	2.98-3.41†	6-8
wet, % of dry	73	75-85
Elongation, dry, %	8	
Elongation, wet, %	9	
Initial modulus g/d	149†-167	170-250

*dry jet-wet spun

†Liu, C. K.; Cho, J. J.; Cuculo, J. A., unpublished data.

Conclusions and Summary

A lyotropic, nematic solution of cellulose was formed in a NH_3/NH_4SCN solvent in what are presumably good solvent compositions. Evidence strongly suggests that the twisted nematic or cholesteric structure that results when solutions of chiral cellulose chains interact may be repressed or compensated so that interactions among chiral centers are minimized. Our reasoning is based a body of experimental evidence which includes:

- Nematic phase patterns observed under a polarizing microscope,
- Low specific rotation of solutions exhibiting those nematic phase patterns and the absence of significant change in that property with increasing storage time,
- The tendency of the pitch of the cholesteric solutions to increase as the conditions for formation of the nematic solutions were approached,
- An increase in minimum cellulose concentration for mesophase formation with increase in NH_4SCN concentration up to 75.5%,
- A solvent composition that would otherwise promote the development of a nematic phase, at higher cellulose concentrations yields solutions containing some cholesteric mesophase.

The fibers spun from nematic phase systems and solvent composition 25.0/75.0 exhibited twice the tenacity and modulus of those extruded from the system of the solvent composition 29.3/70.7. The two fiber types showed significant differences in birefringence, and in the appearance of their external surfaces and fracture surfaces. The comparable moduli of the fiber spun from a nematic solution with those of Fortisan may result from high uniaxial orientation of cellulose molecules in the former fiber.

Literature Cited

1. Bawden, F. C.;Pirie, N. W. *Proc. Roy. Soc., Ser. B* **1937**, *123*, 274.
2. Elliott, A.;Ambrose, E. J. *Disc. Faraday Soc.* **1950**, *9*, 246.
3. Robinson, C. *Trans. Faraday. Soc.* **1956**, *52*, 571.
4. Flory, P. J. *Proc. Roy. Soc., Ser. A* **1956**, *234*, 60.
5. Flory, P. J. *ibid.* **1956**, *234*, 73.
6. Ciferri, A. *Polym. Eng. Sci.* **1975**, *15(3)*, 191.
7. Morgan, P. W. *Macromolecules* **1977**,*.10*, 1381.
8. Kwolek, S. L.; Morgan, P. W.;Shaefgen, J. R.; Gulrich, L. W. *Macromolecules* **1977**, *10*, 1930.
9. Chanzy, H; Peguy, A. *J. Polym. Sci., Polym. Phys. Ed.* **1980**, *18*, 1137.
10. Patel, D. L.; Gilbert, R. D. *J. Polym, Sci., Polym. Phys. Ed.* **1981**, *19*, 1231.
11. Conio, G.; Corazza ,P.; Bianchi, E.; Tealdi, A.; Ciferri, A. *J. Polym. Sci., Polym. Lett. Ed.* **1984**, *22*, 273.
12. Gray, D. G. *J. Appl. Polym. Sci., Appl. Polym. Symp.* **1983**,*37*, 179.
13. Turbak, A. F.; El-Kafrawy, A.; Snyder, F. W.; Auerbach, A. B. U.S. Patent 4 202 252 A, 1982.
14. Johnson, D. L.; U.S. Patent 3 447 939, 1969.
15. Cuculo, J. A.; Hudson, S. M. U.S. Patent 4 367 191, 1983.
16. Hudson, S. M.; Cuculo, J. A. *J. Polym. Sci., Polym. Chem. Ed.* **1980**,*18*, 3469.
17. de Groot, A. W.; Carroll; F. I.; Cuculo, J. A. *J. Polym. Sci., Polym. Chem. Ed.* **1986**, *24*, 673.
18. Chen, Y. S.; Cuculo, J. A. *J. Polym. Sci., Polym. Chem. Ed.* **1986**, *24*, 2075.
19. Demus, D.; Richter, L. *Textures of Liquid Crystals;* Weinheim: NY, 1978, p 17.
20. Adams, J. E.; Hass, W. E. *Mol. Cryst. Liq. Cryst.* **1971**, *15*, 27.
21. Robinson, C. *Tetrahedron* **1961**, *13*, 219.
22. Samulski, T. V.; Samulski, E. T. *J. Chem. Phys.* **1977**, *67*, 824.
23. Toriumi, H.; Minakuchi, S.; Uematsu, Y.; Uematsu, I. *Polym. J.* **1980**, *12*, 431.

24. Toriumi, H.; Kusumi, Y.; Uematsu, I.; Uematsu, Y. *Polym. J.* **1979**, *11*, 863.
25. ASTM, Philadelphia, *ASTM Designation D1795-62 (reapproved 1968), Part 21,* **1974** pp 169-175.
26. de Vries, H. *Acta Crystallogr.* **1951**, *4*, 219.
27. Van, K;. Norisuye, T.; Teramoto, A. *Mol. Cryst. Liq. Cryst.* **1981**, *78*, 123.
28. Werbowyj, R. S.; Gray, D. G. *Macromolecules* **1984**, *17*, 1512.
29. Toriumi, H.,Minakuchi, S.; Uematsu, I. *J. Polym. Sci., Polym. Phys. Ed.* **1981**, *19*, 1167.
30. Uematsu, Y;. Uematsu, I. In *Mesomorphic Order in Polymers and Liquid Media;* Blumstein, A., Ed., ACS. Symposium Series No.74; American Chemical Society: Washington, DC, 1978; p 136.
31. Onogi, Y.; White, J. L.; Fellers, J. F. *J. Polym. Sci., Polym. Phys. Ed.* **1980**, *18*, 663.
32. Tseng, S. L.; Valente, A.; Gray, D. G. *Macromolecules* **1981**, *14*, 715.
33. Terbojevich, M.; Cosani, A.; Conio, G.; Ciferri, A.; Bianchi, E. *Macromolecules* **1985**, *18*, 640.
34. Bianchi, E.; Ciferri, A.; Conio, G.; Cosani, A.; Terbojevich, M. *Macromolecules* **1985**, *18*, 646.
35. Flory, P. J. *Principles of Polymer Chemistry,* Cornell University Press: Ithaca, NY, 1953, p 591.
36. Tesi, A. F. In *Man-Made Textile Encyclopedia,* Press, J. J., Ed.; Textile Book: NY, 1959, Chapter 3.

RECEIVED August 22, 1988

Chapter 12

Cellulose and Cellulose Triacetate Mesophases

Ternary Mixtures with Polyesters in Trifluoroacetic Acid–Methylene Chloride Solutions

Y. K. Hong, D. E. Hawkinson, E. Kohout, A. Garrard, R. E. Fornes, and R. D. Gilbert

Fiber and Polymer Science Program, North Carolina State University, Raleigh, NC 27695

Cellulose and cellulose triacetate (CTA) form mesophases in trifluoroacetate acid (TFA) - CH_2Cl_2 mixtures. Anisotropy is observed as low as 4% (w/w) concentrations for cellulose solutions; well below predicted values. For cellulose triacetate, the critical concentration is 20% (w/w). Both cellulose and CTA are slowly degraded in the TFA-CH_2Cl_2 solvent. Ternary mixtures of cellulose or CTA and polyethylene terephthalate (PET) separate into anisotropic and isotropic phase, conforming Flory's prediction, but there was not complete exclusion of the PET from the anisotropic phase. Surprisingly, fibers spun from anisotropic solutions rich in CTA contained the CTA I polymorph. Block copolymers of CTA and poly(ethylene-co-propylene adipate) (PEPA) formed mesophases. When admixed with PEPA, phase separation did not occur, suggesting molecular composites may be prepared from mixtures of block copolymers, consisting of rigid (or semi-rigid) and flexible block and flexible polymers.

The lyotropic mesophases of cellulose and cellulose derivatives were first observed only relatively recently (*1-3*). It is of interest to note that Flory in his now classical papers (*4,5*) predicted in 1956 that cellulose or cellulose derivatives should exhibit liquid crystal behavior. Since Werbowyj and Gray (*1*) first reported mesophases of hydroxylpropyl cellulose in water, the field has expanded rapidly (for reviews see References 6 and 7). Undoubtedly, the activity in this area originates from a desire to prepare fibers or films of cellulose or cellulose derivatives with superior properties as well as to understand the purely scientific aspects of the systems.

Chanzy and Peguy (2) first reported that cellulose forms a lyotropic mesophase in a mixture of N-methyl-morpholine N-oxide and water. Solution birefringence occurred at concentrations greater than 20% (w/w) cellulose. Patel and Gilbert (*3*) showed mixtures of trifluoroacetic acid (TFA) and chlorinated alkanes (*1,2* dichloroethane, CH_2Cl_2) are excellent solvents at room temperature for cellulose and cellulose triacetate and superior to pure TFA. Lyotropic mesophases were obtained in 20% (w/w) solutions of cellulose in those solvent mixtures. The optical rotatory power of the solutions suggest the lyotropic mesophase is cholesteric, as expected, due to the chirality of cellulose.

0097–6156/89/0384–0184$06.00/0

More recently, Hawkinson (*8*) observed birefringence between crossed polars for solutions of cellulose in TFA-CH_2Cl_2 (70:30 v/v) at concentrations as low as 4% (w/w).

Cholesteric lyotropic mesophases of cellulose in dimethylacetamide-LiCl solutions have been observed by Ciferri and coworkers (*9-11*). While cellulose/TFA-CH_2Cl_2 mesophases have positive optical rotations, the cellulose/ LiCl/DMAC mesophases have negative rotations.

McCormick et al (*12*) observed that cellulose concentrations of 10% (w/w) and above in 9% LiCl/DMAC appear lyotropic after slight shearing, but a pure anisotropic phase was not observed even in 15% (w/w) cellulose solutions.

Chen and Cuculo (*13*) found solutions of cellulose in a mixture of liquid ammonia/NH_4SCN (27:73 w/w) are liquid crystalline at concentrations between 10 and 16% w/w depending on the cellulose molecular weight. Optical rotations of the solutions indicate the cellulose mesophase is cholesteric. As in the case of LiCl/DMAC solutions, the optical rotations were negative.

Recently Yang (*14*) found that cellulose with a degree of polymerization of 210 formed a mesophase at 3.5% (w/w) concentration when the liquid ammonia/NH_4SCN ratio was 30:70 (w/w).

These observations (*8,10,14*) suggests that the Kuhn model (*15*) must apply to cellulose rather than Flory's lattice model which predicts the critical concentration (V_2^C) is given by

$$V_2^C = \frac{8}{p} (1 - \frac{2}{p})$$

where p is the polymer aspect ratio. For semi-rigid polymers persistence lengths (*16*) are used instead of the polymer chain length. The persistence length for cellulose (*7*) depends on the particular solvent employed but is approximately 70 A°. Assuming a width of 8 A° or an aspect ratio of 8.85, the V_2^C value is 0.70 volume fraction. This is much higher than observed.

Typically, the viscosity of a polymer solution containing a polymer that will form a mesophase will exhibit a maximum in the vicinity of V_2^C. Indeed this is often used to determine the V_2^C of a polymer in a particular solvent. The reduction in solution viscosity is due to the reduced polymer-polymer interaction in the ordered phase.

Cellulose triacetate, for example, exhibits this behavior (*17*). However, this has not been observed for cellulose, regardless of the solvent system employed. In the case of LiCl-DMAC, Conio et al (*9*) showed that, due to the close proximity of the cholesteric mesophase to its solubility limit, it is only observed in a metastable condition.

As noted above, liquid crystal solutions of cellulose or cellulose derivatives are of interest because of the possibility of spinning regenerated cellulose fibers with higher strengths and moduli than those obtained from viscose rayon which is spun from isotropic solutions. However, to date there are no published reports of the spinning of fiber from mesomorphic cellulose solutions. O'Brien (*18*) describes the spinning of cellulose triacetate fibers from mesomorphic solutions in either TFA-CH_2Cl_2 or TFA-H_2O mixtures as solvents. Tenacities as high as 13.3 dN/tex were obtained. The triacetate fibers were saponified with NaOMe. The regenerated cellulose fibers had tenacities as high as 16.4 dN/tex with elongations of the order of 10%. These data demonstrate the potential for preparing high strength/high modulus cellulose fibers.

Here we describe some recent work on liquid crystal solutions of cellulose and cellulose triacetate in TFA-CH_2Cl_2 solvent mixtures.

Experimental

Whatman CC 41 microgranular cellulose powder was dried under high vacuum at R.T. for 48 h and then at 110°C at atmospheric pressure for 1 h. It was then stored in a desiccator until use. Cellulose triacetate (CTA; $\overline{M}n$ = 95,000, D.S. 2.8 or $\overline{M}n$ = 110,000, D.S. 3.0) was kindly supplied by Tennessee Eastman. Polyethylene terephthalate was obtained from duPont. Polymethyl methyl methacrylate (PMMA) was obtained from Polysciences. Reagent grade TFA and CH_2Cl_2 were from Fisher Scientific and were used without further purification. Mixtures of TFA and CH_2Cl_2 in various ratios were inverted several times to ensure homogeneity and then stored overnight before use. Stock solutions of cellulose, CTA and PET were prepared separately in calibrated Erlenmeyer flasks with ground glass stoppers that were wrapped with teflon tape. The solutions were allowed to mature 2 weeks with frequent inversions to ensure homogeneity. Ternary solutions of cellulose, or CTA, and PET were prepared by mixing appropriate volumes of stock solutions and mechanically mixed with a magnetic stirrer at 28°C for 5 h and then stored for one week. Some of the ternary systems separated into an isotropic and anisotropic phase. Similar results were observed for mixtures of cellulose triacetate and PMMA. The volume of these phases were measured and the isotropic phase decanted. Each phase was analyzed for cellulose triacetate by extraction with CH_2Cl_2 or 1,1,1-trichloroethane.

A small portion of each solution was cast on a glass slide, a coverslip placed on top and the assembly examined with an optical microscope equipped with crossed polars. A solution was considered to be isotropic if it did not transmit light when viewed between crossed polars. Optical rotations at five different wavelengths were measured at 25°C using an automatic Perkin-Elmer 14 polarimeter equipped with a Hg light source.

Viscosity measurements were made at 25°C and 65% relative humidity with a Wells-Brookfield micro-viscometer equipped with a cone and plate.

Fibers were spun using an air-gap spinning system consisting of an extrusion compartment, three divided compartments for fiber stretching, washing and finishing and a yarn winder. The solutions were extruded using a plunger type extruder, through a 44 hole platinum spinneret at a speed of 2.64 m/min., and passed through the air-gap into a 40 cm long coagulation bath containing MeOH/H_20 (7/3: v/v) (22°C). After winding, the filaments were immersed in H_20 for 5 h and dried in air.

The fibers were annealed at 200°C for 5 min. in N_2, wound parallel on a sample holder and flat plate X-ray photographs taken by exposing the fibers for 1 h (WAXS) to a beam collimated at 90° to the fiber axis. Equatorial diffractometer traces of uniaxially oriented fibers were taken using slit collimation. WAXS patterns were obtained on polaroid film and the diffractometer traces with a Siemens X-ray system. Nickel filtered CuK_α radiation was used. The X-ray unit was operated at 30 KV and 20 mA. The sample to film distance was 71.4 mm.

Tenacity, initial modulus and elongation to break were obtained on fiber samples with an Instron Model 1123 tester using an extension rate of 50.8 cm/min. and a guage length of 2.54 cm at 25°C and 65% relative humidity.

Results and Discussion

Cellulose Mesophases. An anistopropic phase is formed in a 6% (w/w) solution of cellulose in TFA-CH_2Cl_2 (60:40 v/v) and remains anisotropic for at least 16 days (Table I). In TFA-CH_2Cl_2 (70:30 v/v) solutions birefringence was observed as

Table I. Summary of Polarized Light Microscopy Observations for Solutions Containing 4,6,8, and 10% (w/w) Cellulose in TFA-CH_2Cl_2 (60:40 v/v)

Cellulose Concentration	Day 4	Day 7	Day 12	Day 16	Day 42
4.0% (w/w)	I	I	I	I	I
6.0% (w/w)	B/C	B	B	B	I
8.0% (w/w)	B/C	B/C	B	B	I
10% (w/w)	B/C	B/C	B/C	B	B

I = Isotropic; no areas of extinction
I/B = Mainly isotropic; several areas which show extinction remain
B = Birefringent throughout
B/C = Birefringent throughout; undissolved cellulose powder of fiber visible

low as 4% (w/w) (Table II). However, a concentration of 6% (w/w) was required for an anisotropic phase to remain stable for 10 days after complete cellulose dissolution. A 16% (w/w) solution in TFA-CH_2Cl_2 (70:30 v/v) appeared wholly anisotropic after 45 days. The formation of a mesophase at such low concentrations is surprising for a semi-rigid polymer such as cellulose, but similar results have been obtained, as noted above, with the DMAC/LiCl (*9-11*) and liquid NH_3/NH_4SCN (*13,14*) solvents and suggest the Kuhn model is more appropriate for describing cellulose mesophases than the lattice model.

Table II. Summary of Polarized Light Microscopy Observations for Solutions Containing 2-16% (w/w) Cellulose in TFA-CH_2Cl_2 (70:30 v/v)

Cellulose Concentration	Day 4	Day 8	Day 10	Day 20	Day 30	Day 45	Day 60
2.0% (w/w)	I	I	I	I	I	I	I
4.0% (w/w)	B	I	I	I	I	I	I
6.0% (w/w)	B	B	I/B	I	I	I	I
8.0% (w/w)	B/C	B	B	I/B	I	I	I
10.0% (w/w)	B/C	B/C	B/C	B	B	I/B	I
12.0% (w/w)	B/C	B/C	B/C	B	B	B	B
14.0% (w/w)	B/C	B/C	B/C	B/C	B	B	B
16.0% (w/w)	B/C	B/C	B/C	B/C	B/C	B	B

I = Isotropic; no areas of extinction
I/B = Mainly isotropic; several areas which show extinction remain
B = Birefringent throughout
B/C = Birefringent throughout; undissolved cellulose powder of fiber visible

The formation of an isotropic cellulose phase after storage of TFA-CH_2Cl_2 solutions is due to degradation of the cellulose, (Tables III and IV), presumably by attack of TFA at the glycosidic linkages, as the extent of degradation increases with

Table III. Intrinsic Viscosity and DP Data for Whatman CC41 Cellulose and Cellulose Precipitated From 8% (w/w) Cellulose/TFA-CH_2Cl_2 Solutions After 7 Days of Dissolution

Sample	[η] (g/dl)	M.W.	D.P.
Whatman CC41 cellulose	1.08	29,976	185
Cellulose from TFA-CH_2Cl_2 (60:40 v/v)	0.92	25,085	155
Cellulose from TFA-CH_2Cl_2 (65:35 v/v)	0.87	23,574	146
Cellulose from TFA-CH_2Cl_2 (70:30 v/v)	0.79	21,179	131
Cellulose from TFA-CH_2Cl_2 (75:25 v/v)	0.71	18,809	116
Cellulose from TFA-CH_2Cl_2 (80:20 v/v)	0.73	19,399	120
Cellulose from TFA-CH_2Cl_2 (85:15 v/v)	0.65	17,052	105
Cellulose from TFA-CH_2Cl_2 (100:0 v/v)	0.61	15,890	98

Table IV. Intrinsic Viscosity and DP Data for Whatman CC41 Cellulose and Cellulose Precipitated From 8% (w/w) Cellulose/TFA-CH_2Cl_2 Solutions After 12 Days of Dissolution

Sample	[η] (g/dl)	M.W.	D.P.
Whatman CC41 cellulose	1.09	30,285	187
Cellulose from TFA-CH_2Cl_2 (60:40 v/v)	0.69	18,222	112
Cellulose from TFA-CH_2Cl_2 (65:35 v/v)	0.70	18,515	114
Cellulose from TFA-CH_2Cl_2 (70:30 v/v)	0.64	16,761	103
Cellulose from TFA-CH_2Cl_2 (75:25 v/v)	0.65	17,052	105
Cellulose from TFA-CH_2Cl_2 (80:20 v/v)	0.66	17,343	107
Cellulose from TFA-CH_2Cl_2 (85:15 v/v)	0.62	16,180	100
Cellulose from TFA-CH_2Cl_2 (100:0 v/v)	0.56	14,449	89

the TFA-CH_2Cl_2 ratio. However, the decrease in molecular weight after 7 days storage is minimal at 60:40 and 70:30 (v/v) TFA-CH_2Cl_2 ratios and the solutions remain anisotropic. Indeed, from a practical viewpoint, the lowering of molecular weight and the corresponding decrease in solution viscosity (see below) could be an advantage in regards to adjustment of the solution viscosity for fiber spinning. For example, the cellulose molecular weight is deliberately lowered in the viscose rayon process.

The change in cellulose degree of polymerization vs time using TFA-CH_2Cl_2 (70:30 v/v) as solvent is shown in Table V. The decrease in D.P. is nominal, has essentially leveled off after 7 days and, as shown in Table II, the solution is anisotropic for at least 10 days.

Solution viscosities for 2 to 14% (w/w) concentrations of cellulose in TFA-CH_2Cl_2 (70:30 v/v) at three different cone rotation speeds (or shear rates) in the cone/plate viscometer are shown in Table VI and Figure 1.

Table V. Intrinsic Viscosity and DP Data for Whatman CC41 Cellulose and Cellulose Precipitated From 8% (w/w) Cellulose/TFA-CH_2Cl_2 (70:30 v/v) Solutions After 4-20 Days of Dissolution

Sample	[η] (g/dl)	M.W.	D.P.
Whatman CC41 cellulose			
Procedure I	1.09	30,285	187
Procedure II	1.14	31,832	196
Cellulose after 4 days dissolution			
Procedure I	0.86	23,273	144
Procedure II	0.66	17,343	107
Cellulose after 7 days dissolution			
Procedure I	0.67	17,636	109
Procedure II	0.70	18,515	114
Cellulose after 10 days dissolution			
Procedure I	0.72	19,104	118
Procedure II	0.72	19,104	118
Cellulose after 12 days dissolution			
Procedure I	0.74	19,694	122
Procedure II	0.66	17,343	107
Cellulose after 15 days dissolution			
Procedure I	0.74	19,618	121
Procedure II	0.68	17,928	111
Cellulose after 20 days dissolution			
Procedure I	0.71	18,809	116
Procedure II	0.61	15,890	98

Table VI. Viscosity Values for Solutions Containing 2-14% Cellulose in TFA-CH_2Cl_2 (70:30 v/v) After 12 Days of Dissolution. Measurements Were Made at 25°C with a Wells-Brookfield Mode HBT Cone/Plate Viscometer

Cellulose Concentration (% w/w)	Rotation Speed RPM	Viscosity (cps)
2.0	0.5	~ 0.0
	1.0	60.2
	2.5	91.1
4.0	0.5	1,675.5
	1.0	2,631.1
	2.5	1,445.1
6.0	0.5	3,890.4
	1.0	3,063.1
	2.5	2,897.6
8.0	0.5	10,854.2
	1.0	6,571.2
	2.5	5,165.8
10.0	0.5	22,747.6
	1.0	13,693.8
	2.5	7,758.2
12.0	0.5	81,847.9
	1.0	43,745.3
	2.5	20,278.2
14.0	0.5	Off Scale
	1.0	Off Scale
	2.5	Off Scale

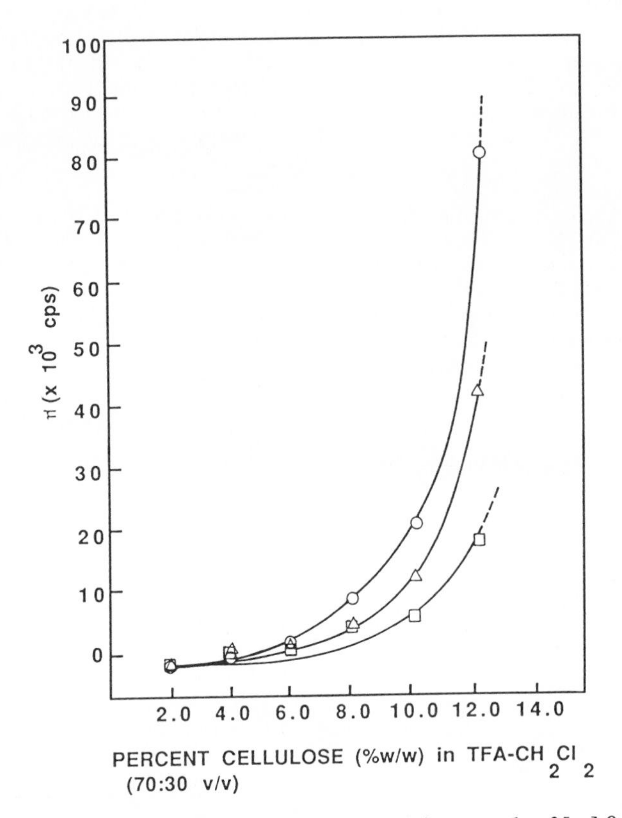

Figure 1. The dependence of η, at cone rotation speeds of [○] 0.5, [Δ]1.0, and [□] 2.5 RPM, on concentration for 2-14% (w/w) cellulose/ TFA-CH_2Cl_2 (70:30 v/v) solutions after 12 days of dissolution.

The above data demonstrates that there is ample latitude in the cellulose-TFA-CH_2Cl_2 system to make adjustments in the spin parameters.

The solution viscosity vs concentration plots do not exhibit a maximum in contrast to typical polymer liquid crystal solutions. As noted in the introduction, this same behavior is exhibited by other cellulose-solvent systems (*9,14*) and, as discussed in the introduction, suggests the mesophase is metastable.

Thin films of the solutions between microscope slides were sheared by applying even pressure on a coverslip while sliding it approximately one cm. The anisotropy appeared to increase as measured by increases in the birefringence. Solutions containing 10-16% (w/w) cellulose developed a threaded texture and the mesophases were stable with time and oriented in the direction of shear. These observations, while not definitive, suggested a cholesteric to nematic transition occurred on shearing.

Cellulose Acetate (CTA) Mesophases. TFA-CH_2Cl_2 mixtures are also excellent solvents for CTA. Patel and Gilbert (*17*) showed the CTA mesophase is cholesteric in nature. They employed various TFA-CH_2Cl_2 ratios. Here, data will be given only for 60:40 (v/v) mixtures.

The rate of degradation is higher in the isotropic solutions (e.g., 14 and 19% (w/w)) than in the anisotropic solutions (Table VII). This would suggest that the mesophases are less accessible and therefore less susceptible to attack by TFA than the isotropic regions.

Table VII. Molecular Weight Data for Original CTA and CTA Precipitated From Solution in TFA-CH_2Cl_2 (60:40 v/v) After 2-12 Weeks of Dissolution

Concentration CTA (w/w)	Molecular Weight [a]						
	Week 0	Week 2	Week 4	Week 6	Week 8	Week 10	Week 12
14%	95,000	58,100	55,300	46,400	38,300	30,900	26,200
19%	95,000	54,600	46,000	37,800	34,900	31,200	26,300
24%	95,000	85,000	77,200	72,300	66,900	57,700	51,600
30%	95,000	82,000	73,700	68,200	60,400	55,600	49,000

[a]Values rounded to nearest hundred

Solution viscosities vs time for 10-32% (w/w) CTA solutions are in Table VIII and Figure 2. The viscosity decreased rapidly at CTA concentrations slightly above 20% and exhibited a minimum at 23% (w/w). This confirms V_2^C is about 20% (w/w), but the minimum does not correspond to the transition of the biphasic phase to a wholly anisotropic phase. A 21% (w/w) solution was completely anisotropic as determined by polarized light microscopy (Table IX). Above 23% (w/w) concentration, the viscosity increased with increasing concentrations. That is, CTA solutions in TFA-CH_2Cl_2 exhibit the expected behavior, contrary to cellulose. However, V_2^C is surprisingly low for a semi-rigid polymer. For example, the V_2^C for poly(phenylene terephthalamide), a rigid rod polymer, is about 9% in 100% H_2SO_4.

Table VIII. Viscosity Data for Solutions of CTA in TFA-CH_2Cl_2 (60:40 v/v) After 2-12 Weeks of Dissolution at Ambient Temperature. Measurements Were Made at 25°C

Concentration CTA (w/w)	Viscosity (Poise)					
	Week 2	Week 4	Week 6	Week 8	Week 10	Week 12
10%	144	121	78	50	53	25
14%	473	281	227	133	105	56
16%	942	664	381	246	175	124
18%	1,229	924	599	375	284	207
19%	1,683	1,123	742	554	352	253
20%	a	1,900	1,280	766	453	314
21%	1,237	a	1,430	1,086	615	362
22%	458	573	1,174	1,034	827	494
23%	383	388	435	463	582	502
24%	407	387	405	361	444	550
26%	512	473	450	422	570	689
28%	665	613	541	643	866	1,080
30%	825	758	682	926	1,756	a
32%	1,012	865	978	1,844	a	a

[a]Viscosity outside range of viscometer

Table IX. Summary of Polarized Light Microscopy for Solutions of CTA in TFA-CH_2Cl_2 (60:40 v/v)

CTA Concentration	Week 2	Week 4	Week 6	Week 8	Week 10	Week 12
14% (w/w)	I[a]	I	I	I	I	I
16% (w/w)	I	I	I	I	I	I
18% (w/w)	I	I	I	I	I	I
19% (w/w)	I	I	I	I	I	I
20% (w/w)	B[a]	B	B	I	I	I
21% (w/w)	A[a]	A	B	B	I	I
22% (w/w)	A	A	B	B	I	I
24% (w/w)	A	A	A	B	B	I
26% (w/w)	A	A	A	A	B	I
28% (w/w)	A	A	A	A	B	I
30% (w/w)	A	A	A	A	B	I
32% (w/w)	A	A	A	A	B	I

[a]I = Isotropic, no birefringence observed
B = Biphasic, areas of birefringence and areas of extinction
A = Anisotropic, birefringence observed throughout

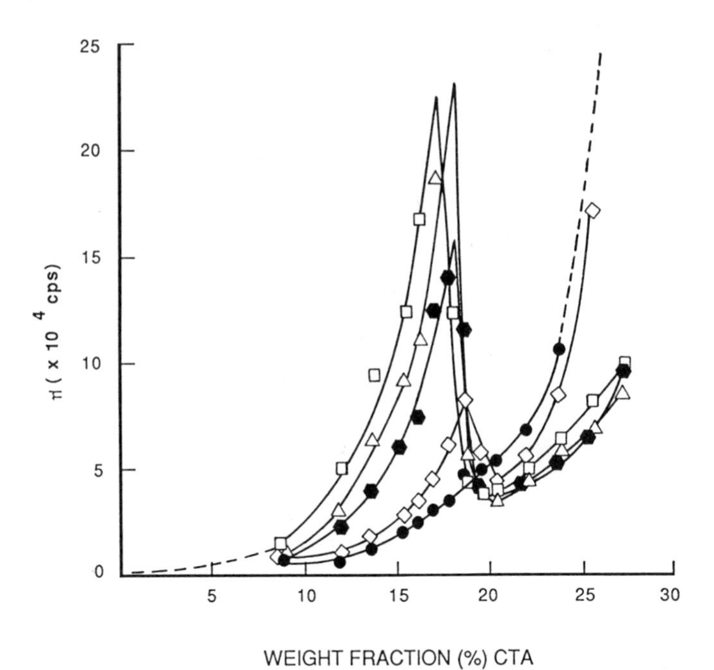

Figure 2. Dependence of viscosity on time of dissolution: Viscosity versus concentration for solutions of 14-32% (w/w) CTA in TFA-CH_2Cl_2 (60:40 v/v) after [□] 2, [Δ] 4, [ ] 6, [◊] 10, and [●] 12 weeks of dissolution.

Figure 2 also demonstrates that the CTA is slowly degraded in the TFA-CH_2Cl_2 mixture. As dissolution time increased, the viscosity peak shifted to higher concentrations. Polarized light microscopy indicated the concentration at which anisotropy occurs also increased slightly with dissolution time (Table IX) and correlates with the shift in the viscosity peaks. The concentration at which the viscosity minimum occurs remained essentially constant with time in solution as did the viscosity minimum (~ 3.5 x 10^4 cps).

Simple shear studies with the CTA solutions showed various textures were obtained depending upon the solution concentration, but they were not stable. Grainy textures were obtained with 22% (w/w) and higher solution concentrations, but no clear evidence of a cholesteric to nematic transition was observed.

Ternary Systems. Flory (19) predicted that in a ternary mixture of rod-like molecules and random coils in an isodiametrical solvent the free energy of the nematic phase would be increased by the presence of the random coil. Consequently, the random coil "is predicted to be virtually excluded" from the nematic phase.

Polyethylene terephthalate (PET) and methyl methacrylate (PMMA) are readily soluble in TFA-CH_2Cl_2 mixtures, but they only form isotropic solutions. The mutual solubility of cellulose (CELLOH) CTA, PET and PMMA in TFA-CH_2Cl_2 offers the opportunity to determine if Flory's prediction applies to ternary systems consisting of cellulose or a cellulose derivative and a synthetic polymer.

Binary solutions of CTA/TFA-CH_2Cl_2 (60:40 v/v), PET/solvent and PMMA/solvent were prepared and mixed in various ratios to obtain ternary solutions. After a lapsed time of 2 weeks, the solutions were observed visually and microscopically. The results are summarized in Table X. In agreement with Flory's prediction, the anisotropic phase is rich in CTA and the isotropic phase rich in PMMA or PET. However, at least within experimental error and in the timeframe of these experiments, the flexible polymer (PMMA or PET) was not completely excluded from the anisotropic phase, contrary to results reported for other combinations of a flexible and a rigid polymer (*20-22*).

The presence of the PMMA or PET lowers the critical concentration of the CTA. For example, a 19.6/1.2/70.2 (w/w/w) ternary solution of CTA, PMMA and TFA-CH_2Cl_2 (6:4 v/v) and an 18.6/1.2/80.2 ternary solution of CTA, PET and TFA-CH_2Cl_2 (6:4 v/v) were biphasic when viewed under crossed polars. Each solution appeared to be one phase but a small isotropic phase may have been present, and required an extremely long time to separate due to the high viscosity of the anisotropic matrix.

The lowering of the critical concentration of the CTA by the presence of the flexible polymer may be explained in terms of a lattice model. The large excluded volume of PMMA or PET decreases the number of lattice sites available to the semi-rigid CTA polymer; thus the CTA molecules are required to pack more efficiently and are more aligned in the ordered phase.

Preliminary data for CELLOH, PET and TFA-CH_2Cl_2 (6:4 v/v) ternary systems indicate they also obey Flory's prediction (Table X), but similar to the results for CTA complete exclusion of the flexible polymers from the anisotropic phase was not realized.

The anisotropy of the phases rich in the semi-rigid polymer (CTA or CELLOH) was demonstrated by optical microscopy and optical rotations. Typical optical rotations for CTA (or CELLOH), PET and solvent ternary mixtures are shown in Table XI.

TABLE X

Ternary Mixtures of PMMA/CTA in TFA-CH_2Cl_2 (6:4 v/v)

Solution No.	CTA / PMMA	Composition CTA/PMMA/Solv. (w/w/w)	Conjugate Phase Volumes Iso/Aniso %	Conjugate Phase CTA/PMMA/Solv. Iso	Composition (w/w/w) Aniso
1	7:3	19.6/3.0/77.3	5.6/94.4	0.9/8.7/90.7	20.7/26/76.7
2	5:5	14.0/5.0/81.0	11.3/80.7	1.1/26.2/72.6	15.6/2.0/82.4

Ternary Mixtures of PET/CTA in TFA-CH_2Cl_2 (6:4 v/v)

Solution No.	CTA / PET	Composition CTA/PMMA/Solv. (w/w/w)	Conjugate Phase Volumes Iso/Aniso %	Conjugate Phase CTA/PMMA/Solv. Iso	Composition (w/w/w) Aniso
3	3:1	14.0/4.6/81.4	43.5/56.5	1.6/11.8/86.4	22.9/0.3/76.7
4	5:1	15.5/3.1/81.4	28.0/72.0	0.13/12.0/86.3	25.7/0.6/73.4
5	6:1	18.1/3.0/78.9	37.0/63.0	1.6/13.0/85.4	27.0/0.1/72.9
6	7:1	18.8/2.7/78.7	21.3/78.7	1.6/15.5/83.0	29.6/0.3/7.02

Ternary Mixtures of PET/CELLOH in TFA-CH_2Cl_2 (6:4 v/v)

Solution No.	CELLOH / PET	Composition CTA/PMMA/Solv. (w/w/w)	Conjugate Phase Volumes Iso/Aniso %	Conjugate Phase CTA/PMMA/Solv. Iso	Composition (w/w/w) Aniso
7	8:1	20.5/2.6/76.9	26.6/73.4	0.9/13.9/85.2	34.6/0.6/64.0
8	2.4:1	9.3/3.9/86.8	17.0/83.0	1.1/20.4/78.5	11.1/0.6/88.3
9	3:1	11.3/3.9/84.8	16.8/83.2	0.7/22.5/76.8	13.6/0.2/86.2
10	1.3:1	8.1/6.5/85.8	34.5/65.5	1.2/17.8/81.0	11.4/0.6/88.0

Table XI. Optical Rotation $\left(\frac{deg.g}{dmml}\right)$ of Ternary Mixtures

CTA, PET, TFA-CH_2Cl_2 (6/4: v/v) Solvent

Solution No.	Composition (CTA/PET/Solvent) (w/w/w)	Wave Length ($1/l^{2}$ x 10^{-6} nm^{-2}) 2.88	2.99	3.35	5.26	7.51
11	23.5/0.9/76.0	667	720	1025	2484	3459
12	25.4/0.6/74.0	1181	1372	2028	2832	3653
13	29.0/0.3/70.6	1109	1263	1701	2788	3824

All Optical Rotations Negative

CELLOH, PET, TFA-CH_2Cl_2 (6/4: v/v) Solvent

	(CELLOH/PET/Solvent) (w/w/w)					
14	12.0/0.3/87.7	14871	18698	21124	51551	76672
15	14.6/0.3/85.1	44047	47244	51924	95495	119254
16	14.4/1.3/84.3	63050	72004	93363	134395	169388

All Optical Rotations Positive

It may be noted that the optical rotations for the solutions containing CELLOH are positive (and higher) compared to those for the solutions containing CTA which have negative rotations. This was observed previously by Patel and Gilbert (*3,17*) for pure CELLOH and CTA solutions in TFA-CH_2Cl_2 mixtures. The sense of the cholesteric helix must be positive in the case of CELLOH and negative for CTA. The reason for the higher optical rotations for the CELLOH solutions compared to the CTA solutions is not obvious based on the number of chiral centers per structural unit, but may be related to the tightness of the pitch. While some data (*17*) exists on the pitch of CTA in TFA-CH_2Cl_2, none is available for CELLOH in TFA-CH_2Cl_2 due to the small domain size of the mesophase, that is, a fingerprint pattern is not observed in the case of CELLOH.

As discussed above, the critical concentration of CTA is lowered in the presence of a flexible polymer (PMMA or PET). However, when the ternary mixtures of CELLOH were prepared (*23*), the very low critical concentration of CELLOH in TFA-CH_2Cl_2 had not been observed (*8*) and presently no results are available for a ternary mixture containing ~ 6% cellulose.

Fiber Formation. A CTA/PET/TFA-CH_2Cl_2; 29.1/0.3/70.6 (w/w/w) solution ($\overline{M}$n of CTA, 110,000 and D.S. = 3.0) was selected for a fiber spinning trial because of its relatively low viscosity (560 poise) and high CTA concentration. Parenthetically, it may be pointed out that the addition of PET lowers the viscosity of CTA solutions (*23*). As a simple laboratory scale spinning system was used, the mechanical properties of the fibers were not optimized. The solution was extruded into MeOH/H_2O (7/3: v/v) coagulation bath using an air-gap between the spinneret and the coagulation bath. The tenacity and modulus of the fiber were determined to be 0.49 and 19.4 GP (4.1 and 164 g/d), respectively. This is significantly higher than for typical CTA fibers and reflects the anisotropic nature of the CTA solution. Higher tenacity and modulus were obtained by O'Brien (*18*) using a more sophisticated spinning apparatus.

The X-ray equatorial diffractogram of the fiber is shown in Figure 3. The diffraction peaks are at spacings corresponding to scattering angles (2Θ) at 7.8, 16.0, 21.2 and 27.6°. These correspond closely to the values reported by Sprague et al (*24*) and to Stipanovic and Sarko's (*25*) predictions for the cellulose triacetate I polymorph. The formation of the CTA I polymorph was surprising as CTA II polymorph is energetically favored. However, Roche et al (*26*) reports the formation of CTA I from solutions of CTA (D.P. ~ 380) in TFA-H_2O (100:8 v/v)

As shown in Figure 1, the viscosity of cellulose solutions in TFA-CH_2Cl_2 do not show a maximum typical of liquid crystalline solutions and are highly viscous above 12% (w/w) concentration. The CELLOH/PET/TFA-CH_2Cl_2 solutions, as noted above, were prepared before it was found that the critical concentration for CELLOH was about 6% (w/w). Due to their high viscosity, it was not possible to extrude satisfactory fibers from CELLOH//PET/TFA-CH_2Cl_2 solutions. Therefore, films were cast. The infrared spectrum (Figure 4) of a film cast from a CELLOH/PET/TFA-CH_2Cl_2, 12.0/0.3/87.7 (w/w/w) solution shows, as expected, a carbonyl peak at 1730 cm^{-1} due to the PET. The broad and strong peaks around 3150-3350 cm^{-1} indicate the cellulose is present as the cellulose II polymorph and strong hydrogen bond are dominant (*27*). The X-ray equatorial diffractogram (Figure 5) indicates the presence of either the cellulose I or IV polymorph in addition to the cellulose II polymorph. It is difficult to distinguish between cellulose I and IV. Interplanar distances (Table XII) shows that the spacing at 5.45 possibly originates from the lattice (101) reflections plane of

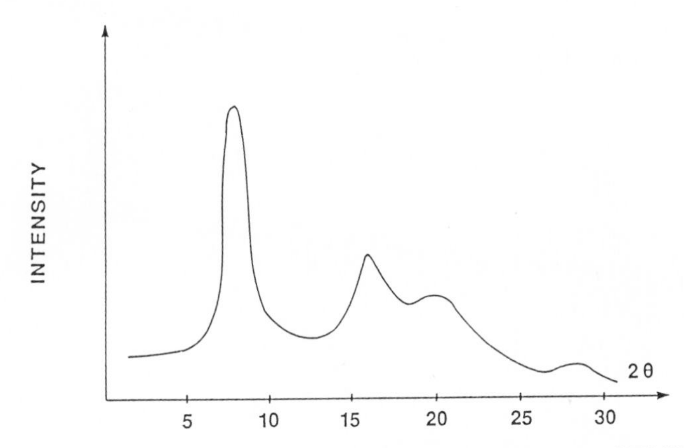

Figure 3. X-ray diffractogram (equatorial scan) of dry spun CTA/PET fiber (annealed) from CTA/PET/TFA-CH_2Cl_2 (29.1/0.3/70.6; w/w/w).

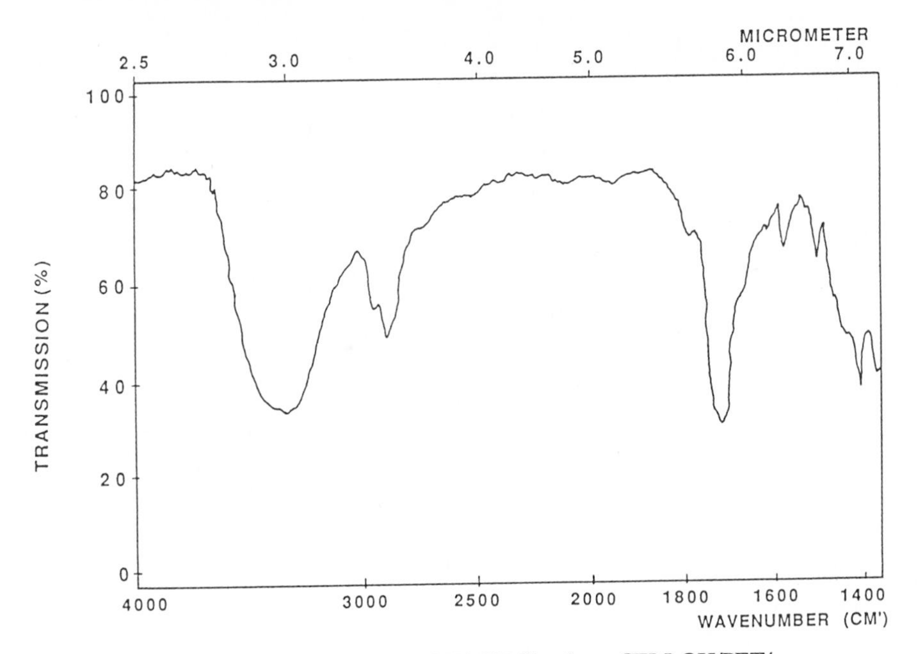

Figure 4. IR spectrum of CELLOH/PET film from CELLOH/PET/TFA-CH_2Cl_2 (12.0/0.3/87.7; w/w) solution.

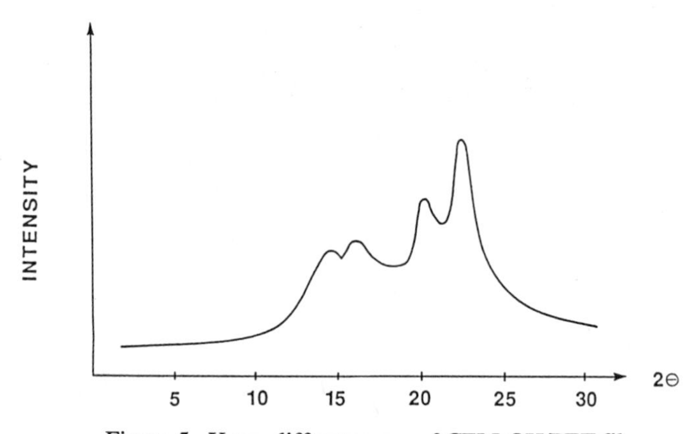

Figure 5. X-ray diffractogram of CELLOH/PET film.

cellulose IV (*28*) and 4.38 and 3.99 from the (101) and (002) reflection planes respectively of cellulose II (*29*).

Table XII. Interplanar Distances

Cellulose From CELLOH/PET/TFA-CH_2Cl_2 (12.0/0.3/87.7; w/w/w/Solution)

Θ°	d spacing (A°)
7.30	6.47
7.95	5.57
10.15	4.38
11.14	3.99

It would be of interest to spin fibers from solutions of cellulose or cellulose-PET at concentrations in the 6-10% (w/w) range to further study the type of polymorph produced and to examine the possibility of obtaining high strength and modulus cellulose fibers and characteristics of intimate blends of cellulose and PET.

Ternary Systems - Block Copolymers. As discussed above, mixtures of cellulose or CTA with PET and CTA with PMMA obey Flory's prediction and separate into isotropic and anisotropic phases.

Ciferri (*30*) suggested that block copolymers, containing rigid (or semi-rigid) blocks and flexible blocks should not phase separate when admixed with the flexible polymer. Gilbert and coworkers (*31*) synthesized block copolymers of CTA and polyesters with the objective of preparing biodegradable polymers. Briefly, the synthesis involves hydrolyzing CTA with hydronium ions to prepare oligomeric, hydroxy-terminated CTA, which is capped with a diisocyanate and then reacted with a hydroxy-terminated polyester. Alternatively, the hydroxy-terminated polyester may be end-capped and reacted with the hydroxy-terminated CTA.

An ABA block copolymer was prepared using the latter route. That is, the A blocks were CTA and the B block an oligomeric poly(ethylene-co-propylene) adipate (PEPA).

Three oligomer hydroxy-terminated CTAs (HCTA) were first prepared. Their molecular weight and A and B points in TFA-CH_2Cl_2 (6:4 v/v) are in Table XIII. That is, these oligomers formed mesophases in TFA-CH_2Cl_2 and the critical

Table XIII. The Influence of Molecular Weight on the A and B Points

Sample	Mv	DP	A% (w/w)	B% (w/w)
HCTA-1	19,232	67	22	33
HCTA-2	15,835	55	27	41
HCTA-5	13,203	46	28	42

concentrations (A point) is dependent on the molecular weight, as expected. The block copolymers (BCOP) formed a biphasic solution at 25% (w/w) and a wholly anisotropic solution at 39% (w/w) based on optical microscopy.

Mixtures of HCTA, PEPA were prepared in TFA-CH_2Cl_2 (6:4 v/v). Visual observations of these ternary mixtures are recorded in Table XIV. The solution coded H-4 phase separated after 2 weeks into isotropic and anisotropic phases.

Table XIV. Results of the HCTA-1/PEPA/TFA-CH_2Cl_2 Ternary Mixtures

HCTA-1 Solution Mv = 19,232 % (w/w)	PEPA Solution % (w/w)	Ratio (w/w)	HCTA-1/PEPA/ Solvent (w/w/w)	Separation	Solution Code
	25.0	10/2.5	32/5/63	N, B	H-5
40.0	20.0	9.4/3.2	30/5/65	N, B	H-6
	16.6	8.8/3.8	28/5/67	Y, I, B	H-4

Y = phase separation was observed
N = no phase separation was observed
I = isotropic
B = biphasic
A = anisotropic

Solutions H-5 and H-6 did not phase separate after 1.5 weeks standing (after standing for a longer time, they did phase separate). Ternary mixture of BCOP/PEPA/TFA-CH_2Cl_2 (60:40 v/v) were prepared. After 6 weeks standing, there was no visible phase separation (Table XV) (phase separation was not visible after 6 months storage). These results indicate that due to the compatibility of the flexible polymer (PEPA) with the PEPA component in the block copolymers, the flexible polymer was not rejected by the mesophase. These data suggest that molecular composites may be prepared from mixtures of block copolymers, consisting of rigid (or semi-rigid) and flexible blocks and flexible polymers.

Table XV. Results of the BCOP/PEPA/TFA-CH_2Cl_2 Ternary Mixtures

BCOP Solution % (w/w)	PEPA Solution % (w/w)	Ratio (w/w)	BCOP/PEPA/ TFA-CH_2Cl_2 (w/w/w)	Separation	Solution Code
	25.0	10/2.5	32/5/63	N, B	B-3
40.0	20.0	9.4/3.2	30/5/65	N, B	B-1
	16.6	7.7/3.3	28/5/67	Y, I, B	B-2

Y = phase separation
N = no phase separation
I = isotropic
B = biphasic
A = anisotropic

Literature Cited

1. Werbowyj, R. S.; Gray, D. E. *Mol. Cryst. Liq. Cryst.* **1976**, *34*, 97 .
2. Chanzy, H.; Peguy, A. *J. Polymer Sci.*, Polym. Phys. Ed. **1980**, *18*, 1137.

3. Patel, D. L.; Gilbert, R. D. *J. Polym. Sci.*, Polym. Phys. Ed. **1981**, *19*, 1231.
4. Flory, P. J. *Proc. R. Soc.* (London), Ser. **1956**, A *234*, 60.
5. Flory, P. J. *Proc. R. Soc.* (London), Ser. **1956**, A *234*, 73.
6. Gray, D. G. In *Polymeric Liquid Crystals* ; Blumstein, A., Ed.; Plenum: New York, 1985.
7. Gilbert, R. D.; Patton, P. A. *Prog. Polym. Sci.* **1983**, *9*, 115.
8. Hawkinson, D., M. S. Thesis, North Carolina State University, 1987.
9. Conio, G.; Corazzo, P.; Bianchi, E.; Teoldi, A.; Ciferri, A. *J. Polym. Sci.*, Polym. Lett. Ed. **1984**, *22*, 273.
10. Bianchi, E.; Ciferri, A.; Conio, G.; Coseni, A.; Terbojevick, M. *Macromolecules* **1985**, *18*, 646.
11. Terbojevick, M.; Coseni, A.; Conio, G.; Ciferri, A.; Bianchi, E. *Macromolecules* **1985**, *18*, 640.
12. McCormick, C. L.; Callair, P. A.; Hutchinson, B. H. *Macromolecules* **1985**, *18*, 2394.
13. Chen, Y. S.; Cuculo, J. A. *J. Polym. Sci.*, Polym. Chem. Ed. **1986**, *24*, 2075.
14. Yang, K. S., Ph.D. Thesis, North Carolina State University, 1988.
15. Kuhn, W. *Kolloid-Z* **1936**, *76*, 258; **1939**, *87*, 3.
16. Peterlin, A. *Polym. Prepr.*, Am. Chem. Soc., Div. of Polymer Chem. **1968**, 9, 323.
17. Patel, D. L.; Gilbert, R. D. *J. Polym. Sci.*, Polym. Phys. Ed. **1981**, *19*, 1449.
18. O'Brien, J. P. U.S. Patent 4 464 323; 4 501 886, 1984.
19. Flory, P. J. *Macromolecules* **1981**, *14*, (4), 1138.
20. Aharoni, S. M. *Polymer* **1980**, *21*, 21.
21. Bianchi, E. A.; Ciferri, A.; Teoldi, A. *Macromolecules* **1982**, *15*, 1268.
22. Marsano, E.; Bianchi, E.; Ciferri, A.; Ramis, E.; Teoldi, R. *Macromolecules* **1986**, *19*, (3), 626.
23. Hong, Y. E., Ph.D. Thesis, North Carolina State University, May 1985.
24. Sprague, B. S.; Riley, J. J.; Noether, H. E. *Text. Res. J.* **1958**, *28*, 257.
25. Stipanovic, A. J.; Sarko, A. *Polymer* **1978**, *19*, 3.
26. Roche, E. J.; O'Brien, J. P.; Allen, S. R. *Polymer Communications* **1986**, *27*, 138.
27. *Instrumental Analysis of Cotton Cellulose and Modified Cotton Cellulose*; O'Connors, J., Ed.; Marcel Dekker: New York, 1972.
28. Howmon, J. A.; Sisson, W. A. In *Cellulose and Cellulose Derivatives*; Part 1, 2nd ed.; Ott, E.; Spurlin, H. M.; Griffin, M. W., Eds.; Wiley-Interscience: New York, 1954; p 23.
29. Aharoni, S. M. *J. Macromol. Sci. Phys.* **1982**, *B21*, 287.
30. Ciferri, A. Private Communication.
31. Gilbert, R. D.; Stannett, V. T. In *Developments in Block Copolymers-2*; Goodman, I., Ed.; Elsevier, 1985.

RECEIVED August 26, 1988

Chapter 13

Polymerization of Lyotropic Liquid Crystals

David M. Anderson and Pelle Ström

Physical Chemistry 1, University of Lund, Lund, Sweden

The polymerization of one or more components of a lyotropic liquid crystal in such a way as to preserve and fixate the microstructure has recently been successfully performed. This opens up new avenues for the study and technological application of these periodic microstructures. Of particular importance are the so-called bicontinuous cubic phases, having triply-periodic microstructures in which aqueous and hydrocarbon components are simultaneously continuous. It is shown that the polymerization of one of these components, followed by removal of the liquid components, leads to the first microporous polymeric material exhibiting a continuous, triply-periodic porespace with monodisperse, nanometer-sized pores. It is also shown that proteins can be immobilized inside of polymerized cubic phases to create a reaction medium allowing continuous flow of reactants and products, and providing a natural lipid environment for the proteins.

This chapter focuses on the fixation of lyotropic liquid crystalline phases by the polymerization of one (or more) component(s) following equilibration of the phase. The primary emphasis will be on the polymerization of bicontinuous cubic phases, a particular class of liquid crystals which exhibit simultaneous continuity of hydrophilic -- usually aqueous -- and hydrophobic -- typically hydrocarbon -- components, a property known as 'bicontinuity' (1), together with cubic crystallographic symmetry (2). The potential technological impact of such a process lies in the fact that after polymerization of one component to form a continuous polymeric matrix, removal of the other component creates a microporous material with a highly-branched, monodisperse, triply-periodic porespace (3).

While there have been efforts to polymerize other surfactant mesophases and metastable phases, bicontinuous cubic phases have only very recently been the subject of polymerization work. Through the use of polymerizable surfactants, and aqueous monomers, in particular acrylamide, polymerization reactions have been performed in vesicles (4-8), surfactant foams (9), inverted micellar solutions (10), hexagonal phase liquid crystals (11), and bicontinuous microemulsions (12). In the latter two cases rearrangement of the microstructure occured during polymerization, which in the case of bicontinuous microemulsions seems inevitable because microemulsions are of low viscosity and continually rearranging on the timescale of microseconds due to thermal disruption (13). In contrast, bicontinuous cubic phases are extremely viscous in general, and although the components display self-diffusion rates comparable to those

0097–6156/89/0384–0204$06.25/0

in bulk, their diffusion nevertheless conforms to the periodic microstructure which is rearranging only very slowly. In fact, recently cubic phases have been prepared which display single-crystal X-ray patterns (14). In the authors' laboratory, experiments are now performed in which bicontinuous cubic phases are routinely polymerized without loss of cubic crystallographic order. The fact that, in spite of the high viscosity and high degree of periodic order, bicontinuous cubic phases have only recently been the focus of polymerization experiments can be traced to several causes, most notably: a) cubic phases cannot be detected by optical textures and usually exist over quite narrow concentration ranges; b) the visualization and understanding of the bicontinuous cubic phase microstructures pose difficult mathematical problems; and c) the focus of research on cubic phases has been on binary systems, in particular on biological {lipid / water} systems, whereas the best cubic phases from the standpoint of straightforward polymerization experiments are ternary {surfactant / water / hydrophobic monomer} systems.

As is clearly discussed in a recent review of polymerized liposomes (15), a distinction must be drawn between *polymerized* and *polymeric* surfactant microstructures. In *polymeric* microstructures, the polymerization is carried out before the preparation of the phase, whereas the term *polymerized* means that the microstructure is formed first, and then the polymerization reaction performed with the aim of fixating the microstructure as formed by the monomeric components. Although this chapter deals mainly with polymerized microstructures, polymeric cubic phases are discussed in a separate section at the end.

The next section and the final section on polymeric cubic phases are intended for those readers who seek a more in-depth understanding of the microstructures involved, including the geometrical aspects as well as the physics behind the self-assembly into these structures. These sections may be omitted by the more casual reader.

The Bicontinuous Cubic Phases: Mathematical Principles.

An understanding of the basic mathematical principles that apply to the physics and the geometry of the bicontinuous cubic phases is necessary for full appreciation of what follows. Since 1976 (1), it has been known that a complete understanding of bicontinuous cubic phases requires an understanding of Differential Geometry and in particular, of a class of mathematical surfaces known as periodic minimal surfaces (often referred to as IPMS, for infinite periodic minimal surfaces; clearly 'infinite' is redundant). A *minimal surface* is defined to be a surface of everywhere zero mean curvature; the mean curvature at a point on a surface is one-half the sum of the (signed) principle curvatures, so that every point on a minimal surface is a balanced saddle point: $\kappa_1=-\kappa_2$. The utility of periodic minimal surfaces of cubic symmetry, and of their constant mean curvature relatives, in the understanding of bicontinuous cubic phases is now well-established, and we begin with a short introduction to these surfaces. There has been considerable confusion in the literature over these complicated surfaces and even of their fundamental basis in the field of surfactant microstructures, but in the last few years this has become considerably clarified.

The first source of confusion was the fact that minimal surfaces represent local minima in surface area under Plateau (or 'fixed boundary') boundary conditions. The importance of this property with respect to cubic phases must be considered to be limited, however, because the surface area of the interfacial dividing surface -- drawn between the hydrophilic and the hydrophobic regions of the microstucture -- is given simply by the product of the number of surfactant molecules, times the average area per surfactant which is strongly fixed by the steric, van der Waals, and electrostatic interactions between surfactant molecules. Therefore this interfacial area does not in general seek a minimum but rather an optimum value, which does not tend to zero because of the electrostatic repulsion between surfactant head groups. Furthermore, the fixed boundary conditions that lead to minimal surfaces are not as appropriate as boundary conditions which result upon enforcement of the *volume fractions* of the hydrophilic and hydrophobic moieties in the unit cell. Minimization of area under such constraints leads to surfaces of constant mean curvature -- or 'H-surfaces' -- which can possess significantly lower interfacial areas than the corresponding minimal surfaces of the same symmetry and topological type (16).

The traditional microstructures -- spheres, cylinders, and lamellae -- all have constant mean curvature dividing surfaces, and, as discussed below, the same appears to be true for bicontinuous cubic phases. However at the same volume fraction, the different competing microstructures give rise to different values of the mean curvature, and a belief that is now firmly embedded in the study of surfactant microstructures is that the structure which is most favorable under given conditions is that which satisfies most closely the 'preferred' or 'spontaneous' mean curvature (17). The spontaneous mean curvature is determined by the balance of forces -- steric, electrostatic, etc. -- between the surfactant head groups, and between the surfactant tails, and thus is sensitive to, e.g., salinity, oil penetration, head group hydration, etc. In the liquid crystals of interest here, the surfactant-rich film is tending toward a homogeneous state in which each surfactant molecule sees the same local environment, regardless of where on the monolayer it is located and, if this monolayer is one-half of a bilayer, regardless of which side of the bilayer it is on. (In certain biological systems there is a significant asymmetry with respect to the two sides of the bilayer, which is of great importance; however, we are dealing for the moment with the symmetric bilayer). Thus each monolayer is driven toward the most homogeneous state which implies a constant mean curvature.

A second source of confusion that still persists to some extent in the literature is the matter of where the interfacial surface is to be drawn. For those cubic phase structures discussed below in which a bilayer is draped over a minimal surface, this minimal surface describes the midplane (or better, 'midsurface') of the bilayer and not the interface between polar and apolar regions; that is, it describes the location of the terminal methyl groups on the surfactant tails, not the dividing point between the hydrophilic head group and the hydrophobic (usually hydrocarbon) tail. The actual polar / apolar dividing surface is displaced from the minimal surface by the length of the hydrophobic tail, on both sides of the minimal surface. While it can be argued as to exactly where in the bilayer profile these two polar / apolar dividing surfaces should be drawn, it is clear than any sensible convention should place them near the first methyl group in the tail and not at the terminal methyl at the tail end. Thus bilayer cubic phases should not be referred to as having a zero mean curvature interface.

Recently, application of geometry and differential geometry to this problem has treated these matters quantitatively. For the case of a cubic phase whose local structure is that of a bilayer, then it has been shown (18) that the requirement of symmetry with respect to the two sides of the bilayer, and therefore of the two aqueous networks lying on the two sides of the surface, leads directly to minimal surfaces as midplane surfaces, and through a construction involving projections of surfaces in four-dimensional space leads to the minimal surfaces which describe the known bilayer cubic phases. Concerning the shape of the polar / apolar interface in such structures, the mean curvature cannot be identically zero, and here two cases must be distinguished. In normal cubic phases, which usually lie between lamellar and normal hexagonal phases, the mean curvature of the interface is on the average toward the hydrophobic regions, and these regions are well-described by interconnected cylinders. The axes of these cylinders are the edges of the two graphs (referred to as 'skeletal graphs' in reference (19); see also Figure 1 below) that thread the two hydrophobic subspaces. These cylinders satisfy both constant mean curvature at the interface and a constant stretch distance for the surfactant tails (except at the junctions of the cylinders). However, in the inverted cubic phases, usually lying between lamellar and inverted hexagonal phases, the constant mean curvature and

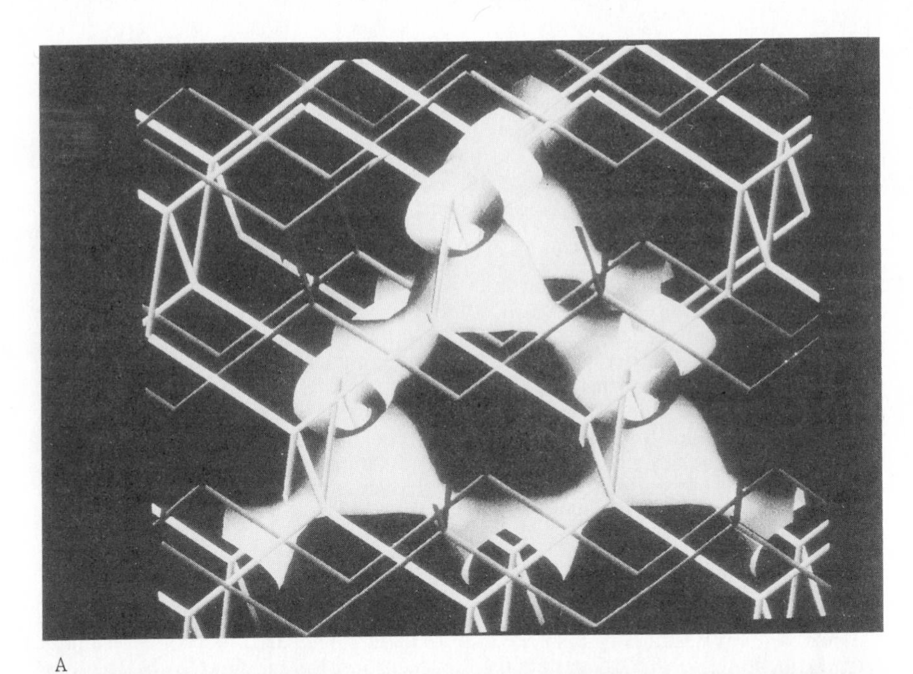

A

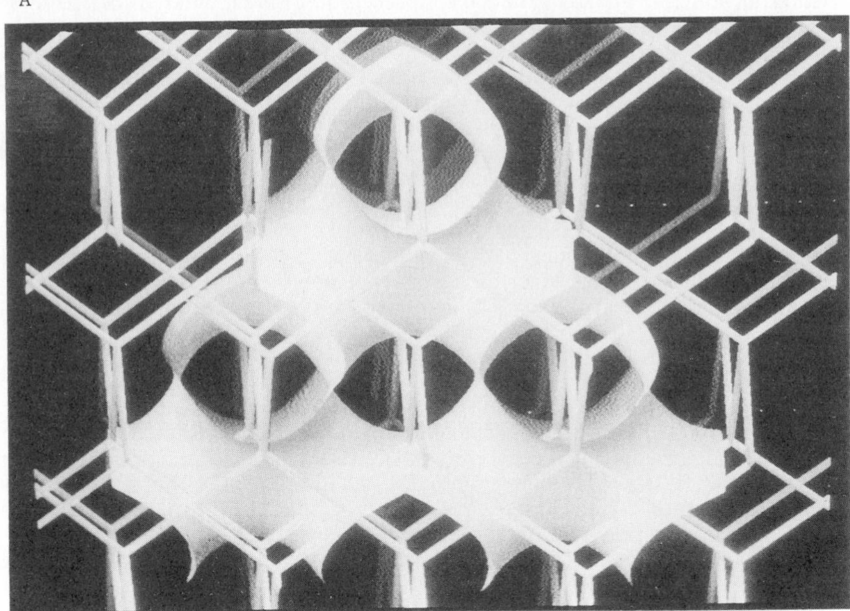

B

Figure 1. A. Computer graphic portion of a periodic surface of constant mean curvature, having the same space group and topological type as the Schwarz D minimal surface. This surface, together with an identical displaced copy, would represent the polar/apolar dividing surface in a cubic phase with space group #224 (Pn3m). The two graphs shown would thread the two aqueous subspaces. B. Computer graphic of a portion of the Schwarz D minimal surface (mean curvature identically zero). In the #224 cubic phase structure, this surface would bisect the surfactant bilayer.

constant distance surfaces do not coincide. This situation has been referred to as 'frustration' (18). Recently, the constant mean curvature configurations have been computed (16), and shown to have rather mild variations in the stretch distance (20), which is measured from the minimal surface to the corresponding point on the constant mean curvature surface.

The Bicontinuous Cubic Phase Microstructures.

We have seen that the balance of forces on the hydrophilic and hydrophobic sides of the surfactant-rich film in a bicontinuous cubic phase determines a 'preferred' or 'spontaneous' mean curvature of the film, measured at the imaginary hydrophilic/hydrophobic dividing surface, so that the optimal shape of this dividing surface is tending toward a homogeneous state of constant mean curvature. In the case where the basic building block of the cubic phase is a surfactant bilayer -- the usual case in binary lipid - water systems -- there is in addition another imaginary surface that describes the midplane (or midsurface) of the bilayer, and this surface must be a minimal surface by symmetry considerations. In this section we discuss each of the known bicontinuous cubic phase microstructures, with the aid of computer graphics that will demonstrate these principles in a visual way.

A Representative Bilayer Structure. An example of a constant mean curvature surface is shown in Figure 1a, together with two skeletal graphs. The surface shown has diamond cubic symmetry, space group F43m. One must imagine an identical copy of the surface shown as being displaced so as to surround the other skeletal graph, leading to double-diamond symmetry, space group Pn3m, #224. One form of inverted cubic phase has this Pn3m symmetry with water located in the two networks lying 'inside' the two surfaces, and the surfactant hydrocarbon tails in the 'matrix' between these two networks, with the two surfaces themselves describing the location of the surfactant head groups, or more precisely, the polar / apolar interface. A triply-periodic minimal surface, known as Schwarz's Diamond (or D) minimal surface (21), shown in Figure 1b, can then be imagined as bisecting the hydrocarbon region. Calculations show that the standard deviation of the stretch distance, from each point on the polar / apolar dividing surface to the minimal surface, is only about 7% of the average distance (20). In the actual cubic phase, the constancy of the mean curvature of the interface might be compromised somewhat in order to achieve even more uniformity in the stretch distance. This would not, however, affect the average value of the mean curvature (Anderson, D. M.; Wennerström, H.; Olsson, U., submitted), which is significantly toward the water.

If, on the other hand, the double-diamond symmetry were found in a normal cubic phase, i.e., with mean curvature on the average toward the hydrocarbon regions, then one would expect to find that the polar / apolar interfacial surface shown in Figure 1a would look instead like interconnected cylindrical rods, because the necks and bulges in Figure 1a would not correspond to water channels but rather to channels occupied by surfactant tails with a preferred stretch distance. Thus far, such a normal cubic phase has not been observed with this symmetry, but has with another symmetry discussed below (#230), and the principles are exactly the same.

It has recently been established (see below) that upon the addition of a protein, for example, to such a structure, a variant of the structure can form in which one of the two water networks is replaced (at least in part) by inverted micelles containing hydrated protein. This changes the space group of the structure, for example #224 changes to #227.

A Monolayer Structure. One of the authors (DMA) has proposed another structure of quite a different nature for a cubic phase occuring in ternary systems involving quaternary ammonium surfactants (16), and this cubic phase is the focus of much of the polymerization work that has been performed. The surfactant didodecyldimethylammonium bromide (DDAB), together with water and a variety of oils, forms a cubic phase whose location is shown in Figure 2 for the case of hexene. Thus the cubic phase exists over a wide range of DDAB / water ratios, but requires a minimum amount of hexene. The same is true for a large number of 'oils' that have been investigated, including alkanes from hexane to tetradecane, alkenes, cyclohexane (22), and monomers such as methylmethacrylate (MMA) and styrene (3). The fact that the cubic phase region extends very close in composition to the L_2 phase region, but not as far as the binary surfactant / water edge, suggests that in this structure the surfactant is locally in the form of a *mono*layer rather than a bilayer.

The model proposed for this cubic phase is shown, for the case of aqueous volume fraction equal to 47%, in Figure 3. One must imagine the oil and the surfactant tails being located on the 'inside' of the dividing surface, water and counterions on the 'outside', and the quaternary ammonium head groups located at or near the depicted surface. The space group is Im3m, #229, which is the same as one of the bilayer cubic structures described below, but these two structures are very different even though the indexing of their X-ray patterns is the same. This structure will be referred to as the 'I-WP' structure (16,19), because the two skeletal graphs are the BCC or 'I' graph (threading the hydrophobic labyrinth) and the 'wrapped package' or 'WP' graph (threading the hydrophilic labyrinth). In Figure 4 are shown three structures in the continuous, one-parameter family (not counting variations in lattice parameter) of I-WP structures, which correspond to aqueous volume fractions of: a) 30%; b) 47%; and c) 65%. This family of constant mean curvature surfaces (16) is proposed to represent the progression in structure as the water / (surfactant + oil) ratio is increased; there is also an increase in lattice parameter with increasing water content, from just under 100Å at low water to about 175Å at the water contents just under 65%. Somewhere around 65% water it appears that there may be a change in symmetry, for lattice parameters at higher water contents are on the order of 300Å. Nonetheless, for water contents less than about 65% the I-WP family of structural models is supported by the following evidence:

1. The indexing and relative peak intensities in SAXS patterns from the cubic phase are fit well by the I-WP model, but not by alternative models (16);

2. TEM micrographs of a polymerized cubic phase match theoretical simulations using the model (3), but not alternative models (see below);

3. Pulsed-gradient NMR self-diffusion data (22) correlate well with theoretical calculations, in which the diffusion equation was solved in the model geometries by a finite element method (Anderson, D. M.; Wennerström, H., in preparation);

4. Values of the area per surfactant head group, calculated from the SAXS lattice parameters assuming the I-WP models, increase from 47Å^2 to 57Å^2 as the water fraction increases from 30% to 65% thus increasing the head group hydration; this compares well with a value of 54Å^2 for the inverted hexagonal phase very near in composition, and with a similar increase in the L_2 phase region (16);

5. The calculated mean curvature of the monolayer goes from toward water at low water content, through zero, to toward oil continuously as the water content increases from less than to greater than 50%; this is well-known in ternary microemulsion systems, and is very hard to reconcile with a bilayer model; futhermore, the mean curvature values in the inverted hexagonal phase at higher oil / surfactant ratio are more strongly toward the water than in the cubic phase assuming the I-WP structure, which fits well with the idea of increased curvature toward water with increasing penetration of oil into the tail region of the monolayer (23);

6. The wide range of hydrophobe / hydrophile ratios in the cubic phase region is also difficult to reconcile with a bilayer model, and in fact has never been observed to this extent in any bilayer cubic phase, but it is readily explained by the progression depicted in Figure 4;

7. The proposed structure at low water content, shown in Figure 4a, ties in very

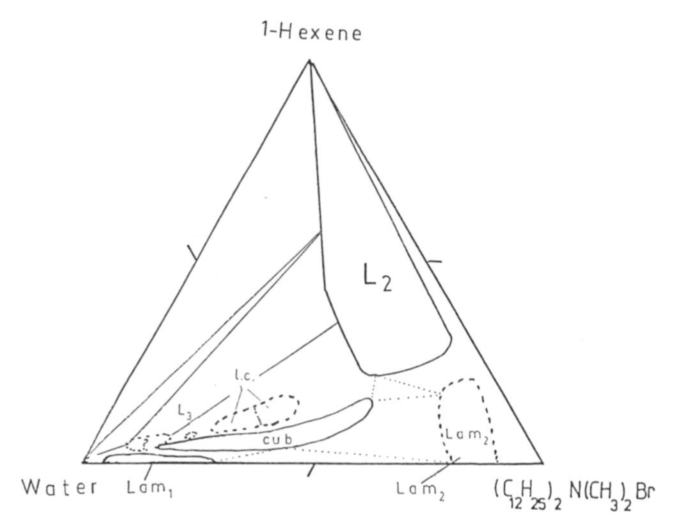

Figure 2. Ternary phase diagram of the system {didodecyldimethylammonium bromide / water / hexene} at 25°C. The nomenclature is cub: cubic phase; Lam_1 and Lam_2: lamellar phases; l.c.: liquid crystalline, inverted hexagonal phase; L_2: microemulsion phase, with curvature toward water. (Courtesy of K. Fontell).

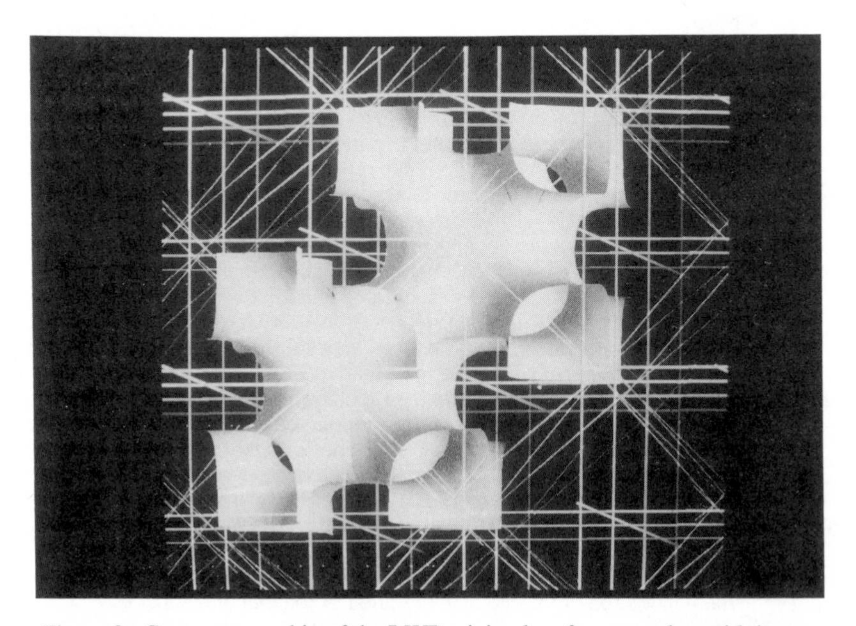

Figure 3. Computer graphic of the I-WP minimal surface, together with its two associated graphs. The lighter graph is the 'I', or BCC, graph, and the heavier graph is the 'WP' or wrapped package' graph. The space group is #229.

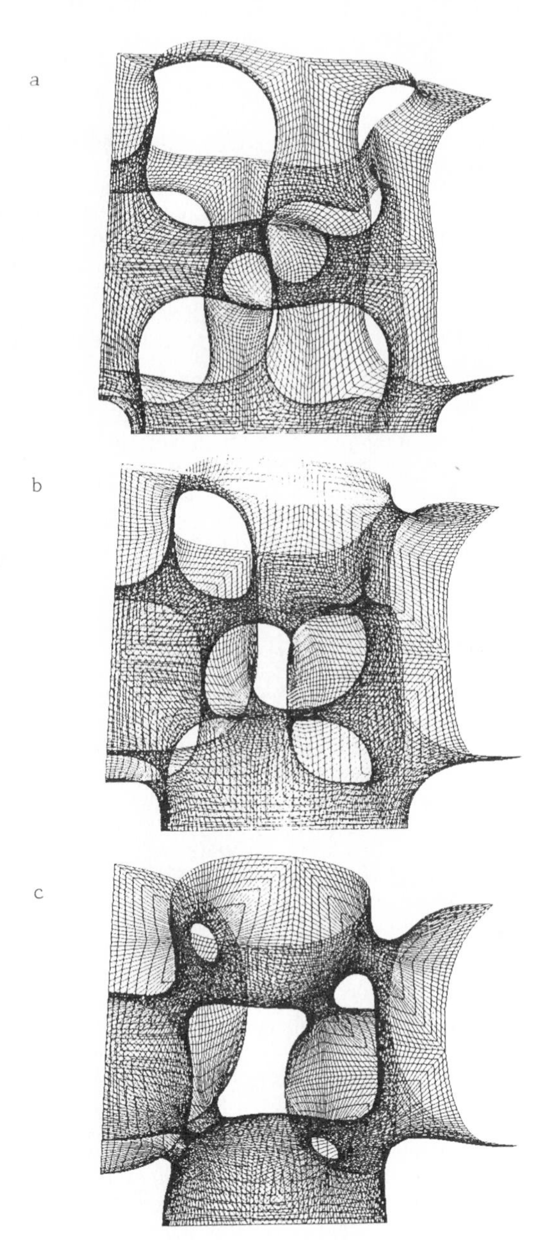

Figure 4. Computer line drawings (without hidden line removal) of three representatives of the I-WP family of constant mean curvature surfaces. These surfaces are invoked to describe the polar/apolar dividing surface in the {DDAB / water / hydrophobe} cubic phases, at water volume fractions of a) 35%; b) 47%; c) 65%.

well with the microstructure that is now generally accepted for the low-water-content microemulsions in the nearby L_2 phase region, namely a bicontinuous, monolayer structure with water lying inside a network of interconnected tubules.

The Known Bicontinuous Cubic Phase Structures. Recording the structures that have been proposed for bicontinuous cubic phases:

>> #224, with the Schwarz Diamond minimal surface describing the midplane of a bilayer; also known as the 'double-diamond' structure, well-established in, e.g., the {glycerol monooleate (GMO or monoolein) / water} system (24), described in detail above; the double-diamond structure is also found in block copolymers (25, 26) (see the final section).

>> #227, obtained from #224 by replacing one of the water labyrinths with inverted micelles; observed when oleic acid is added to monoolein / water at acidic pH (27).

>> #229, the space group of two distinct structures:

a) the *bi*layer structure with the Schwarz Primitive minimal surface describing the midplane of a bilayer; this minimal surface has six 'arms' protruding through the faces of each cube; this structure has been more difficult to establish unambiguously, but appears to occur in, e.g., {monoolein / water} systems and with added cytochrome (27), and in {sodium dodecyl sulphate / water} (28);

b) the I-WP *mono*layer cubic phase described in detail above.

>> #230, with Schoen's 'gyroid' minimal surface (19) describing the midplane of a bilayer (29); the two water networks in this structure are enantiomorphic, and characterized by screw symmetries rather than reflectional or rotational; this appears to be the most common cubic structure, at least in lipids; the normal form of this structure also exists, in which the two enantiomorphic networks are filled with surfactant, and the minimal surface is the midplane of an aqueous network; this normal form occurs in, e.g., some simple soaps (30), and ethoxylated alcohol systems (14).

>> #212, obtained from #230 by replacing one of the water labyrinths with inverted micelles; this is the only known cubic phase with a non-centrosymmetric space group; found in the {monoolein / water / cytochrome-c} system (27), and also by the authors at the same composition but with monolinolein replacing monoolein (see below).

It is interesting to note that, in contrast to the number of bicontinuous cubic phase structures which apparently exist, only one cubic phase structure is now recognized that is not bicontinuous. Furthermore, this structure does not consist of FCC close-packed micelles, but rather a complicated packing of nonspherical micelles (31).

Preparation and Characterization of Polymerized Cubic Phases.

The first bicontinuous cubic phases to be polymerized (3) were the ternary {DDAB / water / hydrophobic monomer} phases described above, which were interpreted as having the 'I-WP' structure. This surfactant was chosen primarily because it was previously known to form bicontinuous phases -- cubic phases and microemulsions -- with many oils or oil-like compounds, including hexane through tetradecane (32), alkenes (23), cyclohexane (33), brominated alkanes (present authors, unpublished), and mixtures of alkanes (33). The location of the cubic phase region in these various systems is rather independent of the choice of hydrophobe, which suggests that the hydrophobe is largely confined to (continuous) hydrophobic channels, having little direct effect on the interactions in the head group region. This makes it an ideal system for investigating polymerization by substituting a hydrophobic monomer.

The composition chosen for the initial experiments was 55.0% DDAB, 35.0%

water, and 10.0% methylmethacrylate (MMA), which had been purified by vacuum distillation and to which had been added 0.004 g/ml of the initiator azobisisobutyronitrile (AIBN). Upon stirring the solution became highly viscous and showed optical isotropy through crossed polarizers, two signs characteristic of the cubic phase (an early name for the cubic phase was in fact the 'viscous isotropic phase'). With other oils such as decane, this composition yields a bicontinuous cubic phase, as indicated by SAXS (16,34) and NMR self-diffusion (34). After equilibrating for one week at 23°C, two samples were prepared for polymerization. The first sample was prepared for SAXS; the phase was smeared onto the end of the plunger of a large syringe, and pushed through an 18 gauge needle into a 1.5 mm i.d. X-ray capillary. The second sample was loaded into a quartz, water-jacketed reaction cell, and nitrogen gas was continually pumped over the sample.

The capillary and the quartz cell were placed in a photochemical reactor having four 340 nm UV lamps, for 36 hours of exposure. At the end of this time the samples were opaque white in appearance. The polymerized material could be rendered clear by the use of a refractive-index matching fluid. To do this, first a large amount of ethanol was used to remove the DDAB, water, and monomeric MMA. Then the sample was dried in a vacuum oven, to yield a solid but highly porous material. Butyl benzene, which has a refractive index (n=1.4898 at 20°C) very close to that of PMMA (1.4893 at 23°C), was imbibed into the porous material, thereby rendering it clear. Upon drying off the butyl benzene, the material once again turned opaque. This is apparently a result of microcrystallites whose sizes are on the order of the wavelength of light; at this low volume fraction of monomer (10.0%), it is easy to imagine that the homogeneity of the polymerized PMMA could be disturbed at the microcrystallite boundaries.

The polymerized sample in the capillary was examined with the modified Kratky Small-Angle X-Ray camera at the University of Minnesota. Due to beam-time limitations (five hours, at 1000 Watts of Cu K_α radiation), the statistics in the data are not particularly good, but (Figure 5) clearly long-range order is indicated by the presence of Bragg peaks. There are not enough peaks for an unambiguous indexing, but the pattern is at least consistent with an indexing to a BCC lattice of lattice parameter 118Å; the first nine theoretical peak positions for this lattice are indicated by vertical lines in Figure 5. The maintenence of cubic crystallographic order through polymerization has also been confirmed recently in K. Fontell's laboratory. The capillary used in the Kratky camera was broken open and the components placed in ethanol, and the insoluble PMMA removed by filter paper and weighed to confirm polymerization of the MMA (monomeric MMA is soluble in ethanol).

The standard method for visualization of microporous polymeric materials is to dry the sample with supercritical drying, which dries the pores without exposing them to the disruptive surface tension forces associated with menisci present during normal evaporation. However, due in part to equipment problems, and in part to the small scale of the pores, this has not yet been performed on a polymerized cubic phase. Transmission electron microscopy has, however, been performed on an air-dried sample. The second sample above was ultramicrotomed at room temperature, and examined in a Jeol 100 CX electron microscope operating at 100KV in TEM mode. Not only the drying process but also, of course, the microtoming procedure have strong disruptive effects on this highly porous material. Nevertheless, the resulting micrograph (Figure 6a; magnification 1,000,000x) indicates regions of periodic order, and in fact the entire field of view in the micrograph gives indications of being a (disrupted) single microcrystallite, in that a (211) direction runs slightly off-horizontal throughout the micrograph. An optical transform of the negative also substantiated the cubic symmetry. Figure 6b is a simulation of the micrograph using the 'I-WP' model structure; a (111) projection of the model structure was calculated by computer, by sending rays through the model and calculating the portion of each ray that lies in void, and in polymer (microscope aberrations were not simulated).

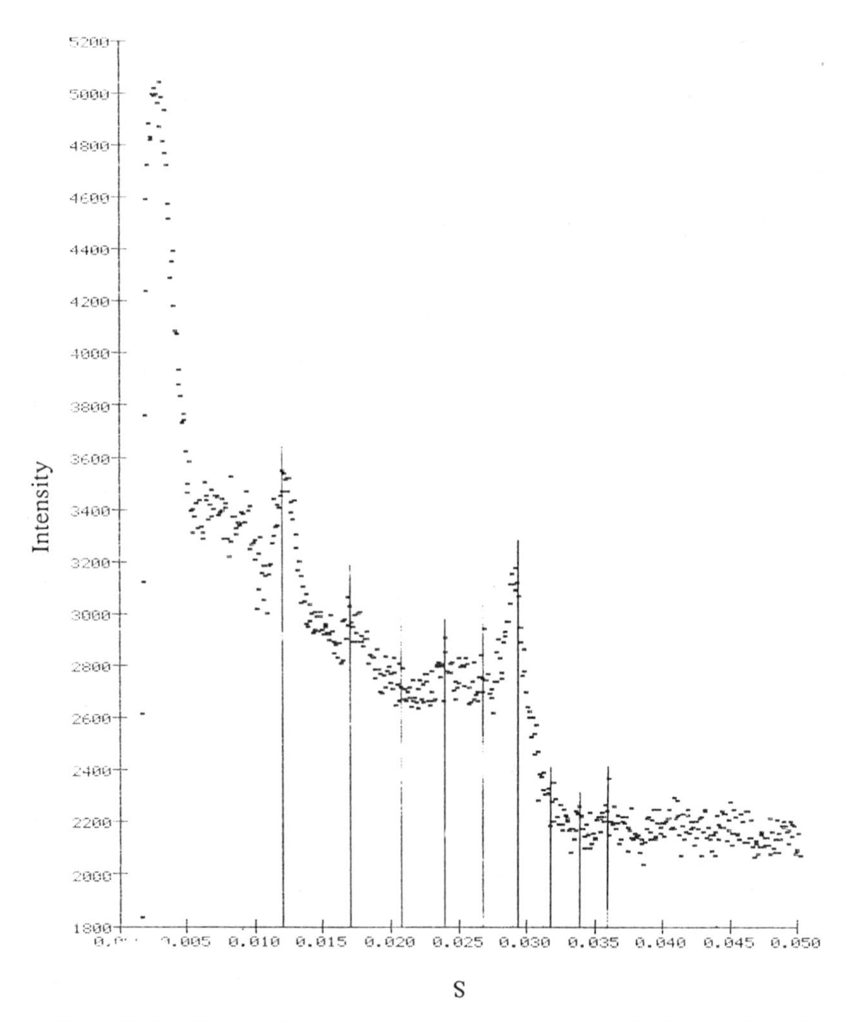

Figure 5. Small-angle X-ray scattering data from a polymerized cubic phase in the {DDAB / water / methyl methacrylate} system. The abscissa is $s=2 \sin \theta / \lambda$ in $Å^{-1}$, where θ is half the scattering angle and λ (=1.54Å) is the wavlength of the radiation used. The vertical lines give the theoretical peak positions for an Im3m lattice with a lattice parameter of 118Å. The maximum at $s \approx 0.0025 Å^{-1}$ is due to the beamstop, and is not a Bragg reflection.

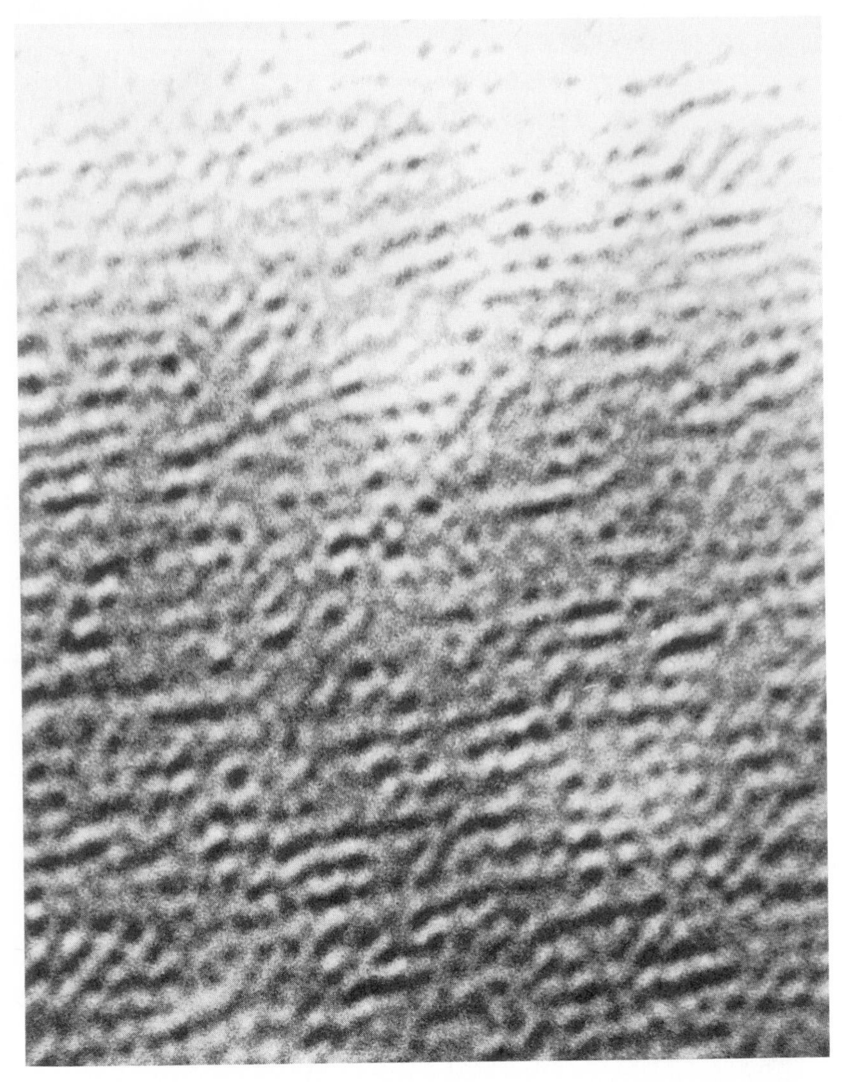

Figure 6a. A transmission electron micrograph of a polymerized cubic phase in the {DDAB / water / methyl methacrylate} system. The dark regions correspond to PMMA and the light regions to void. Magnification is 1,000,000x.

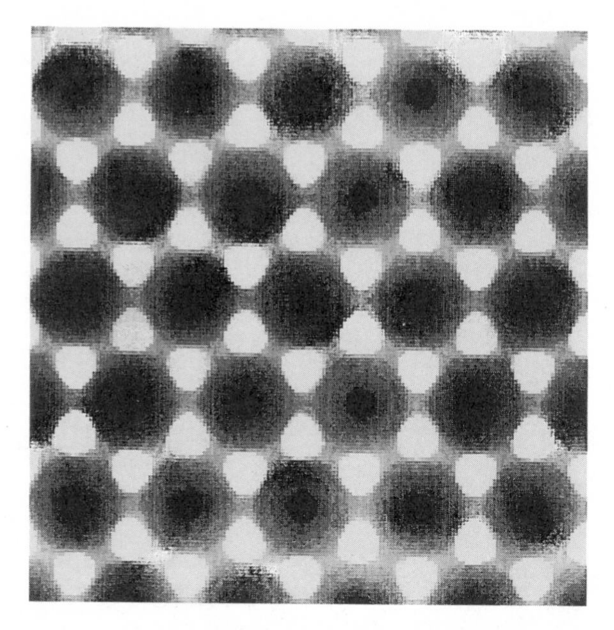

Figure 6b. A computer simulation of the micrograph shown in Figure 6a, computed using the I-WP model structure. The projection is in the (111) direction.

Polymerization in a Nonionic System. Polymerization of the bicontinuous cubic phase in the system {dodecyl hexaethyleneglycol ether ($C_{12}E_6$) / water} has also been performed by the present authors, using acrylamide as the aqueous monomer, and the polymerized phase shown by X-ray to have retained its cubic ordering. The acrylamide made up 19.96wt% of the aqueous phase, and hydrogen peroxide was used as the initiator at 1.1wt% of the acrylamide. This aqueous phase formed 30.30wt% of the total mixture. The polymerization was performed in a nitrogen atmosphere at 23°C, via UV irradiation. Following polymerization, the phase was soaked in ethanol for several weeks, to replace all components except the polymer gel. X-ray was then performed on the polymerized sample, and indexing of the resulting pattern revealed a cubic structure of space group #230, with a lattice parameter of 93Å. At 38wt% water, 62% $C_{12}E_6$, Rançon and Charvolin (14) reported the same space group in an unpolymerized phase, with a lattice parameter of 118Å. In contrast to the latter experiments, no steps were taken to produce a single crystal sample; however, in view of the fact that monodomain cubic phases are relatively easy to produce in this system, a monodomain polymerized cubic phase may become possible.

The successful polymerization of this cubic phase is also of potential importance in that, by keeping the ratio of ethylene oxide to hydrocarbon groups fixed and increasing the molecular weight of the surfactant, it may prove possible to produce polymeric microporous materials with a continuum of pore diameters up toward the micron range (see also the final section). In particular, indexing of X-ray patterns from seven {C_nE_m / water} cubic phases, with $n \approx 17$ and $m \approx 70$ (surfactant mixture obtained from Berol, Inc.) between 25 and 55% surfactant, is consistent with the bicontinuous #230 structure. Pulsed-gradient NMR also proved bicontinuity; the surfactant and water self-diffusion rates were as high as those in low-MW C_nE_m bicontinuous cubic phases (Fontell, K.; Anderson, D.; Olsson, U., in preparation).

Incorporation of Proteins into Polymerized Cubic Phases.

Experiments are now being performed in this laboratory in which proteins, and in particular enzymes, are incorporated into bicontinuous cubic phases and the resulting reaction medium permanented by polymerization. It is well established that the activity and stability of enzymes are generally optimal when the environment of the enzyme is closest to the natural *in vivo* environment, and the lipid bilayer that makes up bicontinuous cubic phases is the normal environment of functioning integral proteins. Polymerization of this continuous bilayer, one example of which is described below, creates by virtue of the bicontinuity a solid, microporous material that allows continuous flow of reactants and products. Furthermore the environment of the protein is precisely controlled sterically and electrostatically, as well as chemically. Control of the geometry of the porespace could be utilized to bias the registry between the enzyme and substrate toward the optimal orientation and proximity, in addition to prividing further control of the chemistry by selection on the basis of molecular size. The electrostatic nature of the pore walls is very homogeneous due to the strong tendency for lipid polar groups to maintain an optimal separation, and it is known that the specificity of many enzymes is sensitive to changes in net charge. Specific surface areas are for enzymatic catalysis are very high: that of the PMMA material described in detail above is approximately 3,500 m^2/gm. In addition the biocompatibility of the presently described materials render them of potential importance in controlled-release and extracorporeal circuit applications.

Immobilization of Glucose Oxidase. The enzyme glucose oxidase was incorporated into the aqueous phase of a cubic phase similar to that polymerized in the previous section, and this aqueous phase made polymerizable by the addition of monomeric acrylamide. Except for a slight yellowish color from the strongly-colored glucose oxidase, the result was an optically clear polymerized material. The concentration of enzyme in the aqueous phase was 10.3 mg/ml, the acrylamide concentration 15.4wt%, and hydrogen peroxide as initiator was present at 0.3 w/w% of the monomer. This aqueous solution was mixed in a nitrogen atmosphere with 24.3 wt% DDAB and 10.93 wt% decane, and the solution centrifuged for one hour to remove any remaining oxygen. This water content, 64.8%, was chosen based on SAXS studies of the cubic phase as a function of water content in similar systems (16; also K. Fontell, unpublished). Above about 63 vol% water, the lattice parameter is larger than 175Å with either decane or decanol, and according to the model shown in Figure 4c the aqueous regions should then be large enough to contain the enzyme while allowing access to it via the porespace left upon removal of the DDAB and decanol.

Two samples were prepared for polymerization. One sample was simply placed in a quartz tube and polymerized for X-ray analysis. The other was smeared onto a nylon backing which had been shaped to fit on the end of a pH probe. Both samples were bathed in nitrogen during UV irradiation. The first sample was about 1.5 mm thick and after polymerization was was a clear solid which could be handled easily; this was loaded into a flat SAXS cell with mica windows. Indexing of the resulting peaks to a BCC lattice was consistent with a lattice parameter of 320Å. The second polymerized sample was soaked for one day in ethanol to remove the DDAB and decane, and then secured over the tip of a pH probe, and the enzyme was shown by the method of Nilsson et al. (35) to have retained its activity in the polymerized cubic phase. This example was intended only for demonstration of a general application, namely in biosensors, and is not particularly impressive in itself because a simple polyacrylamide gel has enough porosity to pass glucose. However, in many potential applications, the substrates to be detected are of higher molecular weight than glucose and the porespace created by the cubic phase microstructure can be tailored to the size of the substrate. In the next example the porosity is due solely to the cubic phase microstructure.

Enzyme Immobilized in a Lipid - Water Cubic Phase. At the time of this report the authors are completing an experiment which demonstrates that proteins can be incorporated, in fairly high concentrations, into bicontinuous cubic phases made with polymerizable lipids that are biocompatible. Glycerol monooleate, or α-monoolein, is an uncharged, biocompatible lipid (e.g., present in sunflower oil), with one fatty acid chain containing a single double bond. A variant of monoolein with a conjugated diene in the chain is *monolinolein*, and the monolinolein - water phase diagram is known to be nearly identical with that of monoolein - water (36). As discussed above, the #212 cubic phase structure has been found in the {monoolein / water / cytochrome-c} system, and the present authors have found the same structure at 6.7wt% cytochrome, 14.8% water, and 78.5% monolinolein, where the monolinolein contained 0.4% AIBN. After equilibration, this cubic phase was placed in the UV photochemical reactor in a water-jacketed cell and bathed in nitrogen in the usual manner. After 48 hours the sample had polymerized and could be held by a tweezers, and was a deep red color, as in the unpolymerized phase, due to the strongly-colored protein. X-ray of the polymerized sample appeared to be consistent with space group #212, with a lattice parameter of approximately 110Å, although the Bragg reflections were very weak. Presently work is under way to further characterize this material; in particular, the model of Luzzati and coworkers (27) implies a stereospecific porespace

Potential Technological Applications.

The polymerization of bicontinuous cubic phases provides a new class of microporous materials with properties that have never before been attainable in polymeric membranes. The most important of these properties are now discussed in turn, and for each an application is briefly dicussed to illustrate the potential importance of the property in a technological, research, or clinical application.

1. All pore bodies and all pore throats are ***identical in both size and shape***, and the sizes and shapes are controlled by the selection of the composition and molecular weights of the components, over a size range which includes pore diameters from 10 to 250Å and potentially into the micron range. Cell shapes cover a range including that from substantially cylindrical to spherical, cell diameter -to- pore diameter ratios cover a range including that from 1 to 5, and connectivities cover a range including that from 3 to 8 pore throats eminating from each cell.

Application: Clearly one important application of microporous materials in which the effectiveness is critically dependent on the monodispersity of the pores is the ***sieving of proteins***. In order that an ultrafiltration membrane have high selectivity for proteins on the basis of size, the pore dimensions must first of all be on the order of 25 - 200Å, which is the size range provided by typical cubic phases. In addition to this, one important goal in the field of microporous materials is the attainment of the narrowest possible pore size distribution, enabling isloation of proteins of a very specific molecular weight, for example. Applications in which separation of proteins by molecular weight are of proven or potential importance are immunoadsorption process, hemodialysis, purification of proteins, and microencapsulation of functionally-specific cells.

2. The porespace comprises an isotropic, ***triply-periodic*** cellular structure. No prior microporous polymeric material, and no prior microporous material of any material with pore dimensions larger than 2 nanometers, has exhibited this level of perfection and uniformity.

Application: Recently the authors have become involved with ***studies of superfluid transitions*** which require microporous materials exhibiting long-range, triply-periodic order. In the Laboratory of Atomic and Solid State Physics at Cornell University, a group lead by Dr. John D. Reppy has been investigating the critical behavior of liquid ^{4}He in microporous media (Chan, M. H. W.; Blum, K. I.; Murphy, S. Q.; Wong, G. K. S.; Reppy, J. D., submitted). Certain theoretical treatments have predicted that the critical exponents characterizing the fluid - superfluid transition are different for disordered than for periodic porous media. The experiments described in the paper now submitted for publication were performed using disordered media: Vycor, aerogel, and xerogel. The group is now proceeding on to a parallel set of experiments using the ordered microporous medium described herein.

3. In certain forms of the material, the microporous polymer ***creates exactly two distinct, interwoven but disconnected porespace labyrinths***, separated by a continuous polymeric dividing wall. This opens up the possibility of performing enzymatic, catalytic or photosynthetic reactions in controlled, ultrafinely microporous polymeric materials with the prevention of recombination of the reaction products by their division into the two labyrinths. These features combine with specific surface areas for reaction on the order of 10^3-10^4 square meters per gram, and with the possibility of readily controllable chirality and porewall surface characteristics of the two labyrinths.

Application: There are in fact at least two distinct biological systems in which Nature uses cubic phases (in unpolymerized form, of course) for exactly this purpose. Electron micrographs of the prolamellar body of plant etioplasts have revealed bicontinuous cubic phase microstructures (37), and lipid extracts from these etioplasts have been shown to form cubic phases *in vitro* (38). The prolamellar body developes into the thylakoid membrane of photosynthesis, which is again a continuous bilayer structure, with the stroma side acting as a cathode and the intrathylakoid side as an anode. Tien (39) states that the chlorophyll dipersed in the lipid bilayer acts as a semiconductor, in that the absorption of light excites an electron to the conduction band and leaves a hole in the valence band. There are at least two reasons why the separation of the aqueous phase into two distinct compartments is important in natural photosynthesis: first, as well as providing an approporiate environment for the pigments, the bilayer acts as a barrier to prevent back-reactions; and second, with the two systems of accessory pigments located in distinct parts of the membrane, each electron/hole pair can be generated by two photons, thus providing an upgrading of photon energy. Second, the endoplasmic reticulum, or ER, which is the site of the biosynthesis of many of the proteins needed by the cell, may also be a bicontinuous cubic phase, for certain electron micrographs indicate cubic order (40). Here the presence of two continuous aqueous labyrinths, one of which is continuous also with the exterior of the cell, creates a very large surface area for reaction and a continuity of 'inner' and 'outer' volumes to prevent saturation of concentration gradients which are the driving force for transmembrane transport. Clearly there is great potential impact in capturing and fixating such systems of high enzymatic activity and fundamental biological importance.

4. The microporous material exhibits in all cases a *precisely controlled, reproducible and preselected morphology,* because it is fabricated by the polymerization of a periodic liquid crystalline phase which is a ***thermodynamic equilibrium state***, in contrast to other membrane fabrication processes which are nonequilibrium processes.

Application: As is well-known in the industry, any microporous material which is formed through a nonequilibrium process is subject to variability and nonuniformity, and thus limitations such as block thickness, for example, due to the fact that thermodynamics is working to push the system toward equilibrium. In the present material, the microstructure is determined at thermodynamic equilibrium, thus allowing ***uniformly microporous materials without size or shape limitations*** to be produced. As an example, the cubic phase consisting of 44.9 wt% DDAB, 47.6% water, 7.0% styrene, 0.4% divinyl benzene (as cross-linker), and 0.1% AIBN as initiator has been partially polymerized in the authors' laboratory by themal initiation; the equilibrated phase was raised to 85°C, and within 90 minutes partial polymerization resulted; SAXS proved that the cubic structure was retained (the cubic phase, without initiator, is stable at 65°C). When complete polymerization by thermal initiation is accomplished, then such a process could produce uniform microporous materials of arbitrary size and shape.

5. ***Proteins, in particular enzymes, can be incorporated*** into the cubic phase bilayer and then fixated by the polymerization, thus creating a permanented reaction medium inheriting the precision of the present material, and maintaining to the highest possible extent the natural environment of the protein. This was illustrated in one of the experiments reported above. Many proteins and enzymes are specifically designed to function in a lipid bilayer, with hydrophilic and hydrophobic regions that match those of the natural bilayer. As shown by K. Larsson and G. Lindblom (41), a very hydrophobic wheat fraction, gliadin, can be dispersed in monoolein, and a bicontinuous cubic phase formed on the addition of water; in this case the protein is in the lipid regions of the cubic phase. Examples of other proteins and enzymes which can be incorporated into bicontinuous cubic phases have been reviewed (42).

Application: Immobilized enzymes offer many advantages over enzymes in solution, including dramatically increased stability in many cases as well as higher activity and specificity, broad temperature and pH ranges, reusability, and fewer interferences from activators and inhibitors. To name a single example in the growing field of ***immobilized enzymes for medical assays***, enzyme tests can distinguish between a myocardial infarction and a pulmonary embolism, while an EKG cannot. The present methods for immobilizing enzymes such as adsorption and covalent bonding have serious drawbacks. Absorbed enzymes easily desorb upon changes in pH, temperature, ionic strength, etc. The covalent bonding of enzymes usually involves harsh chemical conditions which seriously reduce enzymatic activity and cause significant losses of expensive enzymes. Recently a process has been developed for covalently bonding enzymes to collagen in such a way as to avoid exposing the enzyme to harsh chemistry (43). However, collagen is an extremely powerful platelet antagonist, activating fibren and leading to immediate clotting, making it totally unsuitable for applications involving contact with blood. As shown above, enzymes can be immobilized in polymerized bicontinuous cubic phases with the enzyme continually protected in a natural lipid - water environment throughout the process.

6. The components can be chosen so that the material is ***biocompatible***, opening up possibilities for use in controlled-release drug-delivery and other medical and biological applications that call for nontoxicity. It is known that many biological lipids form bicontinuous cubic phases, and many of these have modifications with polymerizable groups, such as the monolinolein case discussed above.

Application: Biocompatible materials of the type described are being investigated as polymerized drug-bearing cubic phases for ***controlled-release applications*** with high stability and versatility. The combination of the biocompatibility and entrapping properties of many cubic phases with the increased stability upon polymerization could lead to new delivery systems, and even the possibility of first-order drug release -- release in response to physiological conditions -- by incorporating proteins and enzymes, as described above, as biosensors. For example, particles with an outer coating of polymerized cubic phase containing glucose oxidase would undergo a decrease in pH with increasing blood glucose levels, which could be used to trigger insulin release.

Polymeric Cubic and Other Liquid Crystalline Phases.

While the primary emphasis of this chapter has been on polymerized liquid crystals, important insight into cubic phases and the driving forces behind their formation can be gained by comparing these with polymeric analogues, in particular with bicontinuous phases of cubic symmetry that occur in block copolymers and in systems containing water and a polymeric surfactant. There are two fundamental reasons why the observation of bicontinuous cubic phases in block copolymers is of tremendous value in helping to understand cubic phases in general: first, the applicability of statistical approaches, and the comparative simplicity of intermolecular interactions (summarized by a single Flory interaction parameter), make the theoretical treatment of block copolymer cubic phases (26) much more straightforward than that of surfactant cubic phases; and second, the solid nature and higher lattice parameters in the copolymer cubic phases make them readily amenable to electron microscopy (25).

The cubic microstructure that has in fact been observed in block copolymers is the #224 structure discussed above, with one of the blocks located in the two channels lying on the 'inside' of the surface, and the other block in the 'matrix' on the 'outside' of the surface, so that the surface itself describes the location of the junctions between the unlike blocks. In the polymer literature this structure has been referred to as the 'ordered bicontinuous double-diamond', or 'OBDD', structure. The structure occurs in medium-MW star diblock copolymers at higher arm numbers, and apparently also in linear diblocks at higher-MW (44), but always at compositions where the matrix component is between 62 and 74 vol%. In early experiments, bicontinuity was indicated by vapor transport, and also by an order of magnitude increase in the storage modulus over that of the cylindrical phase, which occurs at the same composition but lower arm number. TEM tilt-series, together with SAXS measurements, taken at the University of Massachussetts at Amherst, have provided accurate and detailed data on the structure (25). In Figure 7 is shown a split image, with electron microscopy data

Figure 7. A split-screen image, with a TEM micrograph of a cubic structure in a {polystyrene / polyisoprene} star diblock copolymer on the left half, and a computer simulation using the structure indicated in Figure 1 on the right half. The lattice parameter is approximately 300Å, the dark regions correspond to polyisoprene and the light regions to polystyrene.

on the left half, and on the right half a computer simulation using the constant mean curvature dividing surface shown in Figure 1a. The agreement is remarkable.

A theoretical treatment (26) of the OBDD structure, employing the Random-Phase Approximation (RPA), yields accurate predictions of lattice parameters from input data on the two blocks, and rationalizes the occurence of the OBDD at compositions just below 74 vol. % as being due largely to a very low interfacial surface area for the model structure at these compositions, together with a mean curvature that is intermediate between lamellae and cylinders. One important conclusion from the RPA theory is that the interface is very close to constant mean curvature, and this is supported by comparisons of the TEM data with simulations based on various interfacial shapes. However, care must be exercised in carrying over these ideas to the surfactant case, because in the small molecule case there is a higher penalty for variations in end-to-end distances for surfactant tails as compared to polymer chains. Nevertheless, the concepts of interfacial mean curvature, uniformity in stretch distances, and low interfacial areas apply in qualitatively similar ways in the two cases and appear to be the fundamental driving considerations for the occurence of bicontinuous cubic phases in general.

And finally, a word should be said about cubic phases made from polymeric surfactants. Early work was done by Kunitake et al. (45), who produced vesicles from polymeric surfactants. Very recently, polymeric surfactants of the ethoxylated alcohol type were shown to form cubic phases (46). However, those authors were unaware of the notion of bicontinuity in cubic phases, and interpreted their results solely in terms of close-packed micelles. In particular they were unaware of the fact that low-MW ethoxylated alcohol surfactants (such as $C_{12}E_6$) form, in the same region of the phase diagram as their polymeric cubic phase, a bicontinuous cubic phase of the #230 type. With this knowledge in mind, it is quite possible that their polymeric cubic phase was indeed bicontinuous, but unfortunately the authors did little to characterize their phase. Since polymeric surfactants are far from 'typical' polymers, it is difficult to acertain from first principles what the properties of such a phase should be, whether they should have mechanical properties reflective of glassy polymers or closer to those of liquid crystals, for example, or whether they differ from analogous polymer*ized* cubic phases because they are formed by the reversible self-assembly rather than by an (essentially) irreversible polymerization following self-assembly. An experimental complication is the fact that there are no cubic phases in the phase diagram for the monomeric surfactant. This example serves to remind us that the exact relationship between polymeric and polymerized bicontinuous cubic phases is as yet unknown, and many interesting questions remain as to how far the analogy can be carried and whether or not there exists a continuum path between small molecule liquid crystalline and macromolecular bicontinuous states.

Literature Cited

1. Scriven, L. E. Nature 1976, 263, 123.
2. Luzzati, V.; Chapman, D. In Biological Membranes; Academic: New York, 1968; pp. 71-123.
3. Anderson, D. M. U.S. Patent Application #052,713; EPO Patent Application #88304625.2; Japanese Patent Application #63-122193, 1987.
4. Regen, S. L.; Czech, B.; Singh, A. J. Am. Chem. Soc. 1980, 102, 6638.
5. Fendler, J. H. Acc. Chem. Res. 1984, 17, 3.
6. Hub, H.-H.; Hupfer, B.; Koch, H.; Ringsdorf, H. Angew. Chem. 1980, 92 (11), 962.
7. Johnston,D. S.; Sangera, S.; Pons, M.; Chapman, D. Biochim. Biophys. Acta 1980, 602, 57.
8. Lopez, E.; O'Brien, D. F.; Whiteside, T. H. J. Am. Chem. Soc. 1982, 104, 305.
9. Friberg, S.; Fang, J.-H. J. Coll. Int. Sci. 1987, 118, 543.

10. Candau, F.; Leong, Y. S.; Pouyet, G.; Candau, S. J. Coll. Int. Sci. 1984, 101 (1), 167.
11. Thunathil,R.; Stoffer, J. O.; Friberg, S. J. Polymer Sci.1980, 18, 2629.
12. Candau, F.; Zekhnini, Z.; Durand, J.-P. J. Coll. Int. Sci. 1986, 114 (2), 398.
13. Lindman, B.; Stilbs, P. In Surfactants in Solution; Mittal, K. L., Lindman, B., Eds.; Plenum: New York, 1984; Vol. 3, p. 1651.
14. Rançon, Y.; Charvolin, J. J. Phys. 1987, 48, 1067.
15. Regen, S. L. In Liposomes: From Biophysics to Therapeudics; Marcel Dekker: New York, 1987; pp. 73-108.
16. Anderson, D. M. Ph. D. Thesis, University of Minnesota, Minneapolis, 1986.
17. Helfrich, W. Z. Naturforsch. 1973, 28c, 693.
18. Charvolin, J.; Sadoc, J. F. J. Physique 1987, 48, 1559.
19. Schoen, A. H. Infinite Periodic Minimal Surfaces Without Self-intersections; NASA Technical Note D-5541, 1970; Natl. Tech. Information Service Document N70-29782, Springfield, VA 22161.
20. Anderson, D. M.; Gruner, S. M.; Leibler, S. Proc. Nat. Acad. Sci., 1988, 85, 5364.
21. Schwarz, H. A. Gesammelte mathematische Abhandlungen; Springer: Berlin, 1890; Vol. 1.
22. Fontell, K.; Jansson, M. Prog. Coll. Polymer Sci., in press.
23. Ninham, B. W.; Chen, S. J.; Evans, D. F. J. Phys. Chem. 1984, 88, 5855.
24. Longley, W.; McIntosh, T. J. Nature 1983, 303, 612.
25. Thomas, E. L.; Alward, D. B.; Kinning, D. J.; Martin, D. C.; Handlin, D. L. Jr.; Fetters, L. J. Macromolecules 1986, 19 (8), 2197.
26. Anderson, D. M.; Thomas, E. L. Macromolecules, in press.
27. Mariani, P.; Luzzati, V.; Delacroix, H. J. Mol. Biol., in press.
28. Kekicheff, P.; Cabane, B. J. Phys. (Paris) 1987, 48, 1571.
29. Hyde, S. T.; Andersson, S.; Ericsson, B.; Larsson, K. Z. Krist. 1984, 168, 213.
30. Luzzati, V.; Tardieu, A.; Gulik-Krzywicki; T., Rivas, E.; Reiss-Husson, F. Nature 1968, 220, 485.
31. Fontell, K.; Fox, K.; Hansson, E. Mol. Cryst. Liq. Cryst. 1985, 1 (1,2), 9.
32. Blum, F. D.; Pickup, S.; Ninham, B. W.; Evans, D. F. J. Phys. Chem. 1985, 89, 711.
33. Chen, S. J.; Evans, D. F.; Ninham, B. W.; Mitchell, D. J., Blum, F. D.; Pickup, S. J. Phys. Chem. 1986, 90, 842.
34. Fontell, K.; Ceglie, A.; Lindman, B.; Ninham, B. W. Acta Chem. Scand. 1986 A40, 247.
35. Nilsson, H.; Åkerlund, A.-C.; Mosbach, K. Biochim. Biophys. Acta, 1973, 320, 529.
36. Lutton, E. S. J. Am. Oil Chem. Soc., 1966, 42, 1068.
37. Gunning, B. E. S.; Jagoe, M. P. Biochemistry of Chloroplasts, Goodwin, E., Ed.; Academic: London, 1967; Vol. 2, pp. 655-676.
38. Ruppel, H. G.; Kesselmeier, J.; Lutz, C. Z. Pflanzenphysiol., 1978, 90, 101.
39. Tien, H. T. In Solution Behavior of Surfactants; Mittal, K. L.; Fendler, E. J., Eds.; Plenum: New York, 1982; Vol. 1, pp. 229-240.
40. Alberts, B.; Bray, D.; Lewis, J.; Raff, M.; Roberts, K.; Watson, J. D. Molecular Biology of the Cell; Garland: New York, 1983; pp. 335-339.
41. K. Larsson, K; Lindblom, G. J. Disp. Sci. Tech. 1982, 3, 61.
42. Ericsson, B.; Larsson, K.; Fontell, K. Biochim. Biophys. Acta, 1983, 729, 23.
43. Coulet, P. R.; Gautheron, D. C. Biochimie, 1980, 62, 543.
44. Hasegawa, H.; Tanaka, H.; Yamasaki, K.; Hashimoto, T. Macromolecules, 1987, 20, 1651.
45. Kunitake, T.; Nakashima, N.; Takarabe, K.; Nagai, M.; Tsuge, A.; Yanagi, H. J. Am. Chem. Soc. 1981, 103, 5945.
46. Jahns, H.; Finkelman, M. Coll. Polymer Sci., 1987, 265, 304.

RECEIVED September 12, 1988

Chapter 14

Quantitative Small-Angle Light Scattering of Liquid Crystalline Copolyester Films

L. J. Effler, D. N. Lewis[1], and J. F. Fellers

Department of Materials Science and Engineering, University of Tennessee, Knoxville, TN 27996–2200

An IBM personal computer controlled, two-dimensional position sensitive detector (PSD), has been developed for quantitative analysis of small angle light scattering (SALS) patterns. The hardware and software comprising this device are described in detail and compared to the minicomputer PSD developed by Stein and coworkers. The IBM PC controlled PSD developed here was used to study liquid crystalline copolyesters of poly(ethylene terephthalate) and para-oxybenzoic acid. A traditional Guinier based data analysis is used, and then compared to results obtained by approximating the Debye form of the scattering function using regression analysis of a polynomial expansion. Compression molded films were found to be composed of randomly ordered fibrillar domains. This structure would require large scale cooperative motions to achieve global orientation from only locally oriented fibrillar domains. This concept is reinforced by the difficulty in cold stretching these films and no significant orientation being developed in them. However uniaxially stretched melt deformed films were found to possess a high degree of uniaxial orientation.

Various process steps were used to determine their influence on the morphological nature of liquid crystalline copolyester films. Compression molding was used to form quiescent films, while extensional deformation above and below the onset of fluidity, as well as shear deformation above the onset of fluidity was used to make non-quiescent films. It is a basic result that molecular orientation can only be achieved when the deformation is done while the polymer is in a liquid crystalline melt state. Experimental details are given in the subsection 'Materials and Processing,' while an interpretation is offered in the discussion in the subsection 'Morphological and Process Consideration.'

[1]Current address: Michigan Molecular Institute, 1910 West Saint Andrews Road, Midland, MI 48640

0097–6156/89/0384–0225$08.50/0

The basic structural components, and the extent or lack of orientation have been determined through the use of a new SALS device. The basic SALS experimental results are interpreted via standard and a newly developed method for approximating and analyzing the scattering function. The equipment involves the coupling of video and microcomputer technology to create a two-dimensional position sensitive detector (PSD) for small angle light scattering (SALS). Such devices have been available for small angle X-ray, and neutron scattering (SAXS, SANS) for ten years (1), and have more recently been made available for SALS (2-5). The novelty of the device discussed here is the development of technology allowing the use of a microcomputer in the detection and analysis scheme.

Experimental

In addition to the standard discussion of the materials studied, and processing and analytical techniques employed for this work, a rather lengthy presentation on two-dimensional position sensitive detectors for SALS will be presented. A review of prior work performed by other researchers will be discussed, as well as the particulars of the detector developed for this study. Finally, a comparison between the detector presented here, and those developed previously will be made.

Two-Dimensional Position Sensitive Detector, Background. The traditional technique for capturing the scattering patterns from SALS experiments has used photographic film (6-9). This technique generally provides qualitative information such as pattern shape, relative size of pattern, etc. and indirect quantitative information with the aid of a micro-densitometer to determine film emulsion densities (8,9). Photographic films provide high pixel density, approximately 18×10^6 pixels in a 35 mm slide, but have a non-linear intensity response (10,11).

True quantitative SALS requires a photometric technique. Early photometric SALS methods employed mechanically moved photomultiplier tubes (12,13). However, devices of this type have slow scanning rates and lack precision. By 1976 Wasiak, Peiffer and Stein (14) developed a one dimensional PSD for SALS employing a Princeton Applied Research (PARC) Optical Multichannel Analyzer (OMA). This relatively rapid device allows data analysis of the scattering function at a selected azimuthal angle in terms of intensity versus scattering angle plots or variations thereof. This device was superseded by the PARC OMA2 which was adapted for use as a two-dimensional PSD by Tabor, Stein and Long (2-5). The PSD consisted of a vidicon detector or video camera tube, an OMA2 detector control unit which contained the tube power and control hardware, and a 14 bit analog to digital converter (ADC) (15) for digitizing the analog video signal, a microprocessor control unit consisting of a Digital Equipment LSI 11/2 microprocessor, a 64 kilobyte (KB) RAM or user memory, and a flexible disk drive, an image monitor displaying the real time video image, and an intensity monitor displaying the three-dimensional intensity versus x-y location plot (2,5). The analog video signal was collected from the detector in an x-y pixel grid of up to 500 by 500 data points and transmitted to the microprocessor control unit (2,5). Limitations of the manual timing technique and the capacity of each floppy disk, however, reduced the effective number of x-y data points collected in each scan to 2500 or a 50 by 50 x-y grid (5). Each scan could be collected and stored in 4 seconds. Computer control of the timing of the scan and data storage was expected to improve the acquisition rate of a 2500 digitized data point scan to 100 milliseconds (5). Improvements in hardware and software were envisioned to increase the number of digitized data points to a 100 by 100 x-y grid or 10,000 points (2). The collected data were analyzed with the microprocessor and either a Digital Equipment PDP 11/34 minicomputer or a Control Data Cyber 175 mainframe computer (2,5). The digitized data from the OMA2 could then be presented in several forms: (1) an iso-intensity contour plot, (2) a three dimensional intensity versus x-y position plot, or (3) an intensity versus scattering angle plot from a specified azimuthal data slice.

The PSD developed in our laboratories at the University of Tennessee employs a silicon vidicon (SV) image tube as an intensity sensor (10). The SV image tube consists of a silicon photodiode target mounted in the end of an evacuated tube, deflection plates and a cathode type electron gun. The silicon target is a silicon wafer with a microscopic array of about six million photodiode junctions grown upon it, each having a common cathode and an isolated anode (10). The photodiodes are eight micrometers apart and scanned by a 25 μm diameter electron beam (10). This continuously scanning electron beam sets the photodiodes to a default electron voltage (eV) value. Visible light photons strike the silicon target and create electron-hole pairs which deplete the surface charge on the surrounding photodiodes (10). When the electron beam scans the depleted portion of the target, a recharging current flows to reset the photodiodes to the default eV value. The current and corresponding voltage signal (1 volt, 5 MHz bandwidth) are proportional to the number of electron-hole pairs formed in the illuminated 25 μm region of the target, and consequently to the number of light photons striking that region (10,15). Each 25 μm region in the active portion of the target is defined as a picture element or pixel. The maximum number of pixels in a SV imaging tube is about 750,000 or an 860 by 860 x-y array. The effective number of pixels is determined from both the scanning rate and the sampling rate. The position of each pixel on the target can be determined from the scanning rate, the sampling rate and the starting point of the scan (12,15,16).

Two-Dimensional Position Sensitive Detector, Hardware. The video SALS equipment developed and employed in our laboratories (17) is shown schematically in Figure 1. In the current design, all optical components are mounted on a graduated 2 meter optical rail. The light (632.8 nm) from a 2 mW He-Ne laser passes though a vertically set polarization rotator, then through a pair of pinhole collimators to reduce the beam cross-sectional diameter to 0.67 mm when it strikes the sample. The scattered light from the sample then passes through a rotatable analyzer and is projected onto an optically translucent MYLAR film imaging screen. The scattered image on the rear of the screen acts as the imaging source for the MTI-Dage (model SC66) silicon-vidicon (SV) video camera. In the model SC66

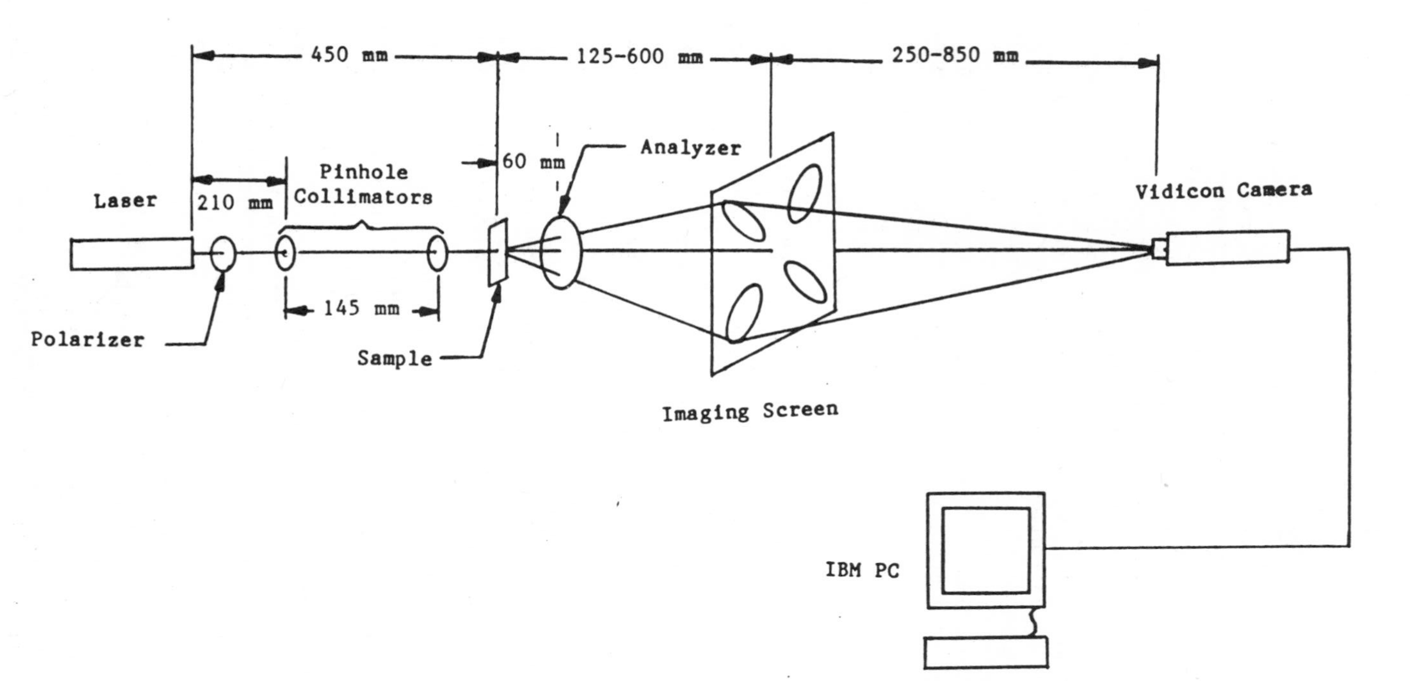

Figure 1: Schematic of Video SALS Setup.

camera, the SV tube is scanned at a rate of 60 frames/s corresponding to the 60 Hz frequency used as the electrical power line standard in the U.S.A. (18,19). Each frame consists of two interlaced fields of 262.5 lines per field or 525 lines per frame (19). The signal sent from the video camera passes over a coaxial cable to an IBM personal computer equipped with a Tecmar VIDEO VAN GOGH (VVG) video digitizer interface card, 640 KB of user ram memory, two 360 KB floppy disk drives, an external 20 MB hard disk drive, and a 5 MB cartridge hard disk drive.

The VVG video digitizer is the heart of the system. The VVG digitizer is the interface between the camera and the personal computer. The analog voltage signal received by the digitizer card is first amplified, then separated into two components, the brightness signal, and the sync pulses (19). The sync pulses (vertical and horizontal) define the beginnings of each field and each line, respectively (19). These pulses are routed to position control circuits on the card to determine the pixel's x-y position. A sample clock is used in conjunction with the horizontal position control to set the sampling rate or the number of pixels sampled per line (19). The brightness signal, a voltage representing the intensity of a single pixel, is processed in two steps. When a sampling pulse is generated by the sample clock to the horizontal position controller, a voltage memory, or Sample/Hold, circuit is activated to store the brightness signal at that point in time (19,20). The brightness signal is then converted by an 8 bit analog to digital converter (ADC) into a binary digit, or bit, pattern representing one of 256 intensity or gray levels (0 = black, 255 = white) (19). The digitized signal of each pixel is stored sequentially as an x-y array in a 64 KB buffer in the IBM personal computer's user RAM. The digitized SALS pattern represented by this array is stored on the 20 MB hard disk for further processing.

Two-Dimensional Position Sensitive Detector, Software. The key step of digitizing the analog video signal is done via the VVG software package. The program digitizes the video image into a grid of 249 by 236 pixel locations. Each pixel is assigned an address in a 64 KB buffer and a shade-of-gray (SG) value that is in proportion to the intensity of the light striking the detector at that location. As such, the SG provides a relative intensity for a monochromatic light source that is dependent upon the gain and contrast setting of the camera, the f-stop of the lens, the sample-to-screen distance, and the screen-to-camera distance. The image in the buffer can be stored on the hard disk in a 64 KB binary array, and can be re-accessed using VVG subroutines. A file can be digitized in 6 s, however, digitizing and storing a file on the hard disk takes approximately 15 s.

Specially written software basically consisted of programs that, with the aid of VVG subroutines, would manipulate the scattering image in some manner. Programs were written that would place the binary image file into a format and size that could be accessed by purchased contouring software. Also, software was generated that would take intensity slices at a specified azimuthal angle. The data from these slices could then be formatted for a number of common methods of data analysis: intensity (I) vs. scattering angle

(θ), I vs. scattering vector (q), I vs. q^2, etc. The programs will also delete the data in a specified Δq around the beamstop if the user so desires.

Two-Dimensional Position Sensitive Detector, Comparisons and Future Improvements. Two advantages exist for the microcomputer controlled PSD reported on here versus the minicomputer controlled OMA2 developed by Stein and coworkers (2,5). The first advantage is the much lower cost of hardware and software for microcomputers compared with minicomputers. The minicomputer controlled OMA2 has a combined hardware and software cost about twice that of the corresponding cost for the microcomputer controlled PSD. A second advantage is the greatly increased x-y resolution compared to the minicomputer controlled OMA2. The latter system has a limitation of 2500 pixels per scan or a 50 by 50 grid (2,5). Even with hardware and software improvements, the capacity of the system would not exceed 10,000 pixels or a 100 by 100 grid per scan (2). The current version of the microcomputer controlled PSD, however, has a capacity of nearly 60,000 pixels per scan or a 249 by 236 grid (19,17). A typical SALS pattern covers some 32,000 pixels or a 160 by 200 grid.

One significant disadvantage exists for our microcomputer controlled PSD when compared to the OMA2 system developed by Stein and coworkers. The current version of our system can distinguish only 256 intensity (SG) levels (19), whereas the OMA2 can distinguish 16,384 intensity levels (15). This difference results from the ADC used in each system. The VVG video digitizer uses an 8 bit ADC (19), whereas the OMA2 detector controller uses 14 bit ADC (15).

The IBM personal computer controlled PSD developed at the University of Tennessee was first made operational in January 1985. In the intervening three years, a number of technological advances have been made in both microcomputers and video digitizing hardware. These advances have been made in three areas: x-y pixel capacity/scan, dynamic range (intensity resolution), and digitizing speed. Currently marketed digitizing boards have x-y pixel capacities per scan up to 307,200 pixels or a 640 by 480 grid and have digitizing speeds of 0.03 seconds (or 30 frames/s) for the entire scan (21-24). These systems use 8 bit flash type ADCs which have conversion rates from 10 to 12.5 MHz and on board RAM memory to store the digitized pixels (21-24). Improvements in dynamic range for PSD's depend on three factors: the microcomputer used, the bit capacity of the RAM memory used and the type of ADC used. The microcomputer and the bit capacity of the RAM memory are interrelated. The IBM personal computer and compatible microcomputers are based upon the 8 bit Intel 8088 microprocessor which manipulates data in 8 bits units. More advanced microcomputers are based upon the 16 bit Intel 80286 or 32 bit Intel 80386 microprocessors which can handle data 16 bits or 32 bits at a time, so the memory depth or capacity can exceed 8 bits. The type of ADC used on the video digitizer card is the most important factor in determining the dynamic range available. Most board level products for IBM PC's use 8 bit ADCs with conversion rates as high as 12.5 MHz. However, we know of only one PC compatible board with an ADC of larger than 8 bits (25). The system can acquire a digitized 256 by 256 pixel array in 0.26 seconds or a conversion rate of 4 microseconds per pixel (25). At the chip level, flash video ADCs exist with a resolution of 10 bits (1024 SG) and conversion rates as high as 50 MHz. A board level product incorporating one of these chips could operate at the real time rate of 30 frames/s for a 512 by 512 pixel array.

Two-Dimensional Position Sensitive Detector, Calibration. A number of tests were performed to assure the accuracy of the data collected by the system. These included testing the detector uniformity to insure that all the photo-diodes on the chip are of equivalent sensitivity. Another calibration of the device was to determine the magnifying affects of the camera. This was done to allow conversion of the x-y pixel grid into the dimensions of reciprocal space. Finally, the contrast setting was adjusted so that a SG value of 0 was produced when the screen was not illuminated, with the gain set at maximum and the f-stop wide open. Further details of the system calibration can be found in Effler's M.S. thesis (17).

Materials and Processing. Copolyesters of poly(ethylene terephthalate) (PET) and para-oxybenzoic acid (POB) were supplied by the Tennessee Eastman Corporation. Past work indicates the copolyesters form thermotropic liquid crystalline phases at compositions containing more than 30 mole% POB (26,27,28). The composition of the copolyester studied here contains 60 mole% POB. Quiescent liquid crystalline films were made by compression molding the copolyester at 210, 230, 255, and 285 °C, and followed by a quench into ice water, ambient air, or cooled in the press with the power off. Film thicknesses ranged between 0.05-0.15 mm. Another sample of the 40/60 PET/POB copolyester was melted at 270 °C in a Mettler hot stage, manually sheared between glass slides, and then ambient air cooled.

Cold stretched films were made by deforming compression molded 40/60 PET/POB films in a T. M. Long biaxial stretching device. Specimens were originally 60 mm x 60 mm. Following a five minute preheat, the films were uniaxially stretched 2 x 1 at 750%/min. with the transverse direction unconstrained. Stretch temperatures were 90, 110, 160, and 170 °C, and were chosen to be below the onset of fluidity for the copolyester at 190 °C, but above the glass transition temperature of 100% PET. Also, highly uniaxially stretched thin film melt extrudates of the 40/60 PET/POB copolyester, and Celanese's VECTRA copolyester, were prepared and kindly supplied by Professor D. G. Baird and T. Wilson of VPI&SU for the case of highly developed uniaxial orientation. The polymers were processed through a 3/4" x 20" single screw extruder, and were collected by a pair of take-up rollers that imparted extensional deformation to the extruded film. For the PET/POB copolyester the extruder barrel and die temperatures were 250, and 200 °C, respectively, while for the VECTRA copolyester the temperatures were 280, and 232 °C (Wilson, T. S.; Baird, D. G., VPI&SU, personal communication, 1988).

SALS was performed using the apparatus described in the Hardware and Software sections. Samples were placed onto the sample holder and illuminated with the laser. The scattered radiation was passed through the analyzer, with the system in the Hv mode (crossed polarizers). The imaging screen was adjusted to maximize the pattern size. The camera position was then adjusted to maximize the video image size. The image was then focused, and the f-stop adjusted to optimize the intensity distribution. Finally, the image was digitized and stored via the IBM PC.

Optical microscopy was performed with an Olympus BH-2 transmitted, polarized light microscope in conjunction with a Polaroid 4" x 5" flat plate film camera. Objective lens of 4x, 10x, 20x and 40x magnification were available, and when coupled with the 10x eye- and photoeye- pieces yield viewed magnifications of 40x, 100x, 200x, and 400x. Samples were cut from the compression molded films and mounted on a glass microscope slide and covered with a cover slip. The samples were placed in an adjustable x-y mechanical stage. This mechanical stage was situated on a larger 360° rotatable stage, that permitted observation of any optical orientation in the film plane. All the work was done between crossed polarizers.

Specimens for SAXS were sandwiches of several small rectangular pieces cut from the compression molded films. These sandwiches were made because the thickness of the molded films did not provide enough mass to scatter X-rays effectively. The average thickness of the sandwiches was 0.5 mm, and were carefully constructed to avoid masking any directional biases. SAXS was performed using the 10 m SAXS camera available at the National Center for Small Angle Scattering Research (NCSASR) at Oak Ridge National Laboratory (ORNL). Power settings on the Rikagu rotating anode X-ray source (Cu Kα) were 40 kV and 50 mA, with the detector 5 m from the sample. The camera uses a two-dimensional position sensitive detector to capture the scattered radiation. Each sample was run for 1.5 hours to obtain approximately 1×10^{6} counts. The collected pattern had the standard corrections for background, transmission coefficient, and thickness made.

Selected stretched films were investigated using wide angle X-ray diffraction (WAXD). Sandwiches from the SAXS experiments were also used for WAXD. Samples were irradiated for 12 hours using a flat plate film camera. A Philips Cu Kα radiation source was used, with power settings of 20 kV and 30 mA.

Scanning electron microscopy (SEM) was also performed on selected films using an Amray SEM at 30 kV. The samples were cut from the compression molded films, subjected to a five minute oxygen plasma etch, glued onto aluminum sample holders, and then sputtered with a thin gold coating. A Polaroid 4" x 5" flat plate camera was used to obtain photo-micrographs from the video screen.

Methods of Data Analysis

A general relationship for the scattered intensity from a statistically isotropic (no global orientation) material is given by Debye (29-30) as:

$$I(q) = K \sum_{i}^{N} \sum_{j}^{N} \sin(qr_{ij})/qr_{ij} \qquad (1)$$

where: $I(q)$ = scattered intensity
K = calibration constant
N = number of scattering elements

r_{ij} = the distance between the ith and jth scattering elements
q = scattering vector, $(4\pi/\lambda)\sin(\theta/2)$
λ = wavelength of the radiation
θ = scattering angle

Prior to the advent of computers, fitting data to equation 1 was virtually impossible. To circumvent this problem various mathematical models and simplifications were performed to develop more tractable analyses. Perhaps the most famous model is the Guinier approximation (29,30,32). If the term $\sin(qr_{ij})/qr_{ij}$ is expanded using a Taylor's series, equation 1 becomes:

$$I(q) = K \sum_{i}^{N} \sum_{j}^{N} \left(1 - \frac{q^2(r_{ij})^2}{3!} + \frac{q^4(r_{ij})^4}{5!} \ldots\right) \qquad (2)$$

Guinier (33) defined the nth moment of r as:

$$\langle r^n \rangle = 1/N^2 \sum_{i}^{N} \sum_{j}^{N} (r_{ij})^n \qquad (3)$$

Equation 2 may now be expressed as:

$$I(q) = KN^2 \left(1 - \frac{q^2\langle r^2 \rangle}{3!} + \frac{q^4\langle r^4 \rangle}{5!} \ldots\right) \qquad (4)$$

Furthermore, the root mean square radius-of-gyration ($\langle Rg^2 \rangle^{1/2}$) can be defined such that $\langle Rg^2 \rangle^{1/2} = (\langle r^2 \rangle/2)^{1/2}$. Also, if the Taylor's series is truncated after the quadratic term $(-q^2\langle r^2 \rangle/3!)$ then the series resembles the Taylor's series expansion for $\exp(x)$ with $x = -q^2\langle r^2 \rangle/3!$ (32). Finally, N can be defined as $N = V(\Delta\rho^2)^{1/2}$, where V is the irradiated volume and $\Delta\rho^2$ is the average scattering contrast. For a two phase material, $\Delta\rho^2$ represents the square of the difference in refractive indices between phases. To be more exact:

$$\Delta\rho^2 = (\rho_1 - \rho_2)^2 x_1 x_2 \qquad (5)$$

where: ρ_i = the refractive index of phase i
x_i = the volume fraction of phase i

Insertion of the above into equation 4 leads to Guinier's final form (32):

$$I(q) = KV^2\Delta\rho^2 \exp(-q^2\langle Rg^2 \rangle/3) \qquad (6)$$

If the experimental data are now plotted in the form of $\ln[I(q)]$ vs. q^2, a straight line is expected, with the slope being a third of $-\langle Rg^2 \rangle$. Note that $I(q)$ is proportional to the contrast factor $(\Delta\rho^2)$. This observation leads to the Babinet principle which states that by scattering methods alone, it is impossible to determine whether phase 1 is dispersed in 2 or vice versa. So, while scattering experiments can yield useful qualitative and quantitative information, they must always be accompanied by some other morphological investigative technique, such as SEM, to more completely determine the nature of the scattering.

The Guinier analysis is truly valid only for dilute systems of uniform size and shape, with no global order. However, the approximation has shown its usefulness for a variety of other systems (34-36). In the case of oriented systems, data for the semi-log plot are taken as data slices at certain azimuthal angles to a reference axis. For densely packed systems application of the Guinier analysis may or may not yield a physically significant value for $\langle Rg^2\rangle^{1/2}$, but is normally useful in differentiating between samples that are composed of the same material, but processed in different manners.

Occasionally, the data on a semi-log plot will not be linear, but will exhibit some curvature (33,34), as shown in Figure 2. In the past, it was common to draw a straight line through the initial part of the curve, and to then disregard those points which did not fall on the line (35,36). Typically, the curvature in the system was interpreted as being due to either polydispersity, or concentration effects. Obviously determining which points to keep and which to discard can be rather arbitrary at times. To alleviate this problem Gethner (37) recently suggested that the scattering data be fitted to a fourth order polynomial. So that:

$$\ln[I(q)] = B0 + B1q + B2q^2 + B3q^3 + B4q^4 \qquad (7)$$

and B0, B1, etc. are the regression coefficients. Thus, $\langle Rg^2\rangle = -3B2$. A problem with this analysis is that it ignores the behavior of the scattering function as indicated by the Taylor's series expansion in equation 4, where odd powers of q were not present. An alternative approach has been given by Stoll et al. (38) in which they modify Guinier's approximation by allowing x in the Taylor's series expansion of exp(x) to equal $-q^2\langle r^2\rangle/3! + q^4\langle r^4\rangle/5!$. So:

$$I(q) = KV^2\Delta\rho^2\exp(-q^2\langle Rg^2\rangle/3 + q^4\langle r^4\rangle/5!) \qquad (8)$$

The fourth order term now allows for curvature in the semi-log plot. The data can now be fitted to a second order polynomial:

$$\ln[I(q)] = B0 + B1y + B2y^2 \qquad (9)$$

where: $y = q^2$

$\langle Rg^2\rangle$ is now given as $-3B1$. It is expected that B2 should yield information regarding $\langle r^4\rangle$; however, Stoll et al. suggested that this is not the case due to the affects of polydispersity. Another approach has been proposed by Effler (17) in which the data are fit-

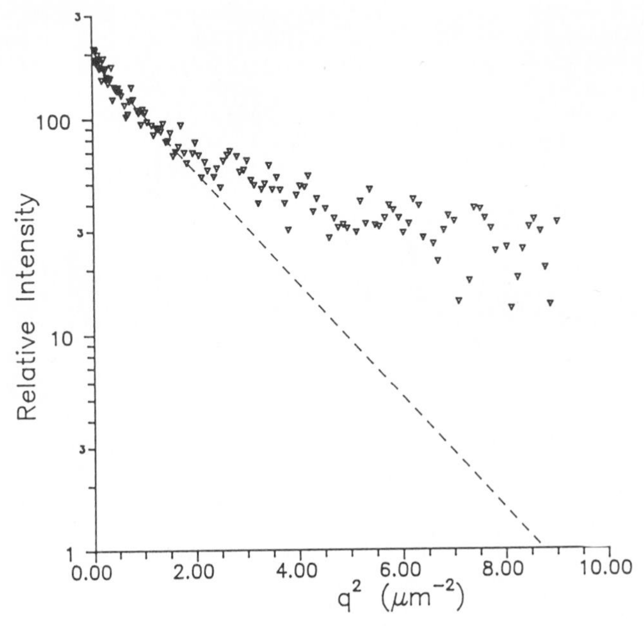

Figure 2: Guinier Plot of a 40/60 PET/POB Copolyester: 255 °C, Ice Water Quench.

ted to a either a sixth or eighth order polynomial in simulation of equation 4. So that:

$$I(q) = B0 + B1y + B2y^2 + B3y^3 + B4y^4 + B5y^5 + B6y^6 \quad (10)$$

Again $y = q^2$ and $\langle Rg^2 \rangle = -3B1/B0$. This method varies from Gethner's in that it allows only for even powers of q in accordance with equation 4. Once more, information about higher moments of r is expected from the analysis. However by studying the use of this approximation with well specified distributions of r, Effler found that the nature of reciprocal space tends to result in an underestimation of the higher moments.

In the above equations, no account has been made for the use of polarized light that is typically used in SALS. In a classic SALS experiment, the incident light is vertically polarized, and the scattered radiation is passed through a second rotatable polarizer referred to as the analyzer. Stein and Rhodes (5) give the relationship for the amplitude of a scattered ray of polarized light (E) as:

$$E = K \int_0^\infty (M \cdot O) r^2 \cos(qr) \sin\alpha \, dV \quad (11)$$

where: K = constant
M = the induced dipole moment of the scattering element
O = a unit vector in the direction of the analyzer
α = the polarizability of the material

The intensity of the scattered radiation is the product of E and its complex conjugate E*. Prud'homme and Stein (39) examined the type of scattering to be expected from a material composed of rod-like aggregates between cross polars (Hv mode). In the case of small rods (<5 µm) with random orientations the scattered intensity (I_{Hv}) is derived as:

$$I_{Hv} = KL^2\delta^2 \int_0^\infty T(r)[\sin qr/qr] r^2 \, dr \quad (12)$$

where: L = the rod length
δ = the average optical anisotropy
T(r) = the correlation function of r

If $T(r) = \exp(q^2/a^2)$, where "a" is the characteristic length, then:

$$I_{Hv} = KL^2\delta^2 a^2 \exp(-q^2a^2/4) \quad (13)$$

Notice that equation 13 is essentially a Guinier type equation, and that equation 12 is simply a special case of the correlation function form of equation 1, given by Debye and coworkers (40,41). As such, when non-linearity in a semi-log plot is observed then the methods of Stoll et al. (38) and Effler (17) should prove useful.

Equations for the scattering behavior of oriented systems of rod-like aggregates quickly become complex. Essentially, the expressions are dependent of the spatial distribution of r and upon the angle that the optical axis of the rod makes with the incident light (39,42-45). In the case of Hv scattering, if the rods pack in some sort of oriented domain structure, a four-point pattern is predicted (41-46). However, if a highly ordered monodomain system is formed a streak-like pattern will form with the streak appearing perpendicular to the direction of the optical axis of the rods (39). Figure 3 shows the various types of patterns that might be observed for a system of rod-like aggregates with various orientations.

Results

Compression molded films of the 40/60 PET/POB copolyester possess a structure of randomly oriented fibrillar domains. The size of these domains are in the range of 1-2 μm. Figure 4 shows a typical isointensity contour plot for these films. The pattern displays cylindrical symmetry, which is consistent with Prud'homme's and Stein's (39) prediction for scattering from a system of randomly oriented rods or disks. Data slices were taken at ±45°, and interpreted using Effler's polynomial analysis. Figure 5 demonstrates a typical intensity vs. q^2 curve, fitted to both the Guinier analysis, and the polynomial analysis. Table I shows a comparison between the values of $\langle Rg^2\rangle^{1/2}$ determined using the polynomial analysis and a standard Guinier analysis on films selected at random, while Table II presents the actual polynomial regression coefficients.

Figure 6 shows a plot of $\langle Rg^2\rangle^{1/2}$ vs. molding temperature for the copolyester films. As can be seen, $\langle Rg^2\rangle^{1/2}$ ranges between 1-2 μm. There is a slight tendency for the size to increase with molding temperature. However, the $\langle Rg^2\rangle^{1/2}$ appears to be independent of the cooling condition. This observation is especially true of the quenched and air cooled samples. Optical micrographs of the copolyesters (Figure 7) reveal a fine "salt and pepper" texture indicating no global orientation, and occasionally fibrils can be seen extending from the edge of a specimen. Under the microscope, these films are highly birefringent, and display an array of colors. The micrographs also confirm that the average domain size is of the order of 1 μm. SEM micrographs (Figure 8) of the films reveals a coarse fibrillar texture which is oriented within the domains, but not between domains.

Cold stretched films were difficult to produce and upon stretching exhibited little, if any, orientation. From Table III it is obvious that of the ten samples stretched only specimens U2 and U8 did not fail upon being loaded. These samples were stretched at 90, and 110 ^{o}C. SALS was performed on all the samples, but only samples U2 and U8 were used in the SAXS, WAXD, and SEM experiments. Figure 9 shows the resulting Hv SALS patterns for some of the cold stretched films. The patterns are circular indicating no orientation. Confirming this lack of orientation are SAXS patterns that showed cylindrical symmetry and WAXD patterns of the films that yielded circular diffraction rings.

However, the SALS pattern (Figure 10) of a film sheared at 270 ^{o}C between microscope slides resembles a streak. The direction of the streak is perpendicular to the shear direction. This result is consistent with Rhodes' and Stein's (42) expression for scattering from an assemblage of highly ordered rods. An optical micrograph

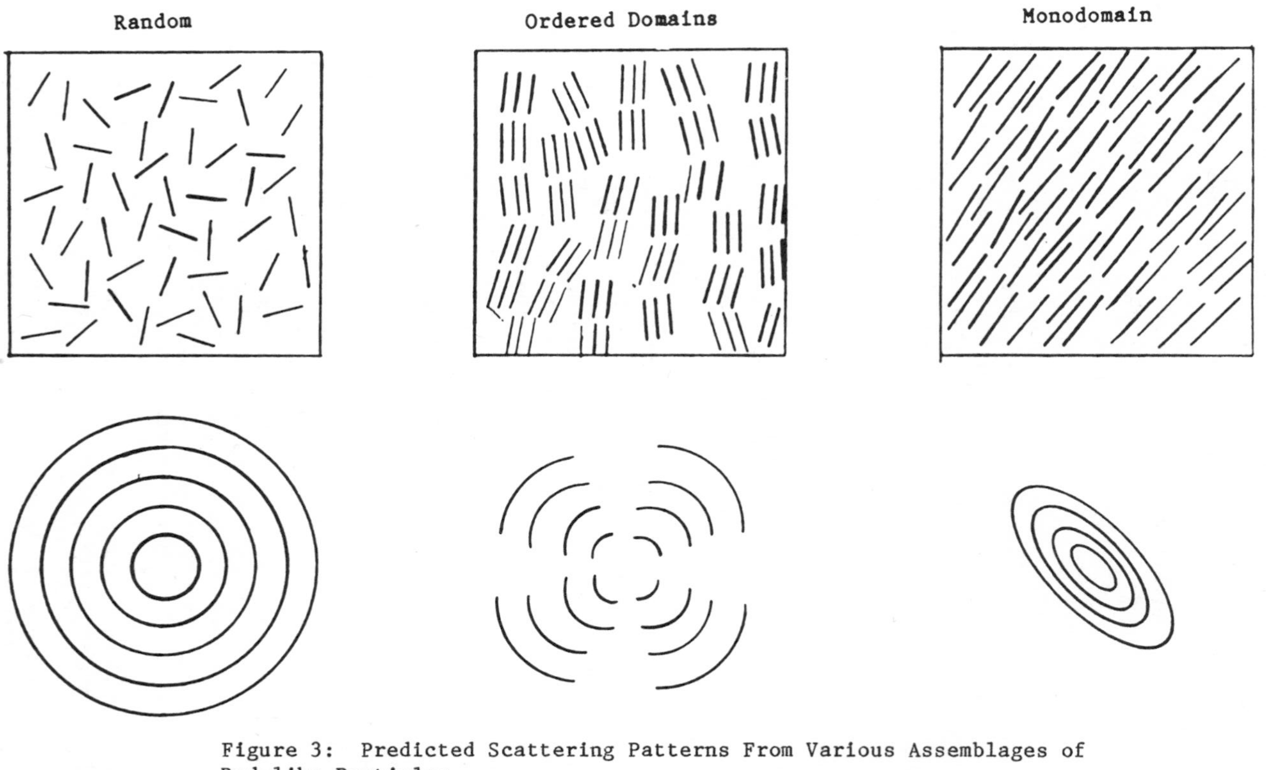

Figure 3: Predicted Scattering Patterns From Various Assemblages of Rod-like Particles.

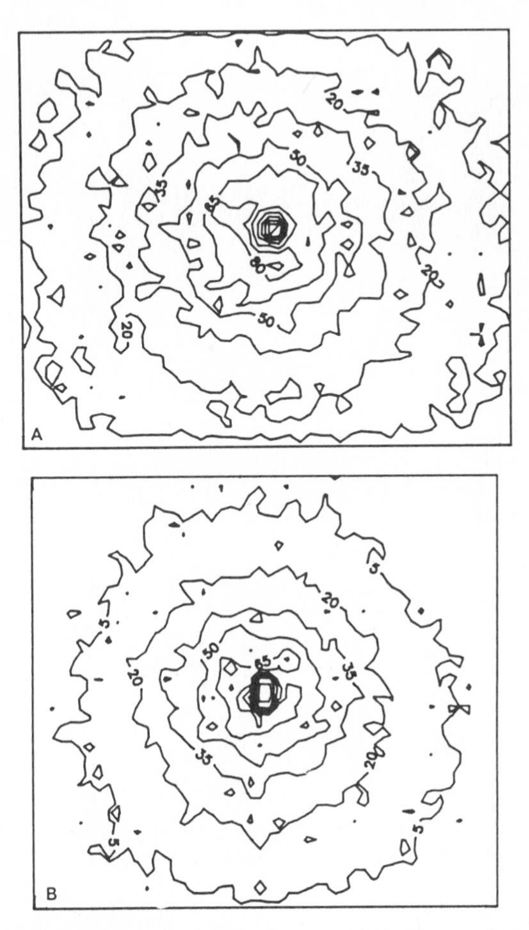

Figure 4: Typical Iso-intensity Contour Patterns of Compression Molded 40/60 PET/POB Copolyester: A) 255 ^{o}C, Press Cooled B) 285 ^{o}C, Ice Water Quenched.

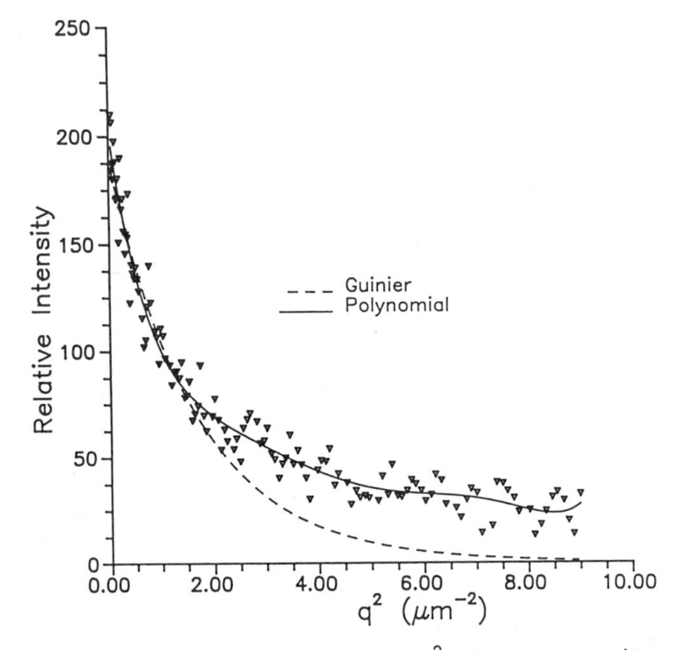

Figure 5: Typical Intensity vs. q^2 Plot for a 40/60 PET/POB Copolyester, 255 ^{o}C, Ice Water Quench.

Table I: Comparison of Results Between the Guinier Analysis and the Polynomial Analysis

Molding Temp. (^{o}C)	Cooling Condition	Guinier Analysis $\langle Rg^2\rangle^{1/2}$ (μm)	Guinier Analysis No. Data Points	Guinier Analysis COD	Polynomial Analysis $\langle Rg^2\rangle^{1/2}$ (μm)	Polynomial Analysis No. Data Points	Polynomial Analysis COD
210	ice water	1.20	56	0.989	1.33	96	0.997
210	press	1.08	59	0.991	1.15	97	0.995
230	ice water	1.39	42	0.970	1.50	75	0.993
230	air	1.12	32	0.963	1.13	102	0.996
255	ice water	1.36	39	0.962	1.53	82	0.989
255	air	1.43	64	0.963	1.68	83	0.987
285	ice water	1.27	71	0.861	1.54	133	0.946
285	air	1.43	50	0.861	1.66	129	0.961

Table II: Representative Polynomial Regression Analysis Results

Molding Temp. (^{o}C)	Cooling Condition	No. Data Pts.	B0	B1	B2	B3	B4	B5	B6	COD
210	ice water	96	97.78	-57.81	18.61	-3.59	0.394	-2.24E-02	5.10E-04	0.997
210	press	97	98.83	-43.55	10.46	-1.50	0.122	-5.12E-03	8.62E-05	0.995
230	ice water	75	100.17	-75.01	33.53	-9.15	1.38	-0.105	3.09E-03	0.993
230	air	102	90.47	-38.43	9.45	-1.41	0.119	-5.10E-03	8.70E-05	0.996
255	ice water	82	91.10	-71.09	33.80	-9.76	1.58	-0.131	4.31E-03	0.989
255	air	83	91.80	-86.12	54.75	-23.27	5.82	-0.753	3.85E-02	0.987
285	ice water	133	72.33	-57.31	29.43	-9.40	1.71	-0.159	5.82E-03	0.945
285	air	129	86.31	-80.28	46.35	-14.94	2.61	-0.231	8.07E-03	0.961

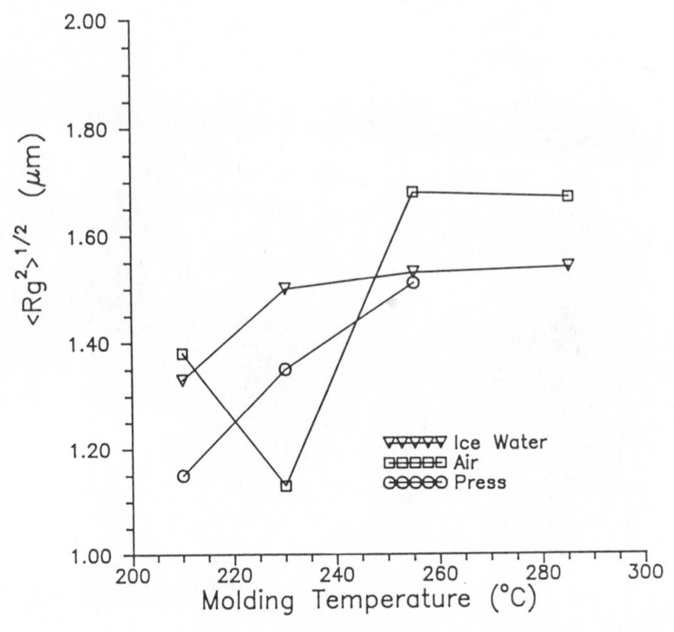

Figure 6: Plots of $\langle Rg^2 \rangle^{1/2}$ vs. Molding Temperature for Compression Molded 40/60 PET/POB Copolyesters.

Table III: 40/60 PET/POB Stretching Results[a]

Sample Code	Unstretched Thick. (mm)	Stretched Thick. (mm)	Stretch. Temp. (°C)	Comments
U1	0.143	0.109	110	Ripped
U2	0.134	0.078	110	OK
U3	0.138	0.114	175	Ripped
U4	0.138	0.121	175	Ripped
U5	0.136	-	110	Ripped in grips
U6	0.142	-	110	Ripped in grips
U7	0.137	0.125	160	Ripped
U8	0.134	0.065	90	OK
U9	0.144	-	90	Ripped in grips
U10	0.137	-	160	Ripped in grips

[a]Attempted Stretch Ratio: 2 x 1
Stretching Rate: 750%/min.

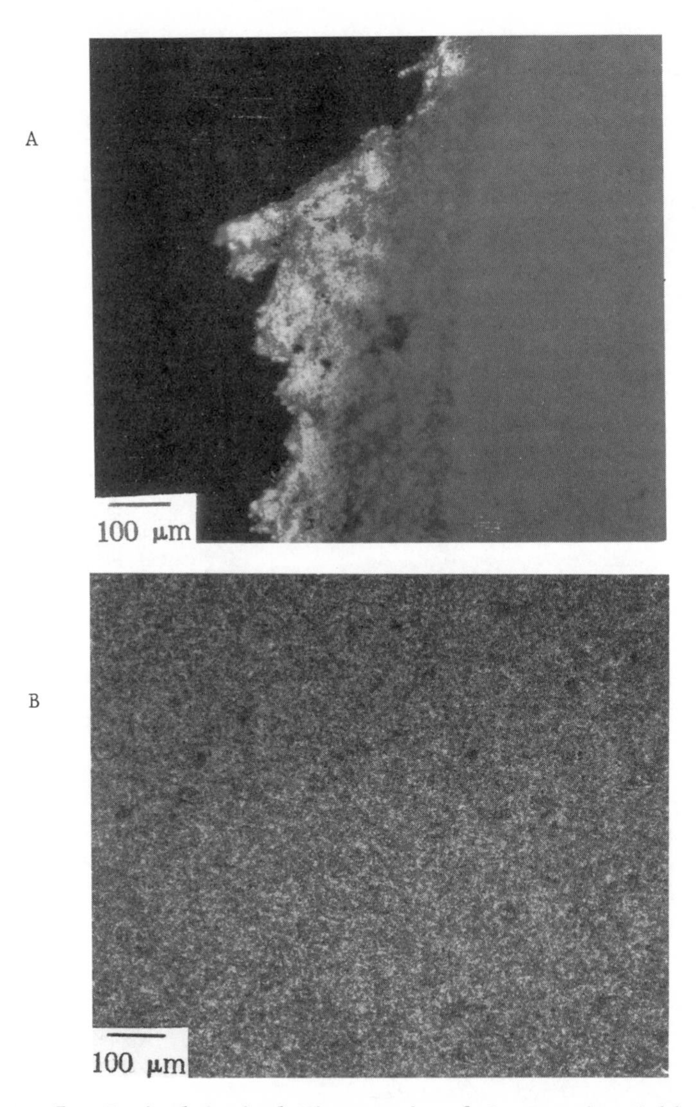

Figure 7. Typical Optical Micrographs of Compression Molded 40/60 PET/POB Copolyesters: A) 255 ^{o}C, Ice Water Quench B) 285 ^{o}C, Air Quenched.

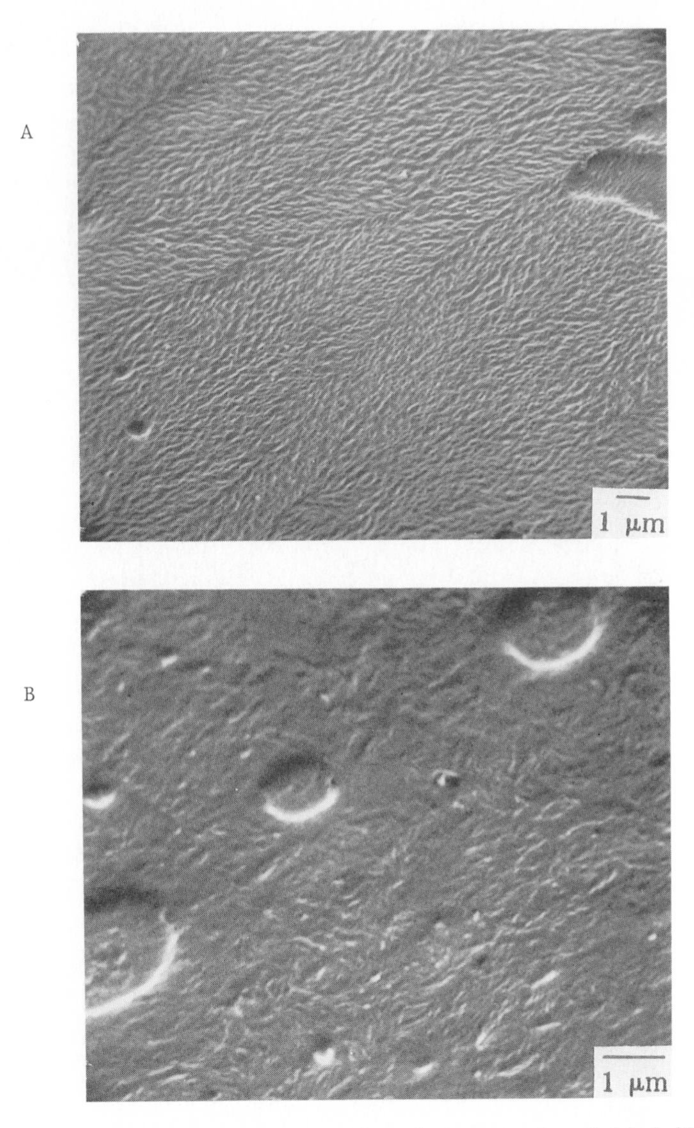

Figure 8: Typical SEM Micrographs of Compression Molded 40/60 PET/POB Copolyesters : A) 255 °C, Ice Water Quench B) 210 °C, Air Quenched.

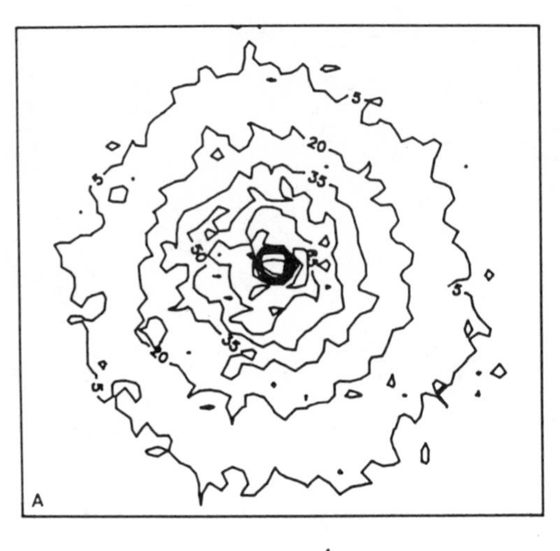

M.D.

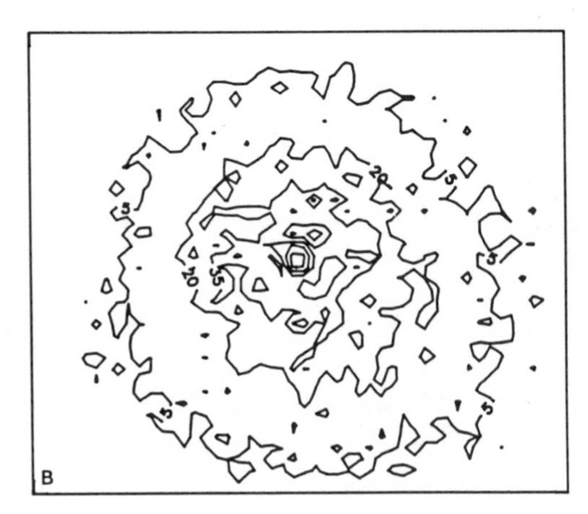

Figure 9: Iso-intensity Contour Patterns of Stretched 40/60 PET/POB Copolyester Films: A) U8 B) U2.

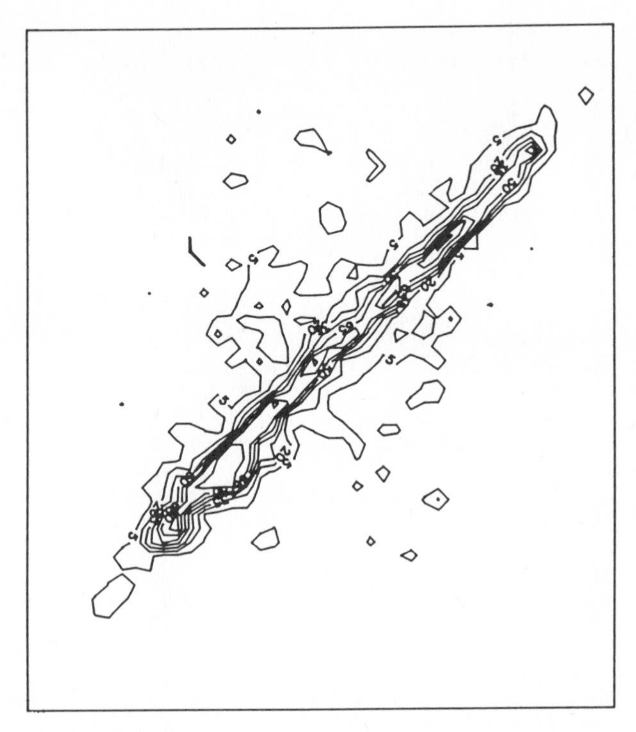

Figure 10: Iso-intensity Contour Pattern of a Sheared 40/60 PET/POB Film.

(Figure 11) of the sheared film reveals a banded structure normal to the shear direction, which fades out when the specimen is aligned in the direction of either the polarizer, or the analyzer. This structure is similar to that reported by Zachariades et al. (47,48) for sheared textures of 20/80, and 40/60 PET/POB copolyesters. Using scanning electron microscopy, and electron diffraction, they found the films to be highly fibrillar, with the fibrils aligned in the shear direction. Shear banding appears to be a common among liquid crystalline polymers and has been observed for poly(γ-benzyl-glutamate) (49), KEVLAR (50), poly(β-thioester) (51), copolymers of chlorophenyleneterephthalate/bis-phenoxyethane (52), and hydroxypropylcellulose (53).

Streak like patterns were also given by the extrusion drawn films supplied by Baird and Wilson (Figure 12). Again, the direction of the streaks are normal to the drawing, or machine direction (MD), as expected (42). Optical micrographs (Figure 13) shows a long fibrillar structure, with the fibrils being located parallel to the MD. While rotating the microscope stage between crossed polars, the field of view would became most intense when the MD was located at 45° to the polarizers, and extinct when the MD was parallel to either polarizer.

Discussion

Guinier vs. Polynomial Analysis. The $\langle Rg^2\rangle^{1/2}$ values obtained from the sixth order polynomial regression analysis are similar to those obtained by standard Guinier analysis, yet they are distinguished in several important ways. Table I reports $\langle Rg^2\rangle^{1/2}$ values, the coefficient of determination (COD), and the number of data points used in each analysis for a variety of samples. Note that while the $\langle Rg^2\rangle^{1/2}$ values from the Guinier and polynomial analyses are similar, the polynomial analysis value is consistently a little bit larger. Two other important characteristic become apparent. Namely, the polynomial analysis has higher valued COD's, where a value of 1 means a perfect fit, and all the data from the slice are used. For example, in the case of the sample molded at 285 ^{o}C and air quenched, the Guinier $\langle Rg^2\rangle^{1/2}$ is 1.43 μm, with a COD of 0.861 from 50 data points. In contrast the polynomial analysis yields an $\langle Rg^2\rangle^{1/2}$ of 1.66 μm, with a COD of 0.991 using 129 data points. In fact a key advantage of this method is that little, if any, data must be disregarded to yield meaningful results.

Further insight into the appropriate behavior of the polynomial regression analysis is found by comparing the results expected from equation 4, to the regression coefficients actually obtained. Table II shows that the coefficients oscillate in sign in accord with the Taylor's series expansion of the scattering function. While this feature, along with the quantitative nature of the improvement in COD, and use of more data points, all help to establish the superiority of this method, Figure 5 makes a fundamentally convincing point. Note the smooth monotonic fitting of the polynomial regression curve to the actual data points. The regression curve looks right, in that it fits the experimental data without doing something perverse to make COD near 1.0 like 'snaking' its way through the data.

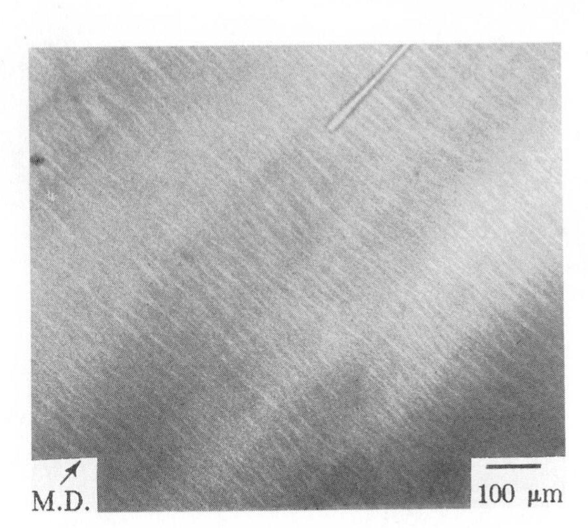

Figure 11: Optical Micrograph of a Sheared 40/60 PET/POB Film.

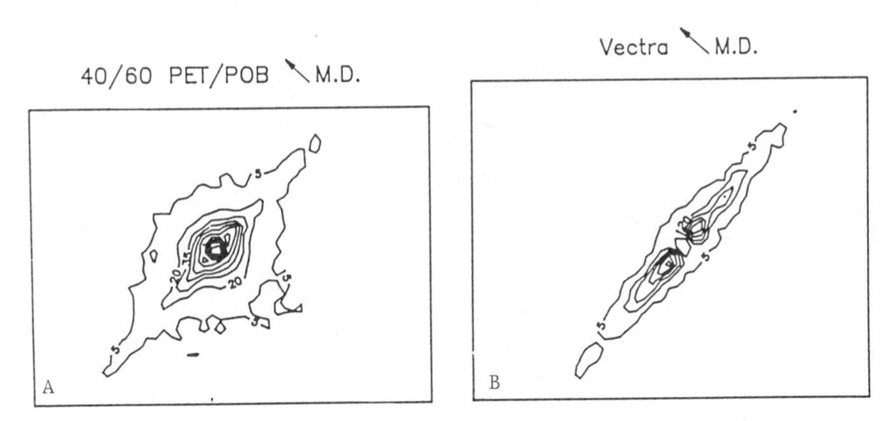

Figure 12: Iso-intensity Contour Patterns of Extrusion Drawn Copolyester Films.

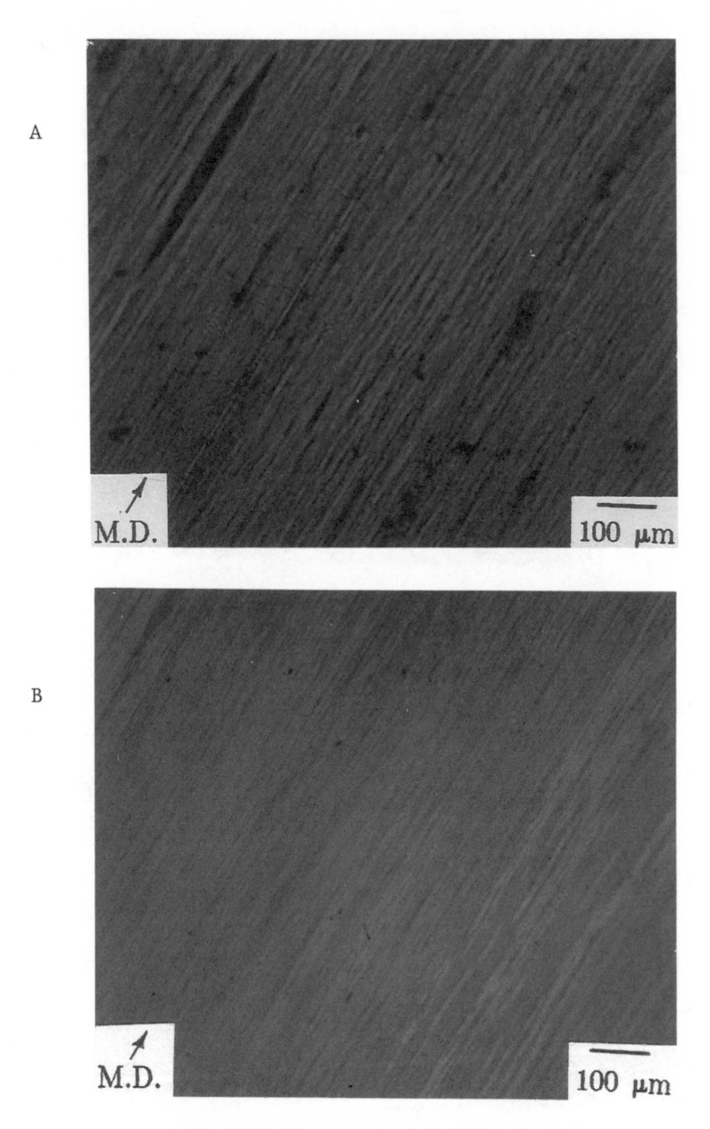

Figure 13: Optical Micrographs of Extrusion Drawn Copolyester Films: A) 40/60 PET/POB B) VECTRA.

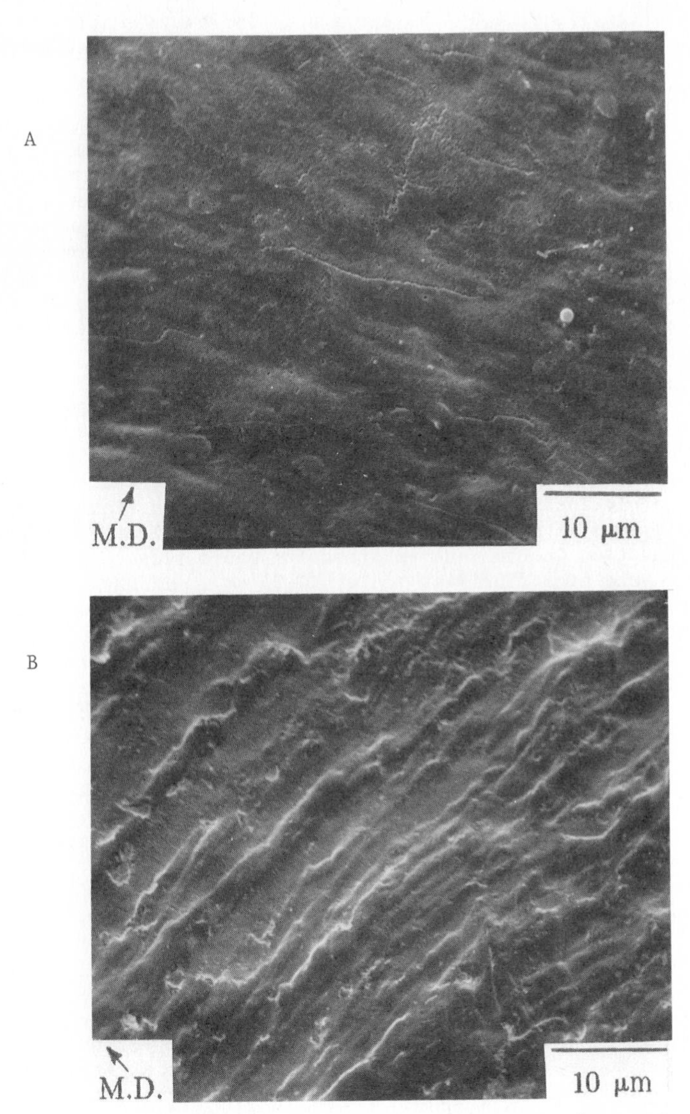

Figure 14: SEM Micrographs of Stretched 40/60 PET/POB Films: A) U8 B) U2.

Morphological and Process Consideration. The insensitivity of domain size to the cooling condition suggests that the structure formed in the melt does not relax rapidly upon cooling. A number of researchers (26,47,48,54,55) have reported for these copolyesters either long structural relaxation times, or structures that do not change when annealed. If the copolyesters are modeled as a system of semi-flexible rods, then this long structural relaxation makes sense. Flory (56) demonstrates that for a system of semi-flexible rods, the equilibrium state is an ordered state below some critical degree of flexibility. With no externally imposed field, there is no preferred orientation direction, so localized order (domains) prevails. The formation of global order requires the cooperative movement of a large number of chain segments, which in turn need more thermal input, time, or an applied external force. In fact, large scale order in these copolyesters is not detected until there is an imposed stress field on these materials while they are in the fluid state as demonstrated in this and other work (17,47,48,54,57,58).

The lack of orientation in the stretched films may again be explained by considering the polymer chains as semi-flexible rods. In flexible coil polymers an applied load is thought to straighten out chain segments in the loading,or machine, direction, and as such, requires mainly localized motions (59). However, in liquid crystalline polymers, large sections of the chain are already extended. Thus, orientation in the machine direction would require motion and rotation on the microfibrillar scale (<0.05 μm). If the surroundings of the chains are rigid enough, slippage or fracture might occur before the necessary movement to obtain orientation. So instead of deforming domains, or aligning fibrils, the load may be causing dislocations of domains via slippage. Figure 14 shows an SEM micrograph of a film stretched at 90 °C. Note the series of parallel bands running normal to the machine direction, reminiscent of optical shear banding. These bands may be the result of slippage of parallel domains past one another while under load. This notion of slippage is also supported by the work of Shih et al. (58), who report only amorphous orientation in 40/60 PET/POB films stretched 2.7 x 1, when plasticized with 1,1,1,3,3,3-hexafluoro-2-propanol. Therefore, if a highly ordered sample is required, the order must be induced while the sample is in the melt.

Acknowledgments

Support for this work was provided by the Center for Materials Processing at the University of Tennessee. We also extend our appreciation to the National Center for Small Angle Research at the Oak Ridge National Laboratory for the use of their facilities.

Literature Cited

1. Schelten, J.; Hendricks, R. W. J. Appl. Cryst. 1978, 11, 297.
2. Tabar, R. J.; Stein, R. S.; Long, M. B. J. Polym. Sci., Polym. Phys Ed. 1982, 20, 2041.
3. Tabar, R. J.; Leite-James, P.; Stein, R. S. J. Polym. Sci., Polym. Phys. Ed. 1985, 23, 2085.

4. Stein, R. S. Polymer J. 1985, 17, 289.
5. Tabar, R. J. Ph.D. Dissertation, University of Massachussetts at Amherst, 1983.
6. Stein, R. S.; Rhodes, M. B. J. Appl. Phys. 1960, 31, 11.
7. Samuels, R. J. J. Polym. Sci., A-2 1971, 9, 2165.
8. Meeten, G. H.; Navard, P. J. Polym. Sci., Polym. Phys. Ed. 1984, 22, 2159.
9. Wissler, G. E.; Crist, B. J. Polym. Sci., Polym. Phys. Ed. 1984, 23, 2395.
10. Talmi, Y. American Laboratory March 1978.
11. Ravich, L. E. Laser Focus 1986, 22, 128.
12. Pakula, T.; Soukup, Z. J. Polym. Sci., Polym. Phys. Ed. 1974, 12, 2437.
13. van Antwerpen, F.; van Krevelen, D. W. J. Polym. Sci., Polym. Phys. Ed. 1972, 10, 2409.
14. Wasiak, A.; Peiffer, D.; Stein, R. S. J. Polym Sci., Polym. Lett. Ed. 1976, 14, 381.
15. OMA 2 Multichannel Spectroscopy Equipment Brochure; EG&G Princeton Applied Research Corporation: Princeton, NJ, 1981.
16. Karasek, F. W. Res./Dev. 1972, 23, 24.
17. Effler, L. J. M.S. Thesis, University of Tennessee at Knoxville, 1987.
18. Ennes, H. E. In Television Broadcasting: Equipment Systems, Operating Fundamentals; Howard W. Sams & Co.: Indianapolis, IN, 1979.
19. PC-Mate Video Van Gogh Installation Manual and Users Guide; Tecmar, Inc.: Solon, OH, 1984, Rev. 1.0.
20. Zuch, E. L. In Data Acquisition and Conversion Handbook; Datel, Inc.: Mansfield, MA, 1979.
21. Maranda, B. American Laboratory April 1987, 41.
22. Low Cost Frame Grabbers DT2803, DT2853 Description Brochures; Data Translation, Inc.: Marlboro, MA, 1986.
23. PC-Eye Video Digitizer Description Brochures; Chorus Data Systems, Inc.: Merrimack, NH, 1986.
24. PC Vision Product Brochures; Imaging Technology, Inc.: Woburn, MA, 1986.
25. Matrix Beam Diagnostic System Product Brochure and Data Sheets; Spiricon, Inc.: Logan, UT, 1986.
26. Jackson, W. J.; Kuhfuss, H. F. J. Polym. Sci., Polym. Chem. Ed. 1973, 14, 2043.
27. McFarlane, F. E.; Nicely, V. A.; Davis, T. G. In Contemporary Topics in Polymer Science 1979, 2, 109.
28. Nicely, V. A.; Dougherty, J. T.; Renfro, L. W. Macromolecules 1987, 20, 573.
29. Kratky, O. In Small Angle X-Ray Scattering; Glatter, O.; Kratky, O., Eds.; Academic: New York, 1982; pp 3-16.
30. Porod, G. In Small Angle X-Ray Scattering; Glatter, O.; Kratky, O., Eds.; Academic: New York, 1982; pp 17-52.
31. Debye, P. Ann. der Phys. 1915, 46, 809.
32. Guinier, A. Ann. Phys. 1939, 12, 161.
33. Kratky, O. In Small Angle X-Ray Scattering; Glatter, O.; Kratky, O., Eds.; Academic: New York, 1982; pp 361-386.

34. Lin, J. S.; Tang, M.-Y.; Fellers, J. F. In The Structures of Cellulose; Atalla, R. H., Ed.; ACS Symposium Series No. 340; American Chemical Society: Washington, DC, 1987; p 233.
35. Statton, W. O. In Newer Methods of Polymer Characterization; Ke, B., Ed.; Interscience: New York, NY, 1964; pp 231-277.
36. Alexander, L. E. In X-Ray Diffraction Methods in Polymer Science; Robert E. Krieger: Huntington, NY, 1979; pp 280-356.
37. Gethner, J. S. J. Appl. Phys. 1986, 59, 4.
38. Stoll, B.; Fellers, J. F.; Lin, J. S. In Scattering, Deformation, and Fracture in Polymers; Material Research Society Symposium Proceedings No. 79, 1987; p 105.
39. Prud'homme, R. E.; Stein, R. S. J. Polym. Sci., Polym. Phys. Ed. 1974, 12, 1805.
40. Debye P.; Bueche, A. M. J. Appl. Phys. 1947, 20, 518.
41. Debye, P.; Anderson, H. R.; Brumberger, H. J. Appl. Phys. 1957, 28, 679.
42. Rhodes, M. B.; Stein., R. S. J. Polym. Sci., A-2 1969, 7, 1539.
43. Stein, R. S.; Rhodes, M. B.; Porter, R. S. J. Coll. Inter. Sci. 1968, 27, 336.
44. Stein, R. S.; Wilson, P. R. J. Appl. Phys. 1962, 33, 1914.
45. Stein, R. S.; Ephardt, P. F.; Clough, S. B.; Adams, G. J. Appl. Phys. 1966, 37, 3980.
46. van Aartsen, J. J. Eur. Polym. J. 1970, 6, 1095.
47. Zachariades, A. E.; Logan, J. A. Polym. Eng. and Sci. 1983, 23, 797.
48. Zachariades, A. E.; Navard, P.; Logan, J. A. Mol. Cryst. Liq. Cryst. 1984, 110, 93.
49. Kiss, G.; Porter, R. S. Mol. Cryst. Liq. Cryst. 1980, 60, 267.
50. Takahashi, T.; Iwamoto, H.; Inoue, K.; Tsujimoto, I. J. Polym. Sci., Polym. Phys. Ed. 1979, 17, 115.
51. Laus, M.; Angeloni, A. S.; Ferruti, P.; Galli, G.; Chiellini, E. J. Polym. Sci., Polym. Lett. Ed. 1984, 22, 587.
52. Graziano, D. J.; Mackley, M. R. Mol. Cryst. Liq. Cryst. 1984, 106, 73.
53. Navard, P. J. Polym. Sci., Polym. Phys. Ed. 1986, 24, 435.
54. Joseph, E. G.; Wilkes, G. L.; Baird, D. G. Polym. Eng. Sci. 1985, 25, 377.
55. Wissbrun, K. F. Brit. Polym. J. Dec. 1980, 163.
56. Flory, P. J. Proc. Roy. Soc. (London) 1956, A234, 60.
57. Muramatsu, H.; Krigbaum, W. R. J. Polym. Sci., Polym. Phys. Ed. 1986, 24, 1695.
58. Shih, H. H.; Hornberger, L. E.; Siemens, R. L.; Zachariades, A. E. J. Appl. Polym. Sci. 1986, 32, 4897.
59. Ward, I. M. In Mechanical Properties of Solid Polymers; 2nd ed., John Wiley and Sons: New York, 1983; Chapter 10.

RECEIVED August 26, 1988

Chapter 15

Dielectric and Electrooptical Properties of a Chiral Liquid Crystalline Polymer

G. S. Attard[1,4], K. Araki[1,5], J. J. Moura-Ramos[1,6], G. Williams[2], A. C. Griffin[3], A. M. Bhatti, and R. S. L. Hung[3]

[1]Edward Davies Chemical Laboratories, University College of Wales, Aberystwyth, SY23 1NE, United Kingdom
[2]Department of Chemistry, University of Swansea, Singleton Park, Swansea SA2 8PP, United Kingdom
[3]Department of Chemistry and Polymer Science, University of Southern Mississippi, Hattiesburg, MS 39406

The dielectric properties of a chiral-nematic liquid crystalline polymer having applications in non-linear optics have been studied over wide ranges of frequency and temperature. Samples of different degrees of macroscopic alignment (homeotropic, planar) are shown to exhibit very different relaxation behaviour, and this is interpreted in terms of the anisotropic motions of the dipolar mesogenic groups. It is shown that fully homeotropic alignment is not achieved and that on removal of the directing electric field a relaxation of alignment occurs, which may be due to the reformation of chiral structures.

There is considerable current interest (1,2) in the synthesis and properties of liquid crystalline (LC) polymers whose side groups contain groups which possess large linear and higher order optical polarizabilities. This interest stems from the fact that films (5 - 100 μm) made from these materials show promise as media for optical information storage and processing, including second harmonic generation (SHG) of laser radiation. Polymeric films may have some advantages over monomeric organic crystals in view of their ease of processing and their resistance to damage by laser beams. In addition, LC polymers may be aligned macroscopically using electric or magnetic fields, giving an enhanced susceptibility and also allowing the macroscopic optical properties to be varied continuously over a wide range. Most of the applications of LC polymers require a film to be aligned macroscopically. In previous papers (3-11) it was shown that acrylate or siloxane polymers could be aligned homeotropically, planarly (or

[4]Current address: Department of Chemistry, The University of Southampton, S09 5NH, United Kingdom
[5]On leave from Science University of Tokyo, Kagurazaka, Shinjuku-ku, Tokyo, Japan
[6]On leave from Departamento de Quimica Engenharia, Technical University of Lisbon, Portugal

0097-6156/89/0384-0255$06.00/0

homeogeneously) or to intermediate extents of alignment by moderate magnetic fields (3-6) or a.c. electric fields (7-11). In the electrical case the alignment process involves the dielectric properties of the polymer, which arise due to the anisotropic motions of dipolar mesogenic groups pendant to the chain (11). It follows that a direct link can be made between the macroscopic alignment behaviour and the molecular properties of the chain (7,10,11). In our earlier studies of siloxane polymers we found (7-13) that the homopolymers could not be aligned by application of the a.c. electric field to the material in its LC state, but copolymers would do so. All materials could be aligned by cooling the melt into the LC state in the presence of the electric field. Homeotropic alignment was achieved readily but planarly aligned material only formed if the cross-over frequency ν_c was below normally accessible power frequencies ($\sim 10^4$ Hz) at the clearing temperature T_c (12,13).

In this paper we describe the behaviour of an LC polymer whose backbone structure is different from that of the conventional acrylate, methacrylate or siloxane polymers (14,15). The polymer I described below was designed for NLO applications in that each repeat unit contained a mesogenic group which has a large second and third order optical polarizability. By incorporating the electroactive group in the side chain of the polymer the problems of low solubility and phase separation which occur in guest/polymer host systems are overcome. Also polymer I has a chiral group in the main chain repeat unit. Using dielectric relaxation spectroscopy we have been able to study the dynamics of the mesogenic groups and the electric-field-induced alignment behaviour. It is shown that this polymer may be aligned directly in the LC state.

Experimental

The polymer had the following structure

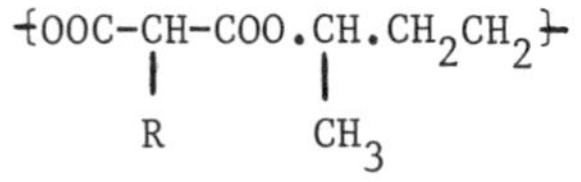

$$\left[\text{OOC-}\underset{\displaystyle R}{\underset{|}{\text{CH}}}\text{-COO.}\underset{\displaystyle CH_3}{\underset{|}{\text{CH}}}\text{.CH}_2\text{CH}_2\right] \qquad \text{I}$$

where R is

$$-(CH_2)_6O-C_6H_4-N = CH-C_6H_4-NO_2$$

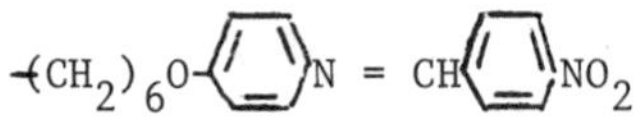

The material was chiral-nematic and had an apparent glass transition at 276 K and a clearing point at 325.2 K, as judged by DSC measurements. The clearing point was very sharp, the biphasic range being $\sim$ 0.5 K as determined by optical microscopy.

Dielectric spectra were obtained in the frequency range 15 to 10^5 Hz using a computer-controlled GenRad 1689 Precision RLC Digibridge. The samples in disc form (1 cm diameter, 120 μm thick) were contained in a three terminal dielectric cell whose electrodes were made in stainless steel. Sample temperatures were controlled to ± 0.01 K by immersing the cell in a thermostatted water bath. Electro-optical measurements were made as follows. A sample was prepared between conducting glass plates (ITO 1 cm^2) at a separation of 24 μm, using a mylar spacer. Polarizers were attached to both glass plates and were crossed. The sample-assembly was mounted in a sealed metal block and was illuminated using light (from a tungsten filament lamp) which was propagated to the sample via an optic-fibre bundle. The light transmitted through the sample was further transmitted along an equivalent fibre-optic bundle to a CdS photodiode, whose output was compared with that from a second photodiode activated by the same bulb. This arrangement ensured that the optical measurements would not be affected by fluctuations in the intensity of the light source.

Samples were aligned by applying an a.c. electric field of 200 V rms (for dielectric measurements) and 50 V rms (for electro-optical measurements), and of variable frequency. It was found that polymer I was two-frequency-addressable such that application of the electric field at 400 Hz to the melt followed by slow cooling gave a homeotropically-aligned sample, while raising the frequency to 10 kHz at this voltage and repeating the cooling from the melt gave a planarly-aligned sample.

Results and Discussion

Figure 1 shows the dielectric loss spectra obtained at 306.2 K for the homeotropic (H), unaligned (U) and planarly-aligned (P) sample, where the H and P materials were prepared by cooling from the melt in the presence of saturating low and high frequency directing electric fields. The H sample has a well-defined low frequency loss peak accompanied by a broad high frequency shoulder. The P sample gives a very broad, featureless absorption whose frequency of maximum loss occurs significantly higher than that for the H sample. The U sample exhibits intermediate behaviour. The isobestic point (at which the loss factor is independent of macroscopic alignment of sample) is seen clearly at 2 kHz. Figure 2 shows the plots of sample capacitance against frequency at 306.2 K for the H and P samples. The crossover frequency ν_c (which is an isobestic point) is seen to occur at 410 Hz at this sample temperature. ν_c determines the nature of alignment if an a.c. directing electric field is able to align a material in its LC state. H or P alignment is obtained for a directing field having $\nu < \nu_c$ or $\nu > \nu_c$ respectively at the fixed temperature (10,16,17). We have determined ν_c from such data (Figure 2) for a wide range of sample temperatures in the LC state of the polymer. Figure 3 shows the plot of log ν_c-vs-$(T/K)^{-1}$ and the rapid variation observed is simply due to the critical slowing-down of the dielectric relaxations in H and P material as the apparent T_g of the polymer is approached. As indicated in the figure, a directing field of $\nu > \nu_c$ will produce planarly-aligned material while that for $\nu < \nu_c$ will produce a homeotropically-aligned material for a given sample temperature.

In contrast with our earlier findings for siloxane homopolymers, we found that polymer I could be aligned directly by application of a strong a.c. electric field to the polymer in its LC state. The kinetics of the alignment behaviour will be described in a future paper, but it should be said that we found that the rate of macroscopic alignment decreased rapidly as the sample temperature was lowered below T_c. As one example of the alignment behaviour, Figure 4 shows data we obtained at 319.2 K. A sample was prepared homeotropically-aligned by cooling from the melt in a saturating low

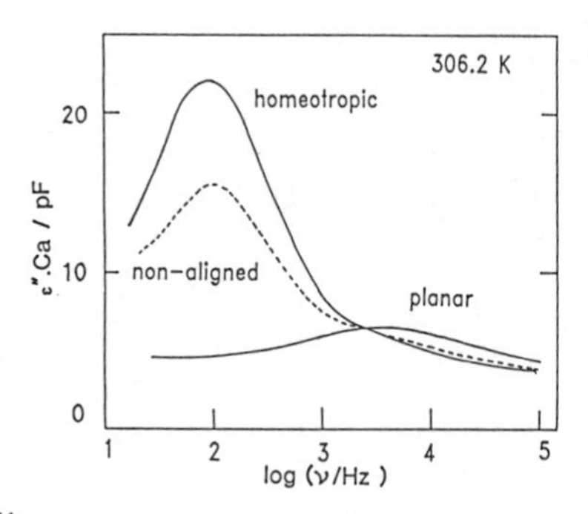

Figure 1. $\varepsilon'' C_a$ against $\log_{10}(\nu/\mathrm{Hz})$ for homeotropic, unaligned and planarly-aligned samples at 306.2 K.

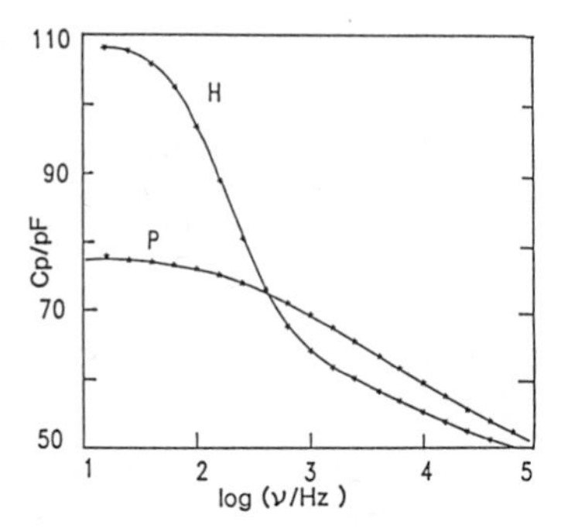

Figure 2. C_p (= $\varepsilon' C_a + C_{edge}$) against $\log(\nu/\mathrm{Hz})$ for homeotropic and planarly-aligned samples at 306.2 K.

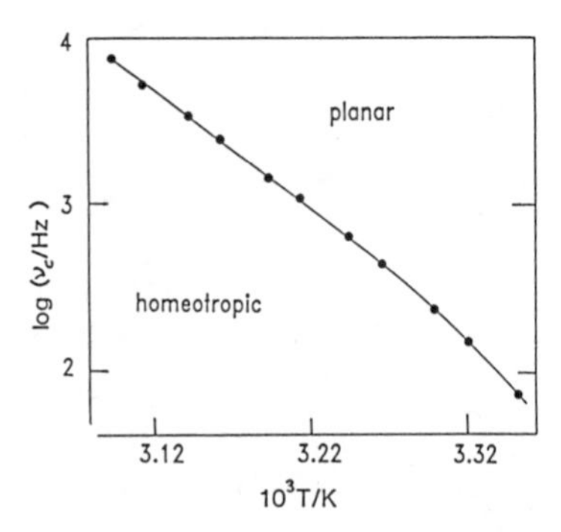

Figure 3. The crossover frequency ν_c as a function of temperature.

frequency electric field. The sample was equilibrated at 319.2 K ($T_{NI} - T = 6$ K) and its loss curve was measured and the value of maximum loss was noted. A saturating field of $\nu = 700$ Hz was applied at this temperature for 30 minutes and the loss curve was again recorded. No change in ε''_{max} was observed showing that $\nu < \nu_c$ at this temperature. A saturating field was applied at a higher frequency for 30 minutes and the loss curve was again recorded. Proceeding in this way the points shown in Figure 4 were obtained. $\varepsilon''_{max}/\varepsilon''_{max}$ (H) is seen to decrease as an S-shaped curve. For $\nu > 5$ kHz the material has transformed to P-alignment. Also shown in Figure 4 is the value of ν_c for T = 319.2 K as determined from the crossover condition (Figures 3 and 4). The value of ν_c coincides with the medium frequency of the S-shaped curve. These data indicate that the two frequency addressing principle applies to this polymer, although it is possible to realign the H-material to some extent by applying a strong electric field for ν slightly less than ν_c. Importantly, we have shown that this polymer may be aligned into the H or P states in the LC phase using electric fields of different frequencies. Having made the P-aligned sample (Figure 4 for $\nu > \nu_c$) we found that application of a strong electric field at 2 kHz transformed the sample back to the H state within 30 minutes at this sample temperature. The material could be switched readily between the states by two-frequency addressing in this way.

Our dielectric data are completed by showing (Figure 5) the values of the static permittivities $\varepsilon_{\parallel}$, $\varepsilon_{\perp}$, $\bar{\varepsilon}$ and ε(iso) for the H,U, P and isotropic samples respectively as a function of temperature. We note that ε and ε(iso) are similar in the region of T_c which is expected if there are no strong angular correlations between dipoles in the LC state. In that case $\varepsilon = \varepsilon$(iso) as may be derived from theory (11). According to the same theory (11)

$$\bar{\varepsilon} = (\varepsilon_{\parallel} + 2\varepsilon_{\perp})/3 \qquad (1)$$

Inspection of the data of Figure 5 and assuming that $\bar{\varepsilon}$ and $\varepsilon_{\perp}$ are correct, i.e. that the U and P samples are fully unaligned and fully planarly-aligned, shows that $\varepsilon_{\parallel}$ (calc) should be near 14.5 whereas the experimental value for $\varepsilon_{\parallel} \sim 12 - 12.5$. This indicates that the H sample is not fully aligned, despite the fact that the values we obtained for the H-sample were the maximum values obtainable as electric field amplitude was increased. In previous studies of a siloxane polymer (8) we found $\varepsilon''_{max}(H)/\varepsilon''_{max}(U) \simeq 2.1$ whereas this ratio (see Figure 1) is $\simeq 1.4$ for the present polymer, again suggesting that the H-sample is not fully aligned. Further evidence comes from the shape of the loss curve and its width at half height $\Delta_{1/2}$ for the H sample. From Figure 1 we obtain $\Delta_{1/2} = 2.0$ for the H-sample. Figure 6 shows comparable data for a siloxane homopolymer

$$Me_3 - Si\text{-}O\left[\begin{array}{c} Me \\ | \\ Si \\ | \\ R \end{array} - O\right]_n Si - Me_3 \qquad \text{II}$$

where $n \simeq 50$ and

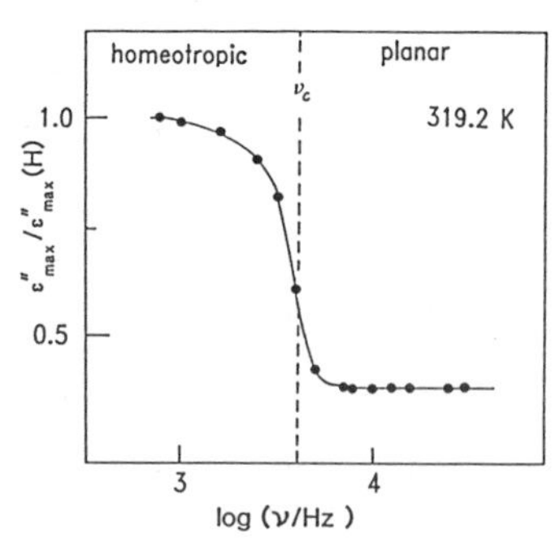

Figure 4. $\varepsilon''_{max}/\varepsilon''_{max}$ (homeotropic) at 319.2 K for the experiment described in the text.

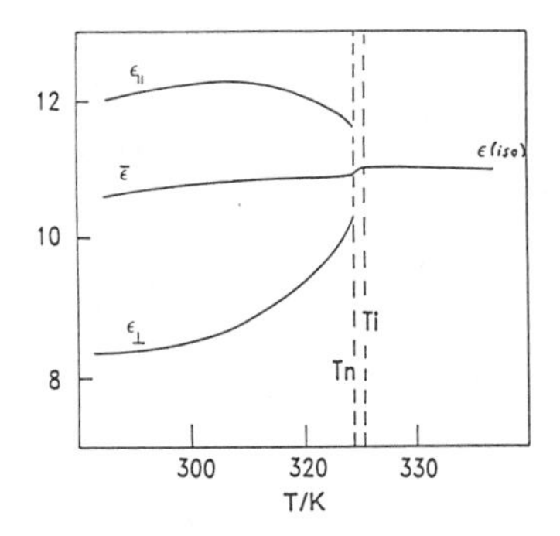

Figure 5. The static permittivities $\varepsilon_{\parallel}$, $\varepsilon_{\perp}$ $\bar{\varepsilon}$ and ε(iso) for the polymer I as a function of sample temperature.

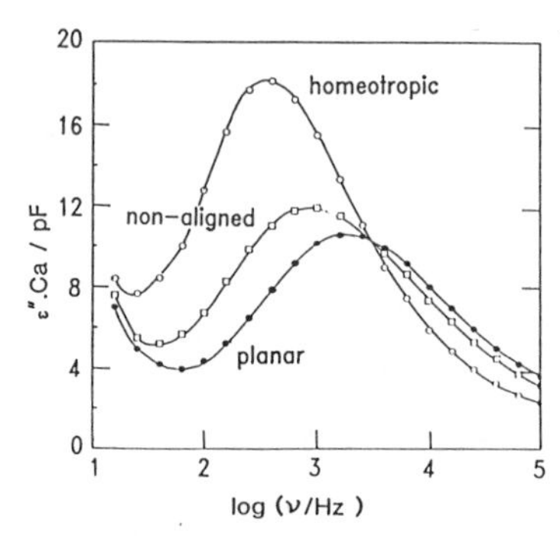

Figure 6. $\varepsilon''C_a$ against $\log_{10}(\nu/Hz)$ for a nematic siloxane LC polymer (II) at 315.2 K in its homeotropic, unaligned and planarly-aligned states.

$$R = (CH_2)_m - O - C_6H_4 - COO - C_6H_3(CH_3) - CN$$

These data are for the polymer in its H, U and P states at 315.2 K. The curve for the H-sample has $\Delta_{1/2} = 1.35$, which is far less than that for the polymer I, also the high-frequency shoulder present in Figure 1 is missing in Figure 6. All these indications suggest that the H-sample for polymer I is only partially aligned. It is interesting to compare Figures 1 and 6. The peaks for the H and U samples for polymer I (Figure 1) are similar, whereas they are very different from polymer II (Figure 6). It might be reasoned that the U sample for polymer I was actually partially-aligned homeotropically, but the continuity of the U and I data in Figure 5 shows that this is not likely to be the case.

At this point it is important to recall the theory of dielectric relaxation in a LC polymer having an alignment which is intermediate between H and P. In an earlier paper (11) we show that if the measuring electric field is uniform within the sample then the measured complex permittivity $\varepsilon(\omega)$ is given by

$$\varepsilon(\omega) = \left[\frac{1 + 2S_d}{3}\right].\varepsilon_{\parallel}(\omega) + \frac{2}{3}(1 - S_d).\varepsilon_{\perp}(\omega) \tag{2}$$

where S_d is the macroscopic director order parameter $(1 \leqslant S_d \leqslant -0.5)$ and

$$\varepsilon_{\parallel}(\omega) = \varepsilon_{\parallel}^{\infty} + \frac{G}{3kT}.\left[\mu_{\ell}^{2}(1 + 2S)F_{\parallel}^{\ell}(\omega) + \mu_{t}^{2}(1 - S)F_{\parallel}^{t}(\omega)\right] \tag{3a}$$

$$\varepsilon_{\perp}(\omega) = \varepsilon_{\perp}^{\infty} + \frac{G}{3kT}.\left[\mu_{\ell}^{2}(1 - 2)F_{\perp}^{\ell}(\omega) + \mu_{t}^{2}(1 + S/2)F_{\perp}^{t}(\omega)\right] \tag{3b}$$

where $\varepsilon(\omega) = \varepsilon'(\omega) - i.\varepsilon''(\omega)$ and $\varepsilon_{\parallel}^{\infty}$ and $\varepsilon_{\perp}^{\infty}$ are limiting high frequency permittivities, S is the local order parameter and the different $F_j(\omega)$, where $i = \ell, t$, $j = \parallel, \perp$ are given by

$$F_j^i(\omega) = 1 - i\omega F\left[\phi_j^i(t)\right] \tag{4}$$

where F indicates a one sided Fourier transform and the $\phi_j^i(t)$ are linear combinations of time correlation functions for the angular motions of the longitudinal and transverse components, μ_{ℓ} and μ_t of the dipole moment of the mesogenic group.

According to these relations, four relaxation modes occur which are appropriately weighted for samples having different degrees of macroscopic alignment. For a fully homeotropic sample $S_d = 1$ and two modes labelled as $F_{\parallel}^{\ell}(\omega)$ and $F_{\parallel}^{t}(\omega)$ contribute to the overall dielectric relaxation. For a fully planar sample $S_d = -0.5$ and two modes labelled as $F^{\ell}_{\perp}(\omega)$ and $F^{t}_{\perp}(\omega)$ contribute to the overall relaxation. For an unaligned sample $S_d = 0$ and all four modes contribute, weighted in accord with Equation 2. It is reasoned that the low frequency peak observed for the H sample in Figures 1 and 6 is the $F^{\ell}_{\parallel}(\omega)$ mode (or the δ relaxation) and is due to the motions of μ_{ℓ} with respect to the local director n giving a dielectric time correlation function $\phi^{\ell}_{\parallel}(t) = \langle \cos \beta (t) \rangle$ where $\beta = \underset{\sim}{\mu}_{\ell} \cdot \underset{\sim}{n}$. The loss peaks for the P-specimens of polymer I and polymer II show no evidence of structure which means that the relaxation functions $F^{\ell}_{\perp}(\omega)$ and $F^{t}_{\perp}(\omega)$ have a similar frequency dependence and are thus not resolved in such frequency plots. The data for the siloxane polymer II can be analyzed quantitatively (11-13) and are found to be entirely consistent with the theoretical description, Equations 2-4. However the data for polymer I studied here are, for the reasons given above, apparently not consistent in the sense that the values for loss factor (Figure 1) and permittivity (Figure 5) for the H-specimen are smaller than are predicted by theory. Several explanations may be sought. First the theoretical relations 1-3 above are derived for the model case where the dipole moment resides in a mesogenic group having a fixed dipole moment $\underset{\sim}{\mu} = \underset{\sim}{z}\mu_{\ell} + \underset{\sim}{x}\mu_{t}$, where $\underset{\sim}{z}$ and $\underset{\sim}{x}$ are unit vectors in the $\underset{\sim}{z}$ ($\equiv \underset{\sim}{n}$) and $\underset{\sim}{x}$ directions in the molecular frame of reference. Polymer I has its main dipole moment in the aromatic head group pendant to the chain but, in addition, there are ester groups along the main chain which are not obviously included in our theoretical description. Motions of the main chain <u>independent</u> of the mesogenic groups would be expected to lead to further dipole relaxation processes which may superpose on the spectra arising from the four relaxation modes characteristic of the LC phase. However it is unclear how <u>all</u> these relaxation processes can be extracted from the data of Figures 1 and 2. Such information could only be obtained by incorporating complementary evidence from related studies of dynamics, e.g. by NMR, ESR or mechanical relaxation, but those methods also suffer in other respects by a lack of frequency coverage or inadequate molecular theory relating macroscopic observations to molecular behaviour. Our main difficulty in understanding our data is almost certainly associated with the response of the chiral polymer I to applied electric fields. Our dielectric measurements were made following the removal of a directing electric field, since the Digibridge could not accommodate making dielectric measurements in the presence of an additional a.c. field. Since the polymer is chiral, we suspected that on removal of the directing field a rapid change of the structure might occur which would not be detectable in our subsequent measurements. As a test of this possibility we measured the optical transmittance of a sample (50 μm) following the application and withdrawal of 50 V rms applied at 400 Hz at different sample temperatures. Any departure from homeotropic alignment (black in crossed polarizers) could be detected as an increased transmittance in time. Figure 7 shows representative data at 313.2 K. When the steady a.c. field is withdrawn there is an immediate change in transmittance followed by a transient which equilibrates after ~ 15 minutes. This texture is only slightly less scattering than the non-aligned material. If the directing field is reapplied a fully-homeotropic film is produced. A surprising feature of these data is

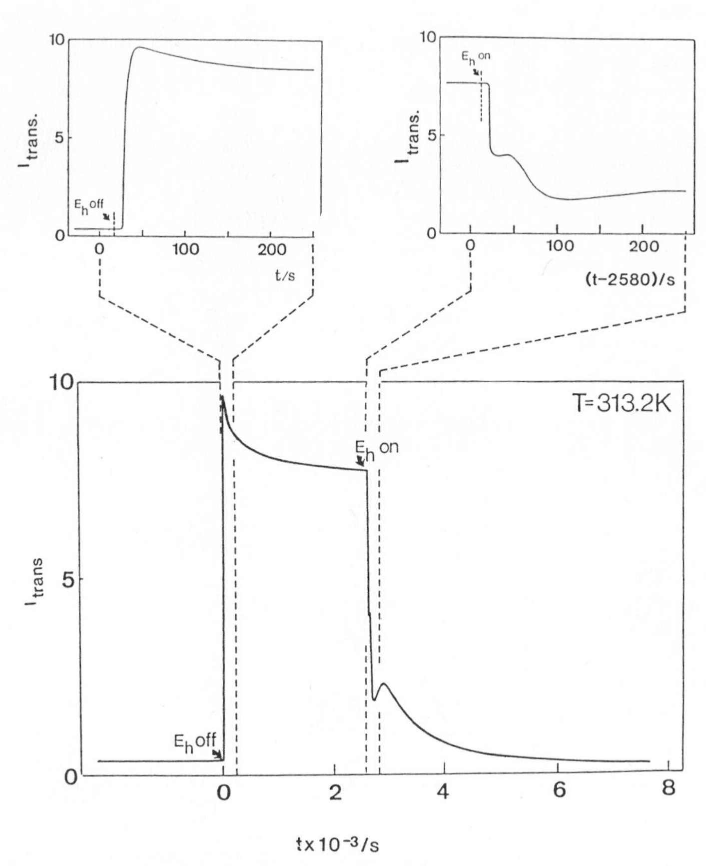

Figure 7. The optical transmittance I_{trans} as a function of time following the withdrawal and subsequent application of a directing electric field as described in the text.

the speed of the initial response (which is of the order of a few seconds). It is apparent that removal of the a.c. field from a homeotropic sample leads to a disalignment of the sample, as we had suspected from the dielectric data discussed above. It seems likely that the disalignment has the following source. In the presence of the steady a.c. field the material is fully-homeotropic and is nematic in nature. On removal of the field, the chiral nature of the polymer leads to the formation of chiral-nematic structures, which has the effect of randomizing, partially, the local directors n which were previously aligned with the directing field, and hence reduces the loss-curve below that expected for a fully-homeotropic specimen, as we have discussed.

Further work is in progress with this polymer to determine its dielectric properties and alignment behaviour in the presence of d.c. directing fields and to compare such behaviour with recent theoretical descriptions of time-dependent alignment induced by steady directing fields (18,19).

Acknowledgments

The Aberystwyth group acknowledge support from the SERC (Electroactive Polymers Specially-Promoted Programme) and from the Treaty of Windsor Anglo-Portuguese Exchange Scheme of the British Council. The USM research was sponsored by the Air Force Office of Scientific Research, Air Force Systems Command, USAF, under Grant Number AFOSR 84-0249.

Literature Cited

1. Nonlinear Optical Properties of Organic Polymeric Materials ; Williams, D.J., Ed., ACS Symposium Series 233; American Chemical Society: Washington, DC, 1983.
2. Nonlinear Optical Properties of Organic Molecules and Crystals ; Chemla, D.S; Zyss, J., Eds.; Academic , New York, 1987.
3. Haase, W., Pranoto, H., Bormuth, F.J., Ber. Buns. Phys. Chem., 1985, 89, 1229.
4. Haase, W., Pranoto, H., Prog. Colloid Polym. Sci., 1984, 69, 139.
5. Pranoto, H., Bormuth, F.J., Haase, W., Kiechle, U., Finkelmann, H. Makromol. Chemie, 1986, 187, 2453.
6. Bormuth, F.J., Haase, W., Zentel, R. Mol. Cryst. Liq. Cryst. 1987, 148, 1.
7. Attard, G.S. and Williams. G. Polymer,1986, 27, 2, 66.
8. Attard, G.S., Williams, G. Liq. Cryst., 1986, 1, 253.
9. Attard, G.S., Williams, G. J. Molec. Electron.1986, 2, 107.
10. Attard, G.S., Araki, K., Williams, G. J. Molec. Electron. 1987, 3, 1.
11. Attard, G.S., Araki, K., Williams, G. Brit. Polym. J.1987,19,119.
12. Attard, G.S., Araki, K., Mol. Cryst. Liq. Cryst. 1986, 141, 69.
13. Araki, K., Attard, G.S., Liq. Cryst.1986, 1, 301.
14. Finkelmann, H., Rehage, G. Adv. Polym. Sci. 1984, 60/61, 99.
15. Shibaev, V.P., Plate, N.A. Adv. Polym. Sci. 1984, 60/61, 173.
16. Clark, M.G., Harrison, K.H., Raynes, E.P. Phys. Technol. 1980, 11, 232.
17. Clark, M.G. Displays, January 1981, 169.
18. Moore, J.S., Stupp, S.I. Macromolecules 1987, 20, 282.
19. Martins, A.F., Esnault, P., Volino, F. Phys.Rev.Lett. 1986,57,1745.

RECEIVED September 12, 1988

POLYMER–POLYMER ASSOCIATIONS

Chapter 16

Phase-Separation Dynamics of Aqueous Hydroxypropyl Cellulose Solutions

Thein Kyu, Ping Zhuang[1], and Partha Mukherjee[2]

Center for Polymer Engineering, University of Akron, Akron, OH 44325

Time-resolved light scattering studies have been undertaken to elucidate the mechanism of thermally induced phase separation in aqueous hydroxypropyl cellulose (HPC) solutions of different molecular weight. The phase diagram was established on the basis of cloud point determination. The phase separation behavior resembles an LCST (lower critical solution temperature), but there appears a discontinuity in the cloud point curve at intermediate compositions. The phase diagram consists of three distinct regions: (1) the isotropic regime at low concentrations, (2) the anisotropic color regime at intermediate concentrations, and (3) the gel regime at high concentrations. The presence of the superstructure in the latter two regimes has complicated the phase separation dynamics; therefore, at this time we will report only the time-resolved light scattering studies on the dynamics of phase separation and phase dissolution in a 10 wt% HPC solution. The time-evolution of scattering curves were analyzed in the context of the linearized theory proposed by Cahn-Hilliard. The linearized theory describes many of the qualitative features of phase separation phenomena, but shows some deficiency in the quantitative comparison. The late stage of spinodal decomposition is dominated by non-linear behavior and the results are interpreted in comparison with recent scaling theories.

The phenomenon of phase separation in liquid crystalline polymer solutions has been the subject of recent interest. This phenomenon was noticed in a number of lyotropic liquid crystal

[1]Current address: Department of Chemistry, University of Kentucky, Lexington, KY 40506
[2]Current address: Institute of Polymer Science, University of Akron, Akron, OH 44325

0097-6156/89/0384-0266$06.00/0

systems, notably in poly-γ-benzyl L-glutamate (PBLG)/dimethyl formamide (DMF) (1), poly-p-phenylene benzobisthiazole (PBT)/methane sulfonic acid (MSA) (2) or sulfuric acid (3), and hydroxypropyl cellulose (HPC)/water systems (4-6). However, the phenomenon of phase separation has been complicated by additional factors such as phase transitions, gelation, and the presence of superstructure.

Miller and co-workers (1) found that poly-γ-benzyl L-glutamate in dimethyl formamide reveals phase behavior (temperature-composition) covering isotropic and anisotropic liquid crystalline phase. When lyotropic solutions are brought into the biphase region by temperature quenching, gelation takes place. This phenomenon is associated with phase separation between polymer- rich and -poor regions and is similar in character with an upper critical solution temperature (UCST). Another rigid-rod polymer system that exhibits gelation associated with phase separation is the poly-p-phenylene benzobisthiazole in methane sulfonic acid (2). Similar observation was also made in the case of PBT-sulfuric acid solution (3), in which the biphase structure changes to a single phase during heating. It returns to a two-phase regime upon cooling, thus is reversible in character.

Aqueous hydroxypropyl cellulose is another type of rod-like material reported to undergo phase separation with heating (4). The phase behavior is similar to that of a lower critical solution temperature (LCST), hence it is different from the above systems. The HPC/water system is an interesting model system because of the rich variety of phase structure of the material. HPC is a semicrystalline polymer in the solid state (7), but exhibits thermotropic liquid crystalline character at elevated temperatures below the melting point (8). It shows isotropic phase in dilute solutions, but forms an ordered liquid crystalline phase with cholesteric structure in concentrated solutions (4).

The phenomenon of thermally induced phase separation in aqueous HPC solutions has been recognized for some time (4), the mechanism by which it occurs was not known until recently (6). According to our previous study (6), the process of phase separation at the isotropic regime (10 wt% aqueous HPC solution) is the spinodal decomposition (SD). In this review article, we continue our efforts to elucidate the dynamics of phase separation as well as the reverse process of phase dissolution using time-resolved light scattering. The technique is fast, non-destructive and provides a good statistical average of phase domains, thus permitting one to follow the rapid spinodal decomposition process. Moreover, the time evolution of scattering function can be analyzed in terms of linear (9, 10) and non-linear scaling theories (11-17) of critical phenomena and related matter in statistical physics.

Materials and Methods

Two different molecular weight HPC specimens, $M_w \sim 60{,}000$ (HPC-E, Hercules Inc.) and $M_w \sim 100{,}000$ (HPC-L, Aldrich Chemical Co.) were used in this study. Various aqueous HPC solutions, typically from 1% to 80% concentrations, were prepared by dissolving them in distilled water. Concentrated solutions were obtained by evaporating dilute solutions and the percent concentration was calculated on the basis of weight change. The solutions were sealed in a demountable cell (Wilmad Glass Co., Model WG-20/C) of 1 to 0.1 mm path length.

A time-resolved light scattering set-up, schematically shown in Figure 1a, was utilized (18). It consists of a laser light source (He-Ne laser with a wavelength of 6328 A°), a set of polarizer and analyzer, a sample hot stage and a screen. Scattered patterns from polymers may be photographed using a Polaroid instant camera (Polaroid Land Film Holder 545). The scattered intensity can be quantitatively monitored by means of a two-dimensional Vidicon camera (1254 B, EG&G Co.) coupled with a detector controller (Model 1216, EG&G Co.). The analogue signal is digitized and analyzed on an OMA III (Optical Multichannel Analyzer) system. The scan rate is typically 30 ms for one-dimensional scan and about 0.5 to 1.5 s for two-dimensional mode depending on the number of pixels chosen for grouping. Various modes of data acquisition are available for selection to commensurate with experimental configurations. The raw data are further transferred to an off-line computer (IBM-XT) for post data treatments such as background correction, data smoothing, rescaling, etc. Generally, a set of sample hot stages is used for temperature jump studies; one is controlled at an experimental temperature and the other is preheated below phase separation points. The schematic drawing of the hot stage is depicted in Figure 1b. Cloud point phase diagrams were established at various heating rates and the equilibrium cloud points were estimated by extrapolating the data to the zero heating rate. Various T-jump experiments were carried out from room temperature to 44 - 48°C at one degree intervals and the reverse T-quench experiments were undertaken from 45 to 43 ~ 40°C.

Phase Equilibria and Phase Morphology

The dilute aqueous HPC solutions were transparent at ambient temperature, but turned milky upon heating to 45°C, indicating the occurrence of thermally induced phase separation. In the light scattering study, a scattering halo developed in the V_v (vertical polarizer with vertical analyzer) configuration, however, no pattern was seen in the H_v (horizontal polarizer with vertical analyzer) polarization. This kind of behavior can be expected if the scattering arises predominantly from concentration or density fluctuations rather than orientation fluctuations. Figure 2 shows two and three dimensional perspective plots of a scattering halo for the 10 wt% aqueous HPC-L solution without using any polarizers. The interconnected biphase structure has been seen under optical microscope (6), but the picture is not shown here. The revelation of scattering halo suggests that the process of phase separation occurs via spinodal decomposition. The alternative

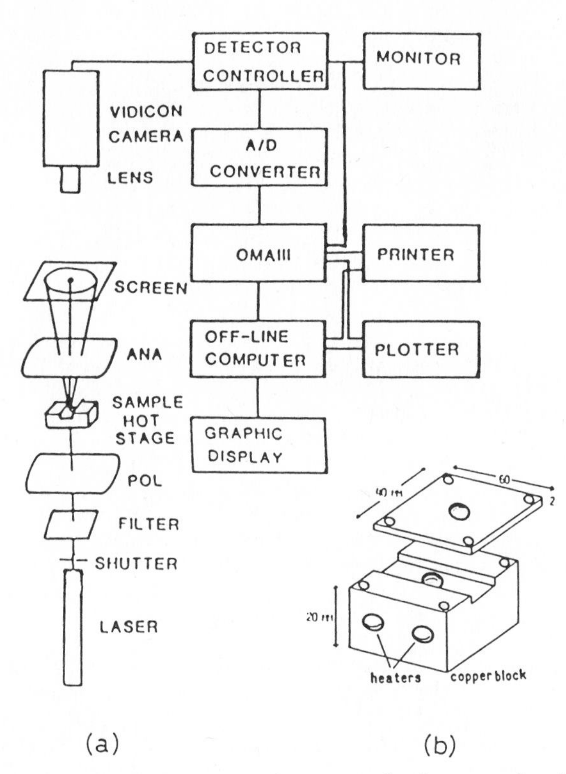

Figure 1. (a) Schematic diagram of time-resolved light scattering and (b) sample hot-stage.

mechanism of nucleation and growth, if it occurs, would have led to the scattering maximum at zero angle which is not seen here.

For the determination of cloud point, the intensity at the scattering wavelength q=3.1 μm^{-1}, at which the scattering maximum first developed, was plotted against temperature in Figure 3. The temperature at which the change of intensity occurs has been regarded as a cloud point. This temperature naturally depends on the rate of heating as can be seen in Figure 4. The equilibrium cloud point was estimated by extrapolating the points to zero heating rate. The cloud point phase diagram obtained for various HPC/water systems is illustrated in Figure 5. The phase behavior is reminiscent of an LCST (lower critical solution temperature), but there is no clear maximum or critical point. Instead, a discontinuity is observed at moderate concentrations of 40 to 60 wt% HPC; i.e., the region where the iridescent color is seen.

The phase diagram is extremely complex, consisting of three distinct regions: (1) the isotropic region at low concentrations, (2) the anisotropic color region at intermediate concentrations, and (3) the gelation region at high concentrations. The complex structure of HPC solutions may be best explained by the evolution of the H_V scattering patterns during the course of solvent evaporation as shown in Figure 6. At low concentrations, there appears no scattering indicative of the single phase structure. The concentration rapidly increases around 40 wt% where the overlapping structure of a four-lobe clover at odd multiples of 45° and a small + type pattern in 0 and 90° directions appears. The four-lobe pattern may arise from the collection of rod-like liquid crystalline entities probably having fairly uniform dimension, whereas the inner + type pattern may be a consequence of inter-rod interference. The four-lobe pattern further moves to low scattering angles with continued drying and eventually merges with the + type pattern to form an incomplete spherulite or sheaf structure. At this stage, the concentration levels off around 85 wt% and the sheaf structure persists until the film is completely dried.

The interpretation of the structural evolution of HPC solutions is by no means straightforward because of the strong scattering power of liquid crystalline entities. The region 40 wt%, at which the dual scattering pattern appeared, corresponds to the color region with the cholesteric structure. The wide-angle scattering may be attributed to the aggregates of rod-like scattering entities, whereas the inner scattering (+ type) pattern may be due to the inter-rod interference. Hashimoto and coworkers (19) have interpreted the evolution of the scattering patterns during solvent evaporation in terms of the changes in the size of the rods, the anisotropy of the rods, the number of rods in the assembly (growth of the assembly leading to an increase of overall crystallinity), and inter-rod interference.

With continued solvent evaporation, the four-lobe (x type) and the + type patterns merge to form a sheaf-like structure. In

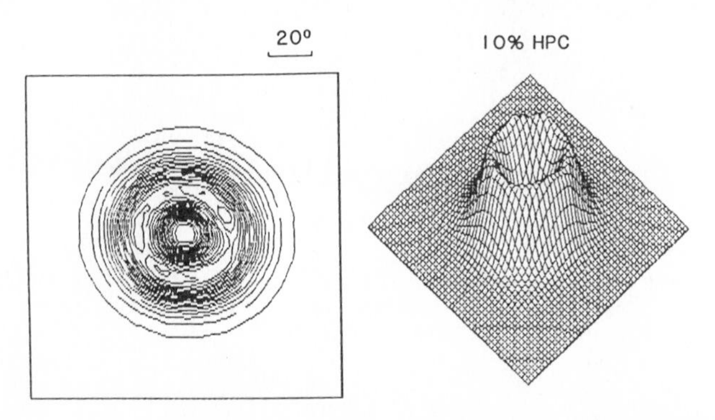

Figure 2. Two- and three-dimensional perspective plots of a scattering halo for the 10% HPC-L solution at 45°C.

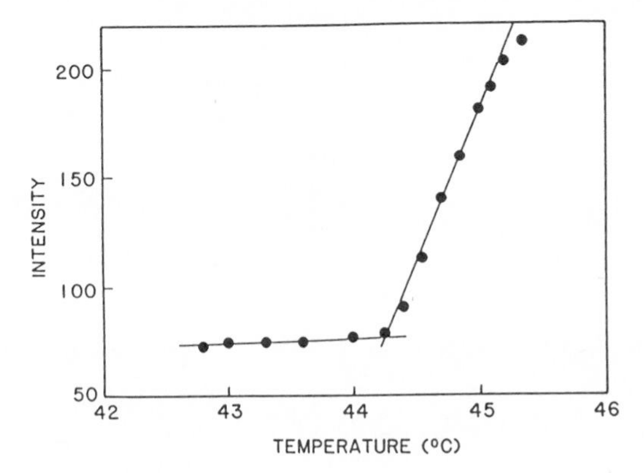

Figure 3. The change of scattered intensity as a function of temperature obtained at 1°C/min.

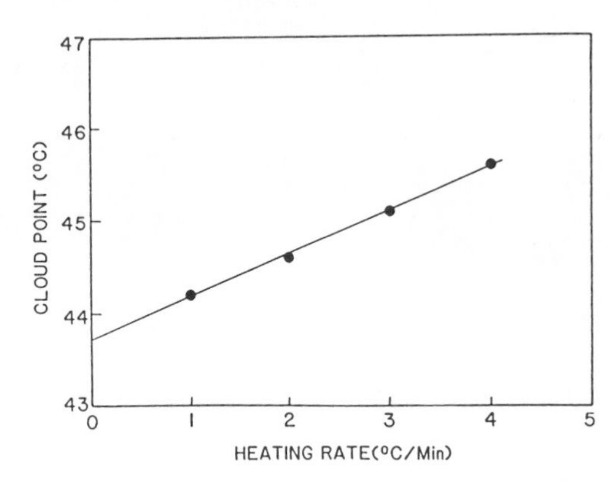

Figure 4. The heating rate dependence of cloud point temperatures.

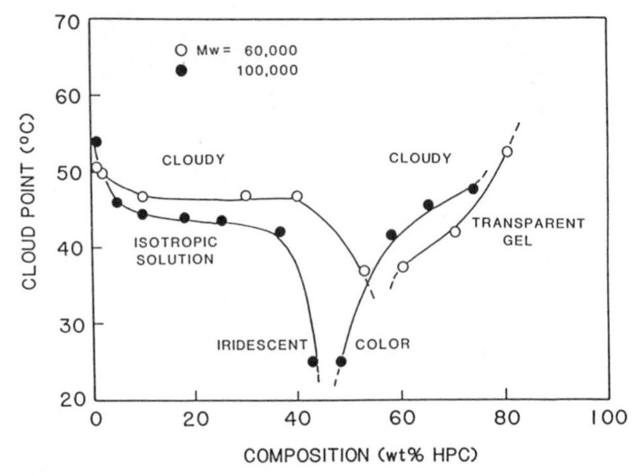

Figure 5. Temperature versus concentration phase diagram for aqueous HPC-E and HPC-L solutions.

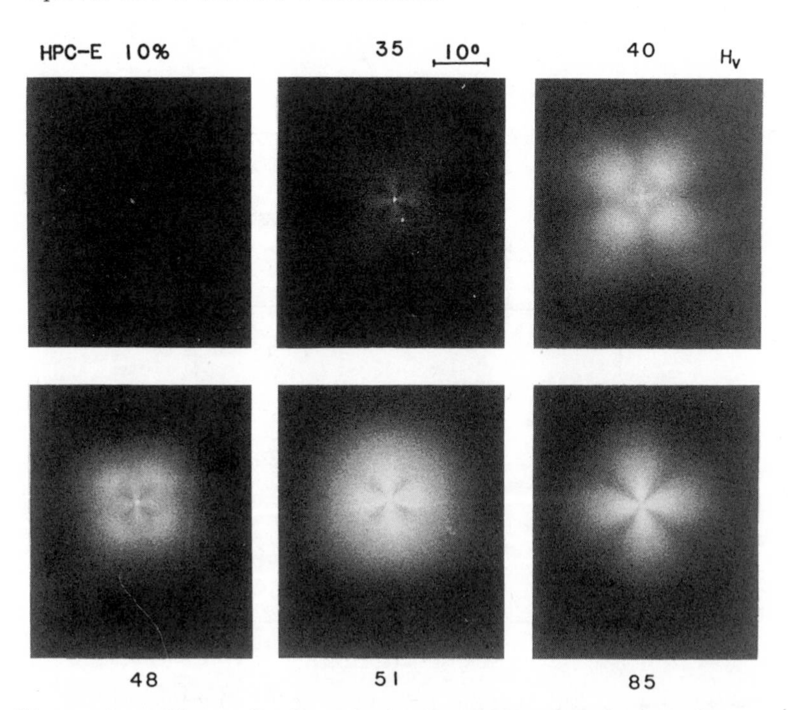

Figure 6. The evolution of the H_v scattering during the solvent evaporation of HPC-L/water. The concentrations were calculated from the weight change of HPC solutions.

this region, the cholesteric structure disappears and the gel structure is formed due to an increase in the overall crystallinity. The optical microscopic investigation revealed the incomplete spherulites. Under appropriate conditions, well-developed spherulites can be observed. When the undried HPC-E film is heated, the typical four-lobe H_v scattering changes to intense isotropic scattering pattern around 128°C and intensifies further for some time, then the intensity reduces with continued heating and completely vanishes around 195°C (Figure 7). It is well established that the HPC exhibits thermotropic liquid crystalline character in the temperature range of 140 to 195°C (8). Therefore, the disappearance of the H_v and V_v scattering around 195°C may be attributable to the nematic-isotropic transition. The transition near 128°C may be associated with the solid-nematic transition. This transition is probably affected by the presence of water, because it is very difficult to completely remove water from HPC specimens even at such high temperatures. The change of four-lobe to an intense isotropic scattering cannot be accounted for exclusively in terms of the spherulite melting, since there is a good chance that the phase separation may be involved (Figure 5). Since the phase behavior is extremely complex, only the time-resolved light scattering results at the 10 wt% concentration will be presented here.

Early Stages of Spinodal Decomposition

Figures 8a and b exhibit the time-evolution of scattering curves obtained during a T (temperature)-jump from room temperature to 44°C. The scattering maximum first appears around q=3.1 μm^{-1} and remains stationary for some period, then shifts rapidly to low scattering wave numbers, indicating the nonlinear nature of phase separation in the 10 wt% HPC solution. The early period of phase separation, at which the peak is virtually invariant, is less obvious at higher T-jumps. Hence, it does not represent the general characteristics of phase separation. In the reverse quench case from 45 to 43°C in Figure 9, the scattering maximum decays with elapsed time without any movement of its position. The invariance of the scattering peak is one of the behavior predicted by the linearized theory of Cahn-Hilliard (9), who proposed the exponential growth of scattered intensity:

$$I(q,t) = I(q,t{=}0).\exp[2R(q).t] \quad (1)$$

which is further modified by Cook (10) by taking into account the effect of thermal fluctuations as follows:

$$I(q,t) = I_s(q) + [I(q,t{=}0) - I_s(q)].\exp[2R(q).t] \quad (2)$$

where $I_s(q)$ is the scattering from the stable solution, t the phase separation (or dissolution) time and q the wave number being equal to $(4\pi/\lambda)\sin(\theta/2)$. Here, λ and θ are, respectively, the wavelength of light and the scattering angle measured in the

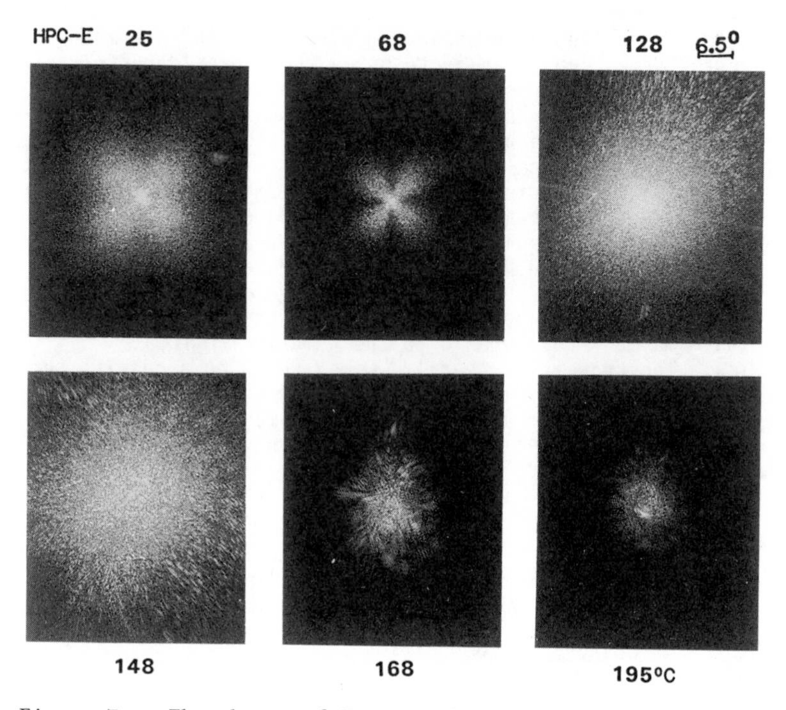

Figure 7. The change of H_v scattering patterns during heating of an undried HPC film. The heating rate is 2°C/min.

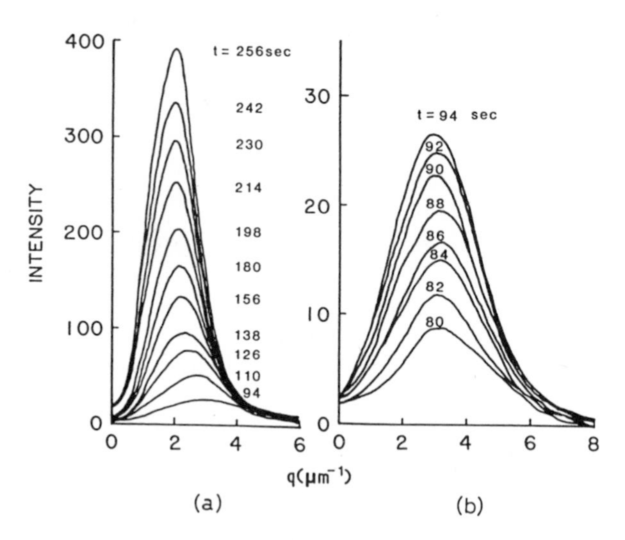

Figure 8. Time-evolution of scattering curves during phase segregation of the 10% HPC-L solution following a T-jump from 23 to 44°C; (a) early period (80 - 94 s) and (b) late stages (94 - 256 s).

medium. The amplification factor R(q) which characterizes the growth (or decay in the reverse quench) process is further related by

$$R(q) = -Mq^2[(\partial^2 f/\partial c^2) + 2\kappa q^2] \quad (3)$$

or

$$R(q)/q^2 = -(D_a + 2M\kappa q^2)$$
$$= -D_a(1 - q^2/q_c^2) \quad (4)$$

where f is the local free energy, c the concentration, κ the composition gradient coefficient, M the mobility and q_c the crossover wave number.

Figures 10a and b show the plots of logarithmic intensity (after Cook's correction) versus elapsed time for the phase separation and phase dissolution, respectively. The plots in the former are hardly linear, and deviations are severe, particularly at late stages of phase separation. This observation is not surprising in view of the fact that the peak invariance is only seen at a very short interval of 80 to 94 s in Figure 8a. Thus, the R(q) can not be determined with good accuracy as it is affected by the non-linear contribution. The data points in the phase dissolution show linear relationships as predicted by the linearized theory. However, it is important to test the linearized theory quantitatively in terms of Equation 4.

As can be seen in Figures 11a and b, the plots of $R(q)/q^2$ versus q^2 in both cases, except for very shallow T-jumps (or quench), are not necessarily linear, pointing to the deficiency of the linearized theory. It is possible that the present system does not follow the mean field prediction of Cahn-Hilliard-Cook simply because the polymer/solvent system is basically not a mean field system. The deviations are severe at small q-region and large T-jumps or large quench depths. However, the data at low T-jumps (or low quench) appear to be fairly linear at large q-region, therefore may be approximated by linear slopes with the help of $q_m^2 = q_c^2/2$. The apparent diffusivities as estimated from the intercepts were plotted against temperature in Figure 12. A negative diffusivity signifies the up-hill and the positive the down-hill diffusion. The temperature at which the diffusivity becomes zero is regarded as the spinodal point (T_s). Near the T_s, the diffusivity D may be scaled as:

$$D = D_0 \cdot \epsilon^{\nu} \quad (5)$$

with

$$\epsilon = (T - T_s)/T_s \quad (6)$$

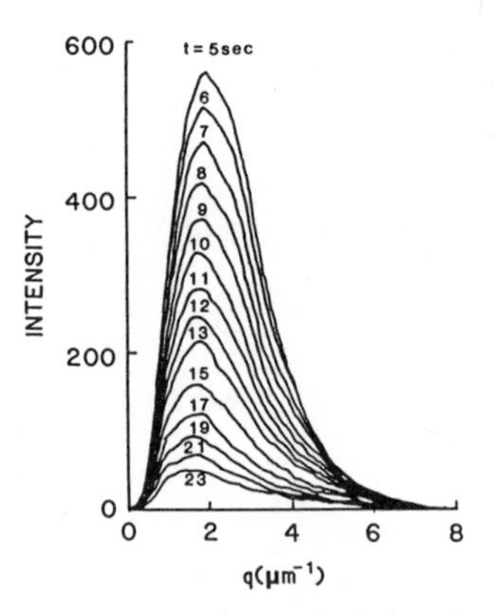

Figure 9. Time-evolution of scattering curves during phase dissolution of the 10% HPC-L solution following a T-quench from 45 to 43°C.

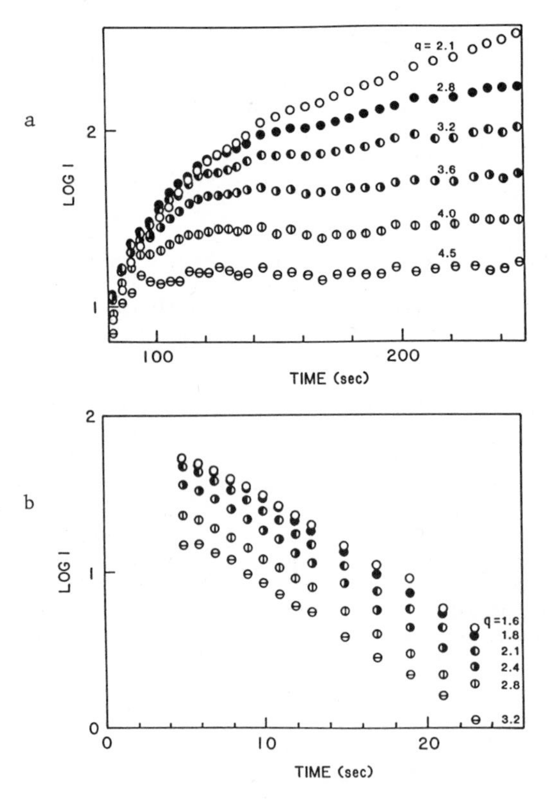

Figure 10. Log I versus t for the 10% aqueous HPC-L solution during (a) phase separation following a T-jump from 23 to 44°C and (b) phase dissolution following a T-quench from 45 to 43°C.

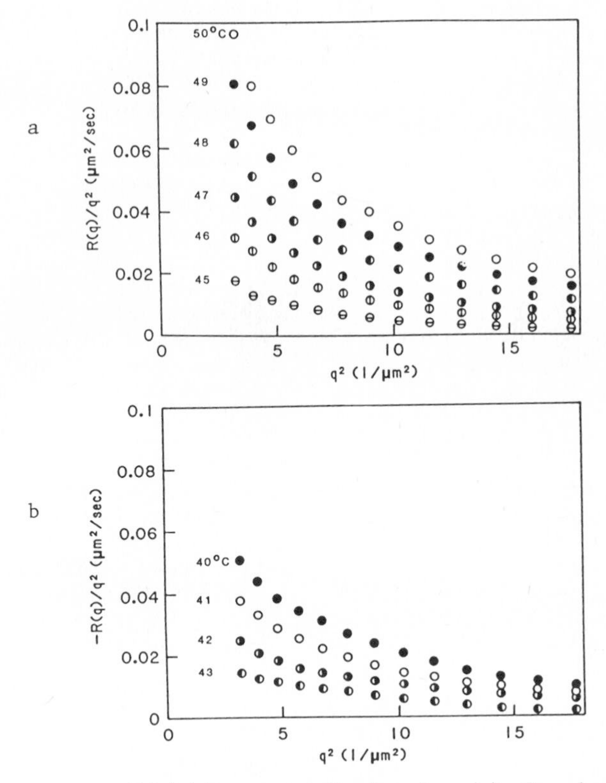

Figure 11. $R(q)/q^2$ versus q^2 plot for (a) the phase separation and (b) the phase dissolution at various temperatures.

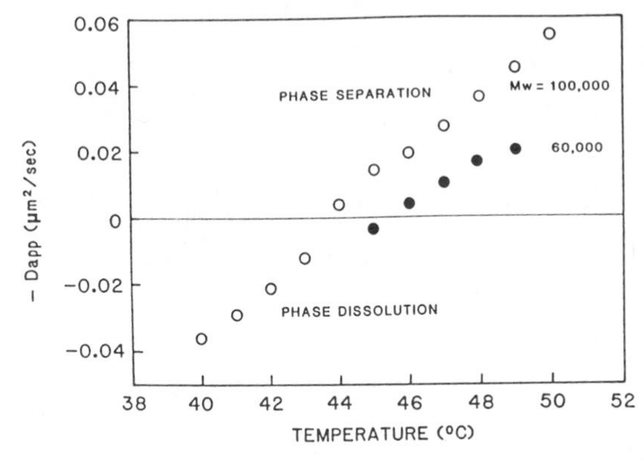

Figure 12. Apparent diffusivity as a function of temperature for phase separation and dissolution of the 10 wt% aqueous HPC-L and HPC-E solutions.

where the critical exponent ν is predicted to be unity by the linearized theory. D_0 is the self-diffusivity and generally considered as constant near the spinodal temperature. However, the temperature dependence of D_0 due to the change in mobility has to be taken into account if the temperature range of the experiment is wide.

The absolute values of D_a for the phase separation and phase dissolution are plotted against ϵ in log-log scale together in Figure 13. In the present case, the value of ν turns out to be unity for the phase separation as well as for the phase dissolution, thereby confirming the prediction of linearized theory. A similar observation was also made by Snyder and coworkers (20) for the polystyrene (PS)/polyvinyl methyl ether (PVME) blends. It is concluded that the diffusion rate is the same for the 60,000 and 100,000 molecular weight HPC specimens, if the diffusivities are referenced to the spinodal temperature.

Late Stages of Spinodal Decomposition

Now, we turn our attention to late stages of spinodal decomposition. Since the phase separation in binary mixture is intrinsically a nonlinear phenomenon, a number of nonlinear theories have been put forward on the basis of statistical consideration, notably the LBM (Langer, Baron & Miller) (11) and BS (Binder & Stauffer) (12) theories. Both theories predicted the power law scheme rather than the exponential growth of the structure

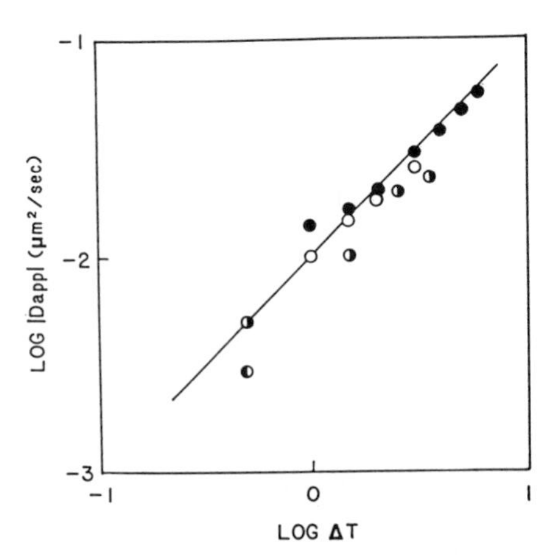

Figure 13. The log-log plot of diffusivity as a function of ΔT ($=T-T_s$) for phase separation and dissolution of the 10 wt% aqueous HPC-L and HPC-E solutions in comparison with the slope of 1. (●) phase separation for $M_w \sim 100K$ specimen, (○) phase dissolution for $M_w \sim 100K$, (◑) phase separation for $M_w \sim 60K$ and (◐) phase dissolution for $M_w \sim 60K$.

function. According to LBM theory, the maximum wave number $q_m(t)$ has a simple form of the power law,

$$q_m(t) \sim r(t)^{-1} \sim t^{-\varphi} \tag{7}$$

with the power $\varphi = 0.21$.

On the basis of the cluster dynamics, BS theory predicted the time evolution of the cluster size r(t) has the same form as Equation 7 and the maximum scattered intensity $I_m(t)$ assumes a similar power law relation, i.e.,

$$I_m(t) \sim t^{\psi} \tag{8}$$

where $\psi = 3\varphi$, with $\varphi = 1/3$ and $\psi = 1$. Later, based on the percolation approach, Siggia (13) reached the same formalism as Equations 7 and 8 with $\psi = 3\varphi$, but the values of φ and ψ are predicted to be 1 and 3, respectively. The plots of log $q_m(t)$ and log $I_m(t)$ versus log t in Figures 14 and 15 show slopes of approximately 1 and 3, which are in comformity with Siggia's prediction for the coarsening process driven by surface tension. This mechanism has been already identified in the off-critical and critical polymer mixtures with large T-jumps (21-22). However, at the very late stages, the growth process appears to slow down as the slope of log q_m versus log t becomes smaller. In this region, phase separated domains coalesce to form larger domains while the interconnectivity breaks down concurrently. It seems the competition between the two opposing mechanisms eventually determines the growth process.

Tests with Scaling Theories

Now we shall examine the time evolution of scattering function in terms of the self-similarity between the phase separated structures (12, 14, 15) The scattered intensity at a given time t for the conserved order parameter system may be given by

$$I(q,t) \sim V\langle\eta^2\rangle\xi(t)^3 s(x) \tag{9}$$

where V is the irriadiated volume, $\langle\eta^2\rangle$ the mean-square fluctuations of refractive indices, s(x) the structure factor and $x = q\xi(t)$. Here, $\xi(t)$ is the correlation length which relates to the wavelength of periodic structure $\Lambda(t)$ by the following equation:

$$\xi(t) = \Lambda(t)/2\pi = 1/q_m(t) \tag{10}$$

From Equations 9 and 10, the structure factor s(x) can be described as

$$s(x) \sim I(q,t)q_m^3(t) \tag{11}$$

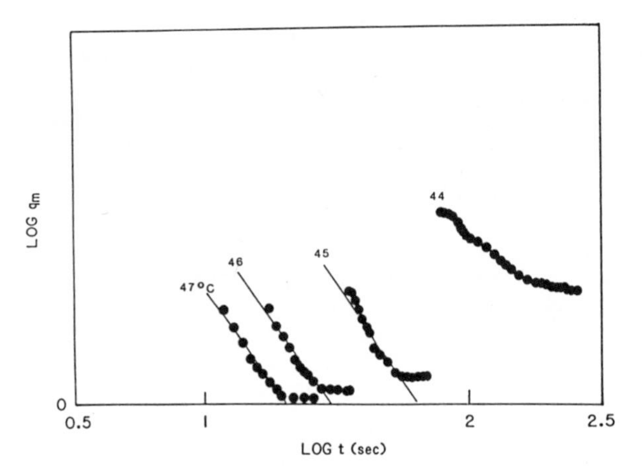

Figure 14. The log-log plot of maximum wave number (q_m) versus time (t) at various isothermal phase separation temperatures.

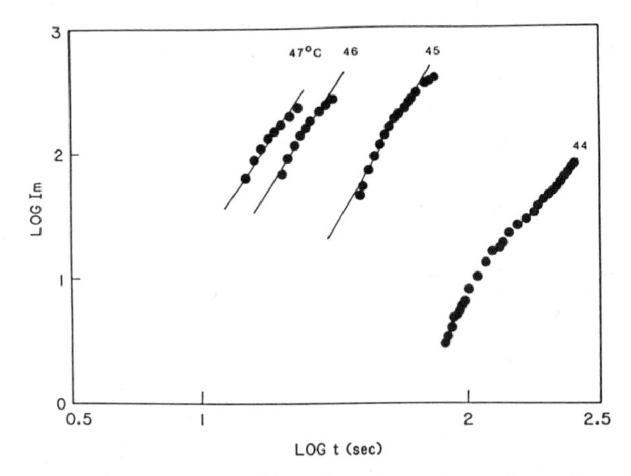

Figure 15. The corresponding log-log plot of maximum intensity (I_m) versus t at various isothermal phase separation temperatures.

In the early stage of spinodal decomposition, $<\eta^2(t)>$ varies with phase separation time, therefore it cannot be scaled by a single length parameter $\xi(t)$. It is necessary to normalize the structure function by the invariant function, i.e.,

$$G(x) = I(q,t)q_m^3(t)/[\int I(q,t)q^2(t)dq] \quad (12)$$

The integration has to be performed in the limit of zero to q_p, i.e., the wave number where the Porod law is satisfied. It should be pointed out that this kind of normalization is not needed if the scaling were to apply only to the late stages of SD. Figure 16 shows the plot of the normalized scaled structure factor G(x) as a function of x for both early and late stages of SD. The superposition of the scattering data is fairly good, suggesting that the self-similarity is preserved in the demixing of aqueous HPC solutions.

The next scaling theory, which is of interest to test with experiments, is that of Furukawa (16) who proposed that the shape of the structure function can be scaled by the following equation:

$$S(\chi) = \frac{(1 + \gamma/2)\ \chi^2}{\gamma/2 + \chi^{2 + \gamma}} \quad (13)$$

where χ = qr and γ is predicted to be equal to 2d for the critical mixture and (d + 1) for the off-critical mixture; d is the dimensionality of growth. In the region of $q < q_m$, the structure function s(q,t) or the scattering function I(q,t) may have the form of

$$I(q < q_m,t) \sim q^2 \quad (14)$$

whereas, in the high q region ($q > q_m$), it may be represented by:

$$I(q > q_m,t) \sim q^{-\gamma} \quad (15)$$

where γ will be 6 for the critical mixture and 4 for the off-critical mixture. Figure 17 shows the log-log plots of I(q,t) versus q for the time-evolution of scattering curves for the T-jump at 45°C. The slopes of 2 and -4 are shown for comparison purpose. As can be noticed from Figure 14, the shape of the intensity curve at the large q-region ($q > q_m$) gives the slope of -4, implying that the kinetics of 10% HPC solution is reminiscent of the behavior of the off-critical mixtures. At low q-region ($q < q_m$), the slope is slightly less than the predicted value of 2. Since the beam stop was used in the experiments to protect the possible damage of the Vidicon detector, the parasitic scattering cannot be removed completely. Hence, the intensity may be affected by this parasitic scattering, especially at low scattering angles. The slopes of 2 and -4 have been experimentally observed for metal alloys (23) and polymeric mixtures (22, 24) and thus are in good agreement with the present study.

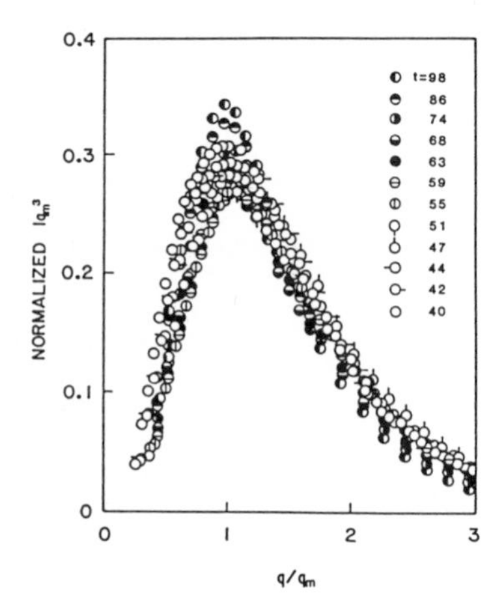

Figure 16. The plot of G(x) against $x(=q/q_m)$ for the 10 wt% aqueous HPC-L solution following a T-jump from 23 to 45°C.

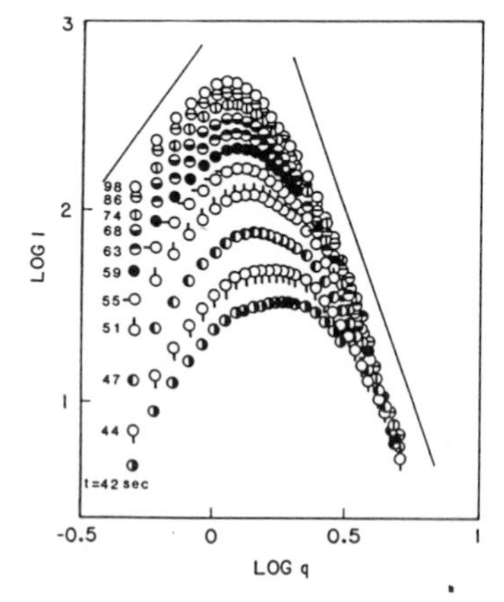

Figure 17. The log-log plot of scattered intensity (I) versus wave number (q) for the 10 wt% aqueous HPC-L solution following a T-jump from 23 to 45°C.

Conclusion

We observed that the phase separation in the 10% aqueous HPC solutions is the spinodal decomposition. The linearized theory of Cahn-Hilliard predicts many of the qualitative features of the SD in the present system, but is not adequate in the quantitative comparison. The phenomenon of SD is non-linear in nature dominated by the late stage of SD. The growth mechanism has been identified to be the coarsening process driven by surface tension. The kinetics of phase separation at the 10% aqueous HPC resembles the behavior of off-critical mixtures.

Literature Cited

1. Russo, P. S.; Magestro, P.; Miller, W. G. In Reversible Polymer Gels and Related Systems; Russo, P. S., Ed.; ACS Symposium Series No. 350; American Chemical Society: Washington, DC, 1988; Chapter 11.
2. Hwang, W. F.; Wiff, D. R.; Benner, C. L.; Helminiak, T. E. J. Macro. Sci. Phys. 1983, B-22, 231.
3. Russo, P. S.; Siripanyo, S.; Saunders, M. J.; Karaz, F. E. Macromolecules 1986, 19, 2856.
4. Werbowyj, R. S.; Gray, D. G. Macromolecules 1980, 13, 69.
5. Nishio, Y.; Yamane, T.; Takahashi, T. J. Polym. Sci. Polym. Phys. Ed. 1985, 23, 1043.
6. Kyu, T.; Mukherjee, P. Liq. Cryst. 1988, 3, 631.
7. Samuels, R. J. J. Polym. Sci. 1969, A2, 1, 1197.
8. Shimamura, K; White, J. L.; Fellers, J. F. J. Appl. Polym. Sci. 1981, 26, 2165.
9. Cahn, J. W.; Hilliard, J. J. Chem. Phys. 1958, 28, 258; Cahn, J. W. J. Chem. Phys. 1965, 42, 93.
10. Cook, H. E. Acta Metall. 1970, 18, 297.
11. Langer, J. S.; Bar-on, M.; Miller, H. D. Phys. Rev. A 1975, 11, 1417.
12. Binder, K.; Stauffer, D. Phys. Rev. Lett. 1973, 33, 1006; Adv. Phys. 1976, 25, 343.
13. Siggia, E. D. Phys. Rev. A 1979, 20, 595.
14. Goldburg, I. W. NATO Adv. Study Inst. Ser. B 1981, 73, 383.
15. Furukawa, H. Phys. Rev. A 1981, 23, 1535.
16. Furukawa, H. Physica 1984, 123A, 497.
17. Gunton, J. D.; San Miguel, M.; Sahri, P.S. In Phase Transitions and Critical Phenomena; Domb, C.; Lebowitz, J. L., Eds.; Academic: New York, 1983; Vol. 8, Chapter 3.
18. Kyu, T.; Saldanha, J. M. J. Polym. Sci. Polym. Lett. Ed. 1988, 26, 33.
19. Hashimoto, T.; Ebisu, S.; Kawai, H. J. Polym. Sci. Polym. Phys. Ed. 1981, 19, 59.
20. Snyder, H. L.; Meakin, P.; Reich, S. Macromolecules 1983, 16, 757.
21. Sato, T.; Han, C. C. J. Chem. Phys. 1988, 88, 2057.
22. Kyu, T.; Saldanha, J. M. Macromolecules 1988, 21, 1021.
23. Komura, S.; Osamura, K.; Fujii, H.; Takeda, T. Phys. Rev. B, 31, 1278.
24. Takahashi, M.; Horiuchi, H.; Kinoshita, S.; Ohyama, Y.; Nose, T. J. Phys. Soc. (Japan) 1986, 55, 2689.

RECEIVED August 22, 1988

Chapter 17

Kinetic Study of Association Processes between Polymer Latex Particles

Hiromi Kitano and Norio Ise

Department of Polymer Chemistry, Kyoto University, Kyoto, Japan

The association processes between polymer latex particles were studied. Advantage was taken of latex particles being large enough to be seen under an ultramicroscope connected to an image-processing system. The kinetic laws were examined directly by visual imagery. The forward rate constants (k_f) of the association of oppositely charged latex particles obtained were unexpectedly close to, although slightly smaller than, the theoretical values for the diffusion-controlled association process of neutral species. The k_f values of the association of antigen- and antibody-carrying latex particles were evaluated to be several times larger than that of the free antigen-antibody association system. The introduction of a fluorescence dye into the latex particles enabled us to confirm that the association took place specifically between different kinds of latex particles.

The kinetic studies of chemical reactions in general have been performed by measuring the time changes of parameters associated with the reaction. It would be quite interesting to study kinetic laws without introducing the assumption that the absorbance in spectrophotometry is strictly proportional to the concentration of the species in consideration.

This study also has another significance in colloid science. Polymer latices have been widely used in the diagnosis of various kinds of diseases(1). Agglutination methods can be used to recognize and estimate specific compounds in biological samples in the actual diagnosis of diseases. Agglutination of latex particles by the increase in ionic strength has been studied extensively (2,3). However, the basic nature of the association of latex particles (colloidal particles in general), and hence the rate constant of the association at the very initial stage, has not been studied rigorously.

It is of course desirable to determine the association rate at a much earlier stage to examine precisely the kinetic equation of

0097–6156/89/0384–0284$06.00/0

coagulation, for example, Smolukowski's equation, since this equation is derived for "binary" collision. Herein, we report the initial rate of association of polymer latex particles by counting dimer particles (4-6).

Experimentals

Materials

Electrically Charged Latex Particles

Three kinds of anionic latex particles were purchased from the Japan Synthetic Rubber Co., Tokyo (Immutex G-5301) and the Sekisui Chemicals, Osaka (N-200 and N-700), respectively. The other anionic latices SS-30 and SS-40 were prepared by emulsion polymerization of potassium p-styrenesulfonate and styrene using potassium peroxydisulfate as an initiator. A cationic polymer latex MATA-2 was prepared by emulsion polymerization of 3-methacryloxylaminopropyltrimethylammonium chloride and styrene using a cationic initiator 2-azobis(2-methylpropamidinium) dichloride. Fluorescent latex particles such as SMC-3 and SSH-10 were prepared by mixing fluorescent dyes such as coumarin-6 and Hostalux KCB with styrene before the emulsion polymerization. The anionic latices were converted to the H^+ form using a mixed-bed ion-exchange resin, Amberlite MB-3. The cationic latices were purified by repeated washings with pure water using an Amicon ultrafiltration apparatus. Table I shows the diameter of these latices estimated from electron micrographs using a JEM-100U TEM (Nihon Denshi, Tokyo, Japan), and the charge number on the surface of the polymer latices. The charge density of the latex particles remains constant at the neutral pH examined here, because sulfonate and quaternary ammonium groups on the latex surfaces are highly ionizable.

Latex Particles For Immobilization of Antibody and Antigen

Acrolein-containing latex particles were used for immobilization of proteins. The latex particles were prepared by emulsifier-free polymerization of acrolein and styrene (Table I). Fluorescent latex particles were prepared by mixing fluorescent dyes such as Hostalux KCB (Hoechst) or coumarin-6 into styrene before the emulsion polymerization (Table I). To increase the immunological sensitivity of the protein-conjugated latex particles, a hexyl group was introduced between the latex particle and the protein (Protein-spacer-latex). Proteins such as human serum albumin (HSA), anti-human serum albumin (anti-HSA-IgG), and a fragmented antibody (anti-HSA-IgG-F(ab')$_2$ were immobilized onto the spacer-containing latex particles using a water-soluble carbodiimide. The numbers of protein immobilized onto the latex particle were 16,000(HSA-spacer-AL-2), 16,000(anti-HSA-IgG-spacer-AL-2), 22,000(anti-HSA-IgG-F(ab')$_2$-spacer-AL-3), and 19,000 (HSA-spacer-AL-4). The latex particles modified were purified by repeated washing with distilled water using an Amicon ultrafiltration apparatus with a Millipore membrane.

Methods

Microscopic Observation

The association process was directly observed using a Carl Zeiss microscope (Axiomat) or a fluorescence microscope (DIAPHOT-EF, Nikon, Tokyo) connected to a Carl Zeiss image processing system (IBAS)(7). The microscopic observation was started just after the addition of the latex suspension to another kind of latex suspension in the observation cell using a small polyethylene mixing rod. The suspensions were mixed slowly in a constant manner to avoid the enhancement of the association by the vigorous mixing. The progress of the process was recorded on video tape. By replaying the tape the number of dimeric particles in the visual field was counted at appropriate intervals until it levelled off. The percent of dimers was estimated by assuming the uniform distribution of the dimeric and monomeric particles throughout the suspension. The number of particles used for the calculation was 200-300 at one time, and the uncertainty was within 10 %.

Spectrophotometric Measurements

The association between different kinds of polymer latex particles was also studied spectrophotometrically by the absorbance change at 800 nm using a high-sensitivity spectrophotometer (SM-401, Union Engineering, Hirakata, Japan).

Results and Discussion

Association Processes Between Oppositely Charged Latex Particles

The association in water was too rapid for visual observation. Therefore, the microscopic observation was carried out in a 30(w/v)% aqueous sucrose solution at 25°C, in which the association was slowed down because of the enhanced viscosity. The results are as shown in Figure 1, where a 1 M latex suspension denotes 6.02×10^{23} latex particles in 1 l of the suspension. The reciprocal of the relaxation time of the association of latex particles (9.1×10^{-3} s^{-1}) was obtained from the semilogarithmic plots of the concentration of dimeric particles in the suspension converted from the percent of dimers in the visual field against time.

Clear curves with a single relaxation time could also be observed spectrophotometrically by mixing the anionic latex with an excess amount of cationic latex. The reciprocal relaxation time measured spectrophotometrically under the same experimental condition as in Figure 1 (0.0094 s^{-1}) is in good agreement with the value obtained by microscopy (0.0091 s^{-1}). This shows that the relaxation phenomena observed spectrophotometrically correspond to the binary association of particles of opposite charges provided that the spectrophotometry can be carried out at the very initial stage.

Long after the start of the association, further association of dimeric particles with other particles to form trimers, tetramers and so on, was confirmed by the microscopic observation. The absorbance curve changed in a complicated manner. Thus, the kinetic data obtained long after the onset of the association should not be considered as representing the binary association.

We could obtain the first-order rate constant k_{obs} ($=1/\tau$) at various concentrations of the cationic latex by the first-order analysis. We will subsequently discuss the association process between oppositely-charged latex particles using spectrophotometric data obtained in water. The second-order rate constant k_f for the

Table I. Properties of the Polymer Latices

Latex	Diameter (Å)	Charge Number (per particle)	Charge Density ($\mu C/cm^2$)
MATA-2	3000	7.2×10^4	+4.0
G-5301	3700	2.7×10^5	-10
SS-40	3000	2.0×10^5	-11
SS-30	1250	2.3×10^4	-7.6
N-300	3000	2.3×10^4	-1.3
N-700	7000	4.0×10^5	-4.2
SMC-3[a]	6100	3.3×10^5	+4.7
SSH-10[b]	6000	1.7×10^5	-2.3
AL-2[c]	3750	-	-
AL-3[b,c]	5000	-	-
AL-4[a,c]	4100	-	-

(a) Coumarin-6 is contained.
(b) Hostalux KCB is contained.
(c) Acrolein-containing latex particle.

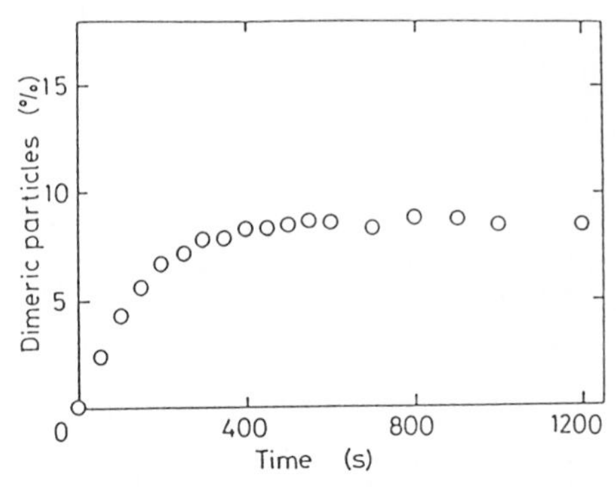

Figure 1. Time dependence of percent of dimeric latex particles in 30 (w/v)% sucrose-water at 25 °C. [MATA-2]=81 pM, [G-5301]=8.1 pM.
Reproduced from Ref. 4. Copyright 1987 American Chemical Society

association of the cationic latex with the anionic latex could be estimated from the slope of the plots of $1/\tau$ vs. concentration of the cationic latex. We could not estimate k_b from the Y-intercept of the plots, and we can only say that k_b is very small.

Table II shows the k_f and k_b values obtained. The table shows the theoretical rate constants for diffusion-controlled association reactions between neutral particles of the same sizes as those of the latex particles calculated by equations (1)-(4)(8-10).

$$f = 6\pi\eta R \tag{1}$$

$$D = kT/f \tag{2}$$

$$k_f = 4\pi N_A(D_a+D_b)R_{ab}/1000 \tag{3}$$

$$k_b = 3(D_a+D_b)/R_{ab}^2 \tag{4}$$

where f, η, R, D_a, k, T, N_A and R_{ab} are the friction coefficient (in g/s), the viscosity of the solvent (g/cm.s), the radius of the particle specified by the subscript (cm), the diffusion coefficient of the particle specified (cm^2/s), the Boltzmann constant, the absolute temperature, the Avogadro number, and the distance of closest approach between particles "a" and "b" (=R_a+R_b in the case of rigid particles. R_a and R_b are the radii of particles "a" and "b" (cm)), respectively. Note that the theoretical and experimental k_f's are unexpectedly in the same order of magnitude.

As mentioned above the increase in the viscosity of the reaction suspension reduced the association rate constant (Figure 2), which suggests that the association process observed here is mostly diffusion-controlled.

If we evaluate the influence of the electrostatic interaction between particles by equations (5)-(7) (11,12) the theoretical k_f value becomes much larger than the experimental value (3.3×10^{17} $M^{-1}s^{-1}$ was obtained for the theoretical value of k_f in the case of association of MATA-2 with G-5301).

$$k_f = f_{el}k_{fo} \tag{5}$$

$$f_{el} = Z/[\exp(Z)-1] \tag{6}$$

$$Z = Z_aZ_be^2/\varepsilon R_{ab}kT \tag{7}$$

where f_{el}, k_{fo}, Z, Z_a, Z_b and ε denote the electrostatic factor, the reaction rate constant without electrostatic effects, Z=U/kT (U;potential energy), total analytical charges of particles "a" and "b" and the dielectric constant of the medium, respectively.

Recent transference experiments showed that the so-called counterion association by latex particles was surprisingly much larger than that by linear macroions; in other words, the net charge number is much smaller than the analytical number(13). Only eight percents of the total analytical number of counterions (H^+) were free for latices having about the same analytical surface charge density as one of our latices, G-5301. Taking this experimental fact into consideration, and by assuming that the fraction of free counterions is 0.08, Z_a and Z_b of the G-5301 - MATA-2 system were assumed to be

Table II. Rate Constants at Various Temperatures in Water [a]

Temp. (^{o}C)	MATA-2 + SS-40 10^{-9}xk_f($M^{-1}s^{-1}$)	MATA-2 + SS-40 k_b(s^{-1})	MATA-2 + G-5301 10^{-9}xk_f($M^{-1}s^{-1}$)	MATA-2 + G-5301 k_b(s^{-1})
15	1.9 (5.6)[b]	0 (82)[b]	1.2 (5.7)[b]	0 (60)[b]
20	2.4 (6.5)[b]	0 (95)[b]	1.4 (6.5)[b]	0 (68)[b]
25	3.0 (7.4)[b]	0 (108)[b]	1.9 (7.5)[b]	0 (79)[b]
30	-	-	2.5 (8.5)[b]	0 (89)[b]

a. spectrophotometric method
b. theoretical values for association of neutral species

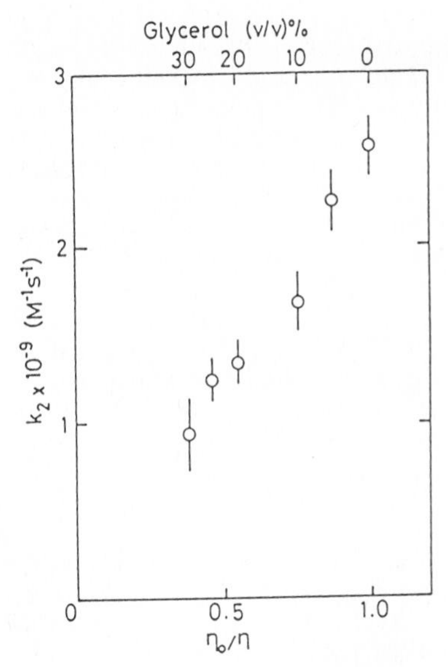

Figure 2. Effect of viscosity on $1/\tau$ of the association process of SMC-3 with SSH-10 at 25°C. [SMC-3]=0.18 pM, [SSH-10]=5.8 pM.

about 22,000 and 5,700, respectively. Therefore, the theoretical value of k_f was estimated to be $2x10^{15}$ $M^{-1}s^{-1}$, which is still much larger than the experimental result. This is probably because the point charge model assumed in the theoretical consideration is far from reality for latex particles. The number of effective charges responsible for the attractive interaction between the latex particles may even be much less than the number of net charges the latex particles have. One of the simplest interpretations is that, in the association processes, only the charges in the relatively narrow surfaces near the collision center play a decisive role; the charges on the surfaces diametrically opposite from the collision center may not be influential because of the relatively large dimension of the latices.

Table III shows the effects of charge density and the diameter of the anionic particles on the forward rate constants (k_f) of the association with MATA-2. The charge density on the latex surface clearly affects the k_f values, which supports the importance of the electrostatic interaction on the association of oppositely charged latex particles. The claim, that the observed rate constant and the theoretical rate constant calculated for neutral particles (by equations (1)-(4)) agreed, would not be physically sound.

Figure 3 shows the influence of sodium chloride on the $1/\tau(=k_f[\text{Cationic latex}]+k_b)$. Judging from the figure, we can claim that the ionic strength does not have a substantial effect on the association of latex particles, because the latex has a large diameter and the number of its effective charges is much smaller than that of net charges. Harding reported a similar retardation effect of the increase in ionic strength on the agglutination of Al_2O_3 with SiO_2 when the two colloidal particles are oppositely charged (14).

We could estimate the activation parameters, $\Delta G^{\ddagger}$, $\Delta S^{\ddagger}$ and $\Delta H^{\ddagger}$, which are listed in Table IV from the temperature dependence of k_f. Theoretical values calculated for the diffusion-controlled reaction are also shown in the table. In the case of $\Delta G^{\ddagger}$, the experimental results are in good agreement with the theoretical values. As for $\Delta H^{\ddagger}$ and $\Delta S^{\ddagger}$, however, the experimental values are larger than the theoretical values. The $\Delta H^{\ddagger}$ value is larger because there is a rate-determining factor which needs a large activation energy in the association reaction. Dehydration of latex particles in the association process as suggested by previous authors (15,16) and conformational changes of charged side chains on the latex surface are possible reasons for the larger experimental values of $\Delta H^{\ddagger}$ and $\Delta S^{\ddagger}$.

A similar effect was observed for the salt-induced coagulation of highly charged latex particles using the microscopic measurements (17). Figure 4 shows the time dependences of monomer, dimer and trimer particles of N-700 at the very initial stage of the salt-induced coagulation in a 1 M NaCl solution at 25°. The binary association constant of the latex particles was evaluated to be $2.8x10^{-12}$ $cm^3.p^{-1}.s^{-1}$ ($1.7x10^9$ $M^{-1}s^{-1}$). From the plots of salt concentration vs. initial turbidity change, the experimental value obtained was concluded to correspond to the rapid coagulation rate constant of the latex particles. The experimental value is, however, much smaller than the theoretical value evaluated for a diffusion-controlled binary association by the equation (3) ($12.3x10^{-13}$ $cm^3.p^{-1}.s^{-1}$, $7.4x10^9$ $M^{-1}s^{-1}$). Such a slow binary association can also

Table III. Rate Constants of Association Process of MATA-2 with Various Anionic Latices at 25 °C [a]

Latex	Diameter (Å)	Charge Density ($\mu C/cm^2$)	k_f ($10^9 \times M^{-1} s^{-1}$)
G-5301	3700	10	1.9
SS-40	3000	11	3.0
SS-30	1250	7.6	1.0
N-300	3000	1.3	0.62

a. in H_2O

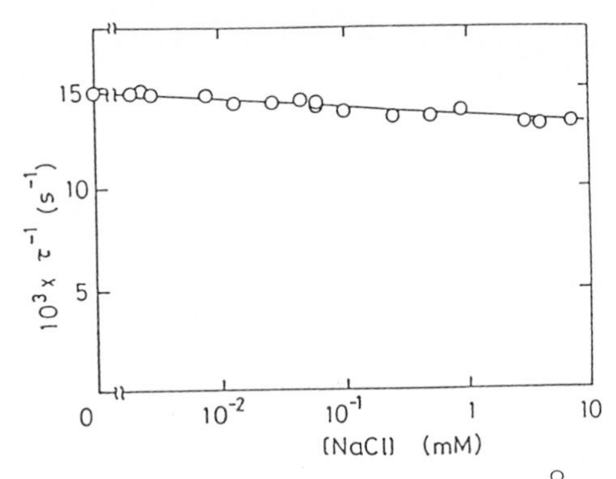

Figure 3. Influence of [NaCl] on $1/\tau$ at 25 °C. [MATA]=8.0 pM, [G-5301]=0.80 pM. In H_2O.
Reproduced from Ref. 5. Copyright 1987 American Chemical Society

Table IV. Activation Parameters of Association Processes Between Latex Particles

	$\Delta G^{\ddagger}$ (kcal.mol^{-1})	$\Delta H^{\ddagger}$ (kcal.mol^{-1})	$\Delta S^{\ddagger}$ (eu)
MATA-2 with G-5301	4.8	7.4	9
Antibody-Latex with Antigen-Latex	7.7	9.3	5
Theoretical values for diffusion-controlled association	3.9	4.4	1.5

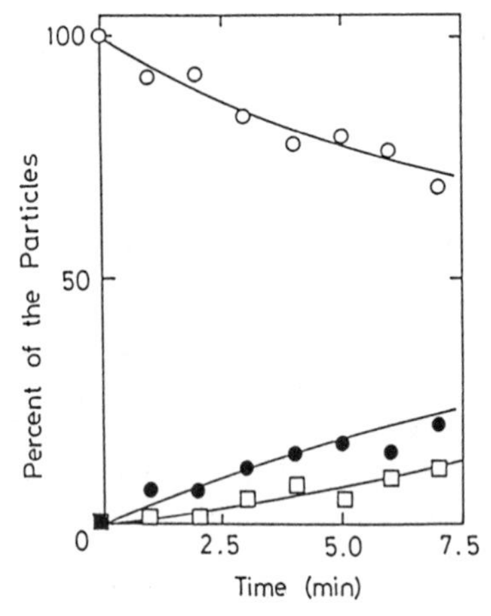

Figure 4. Time dependence of monomer(o), dimer(●) and trimer(□) of N-700 latex particles at 25°C. [NaCl]=1 M.

be attributed to the dehydration and the conformational change of charged side chains on the latex surface in the association process.

Association Processes Between Antigen- and Antibody-Carrying Latex Particles

The interparticle reaction between the Anti-HSA-IgG-spacer-latex and the HSA-spacer-latex was also examined by direct visual observation of latex particles using an ultramicroscope(7). The percent of monomeric latex particles in the visual field was determined with the lapse of time. At time=0, only monomeric particles of the anti-HSAIgG-spacer-AL-2 were found to be present in the visual field and by the addition of a small amount of HSA-spacer-AL-2 suspension into the observation cell, the number of monomeric particles decreased with time and almost levelled off (Figure 5). Although not shown in the figure, the number of dimeric particles in the visual field increased with time correspondingly to the decrease in the number of monomeric particles.

The observed reaction rate constant (k_{obs}) for the association of latex particles could be estimated by plotting log [dimer] versus time by the microscopic measurements. The second-order rate constant, k_f, for the association of latex particles modified with antigens and antibodies by equation (8), could be obtained from the slope of the plots of k_{obs} versus [Anti-HSA-IgG-spacer-AL-2].

$$k_{obs} = k_f \text{ [Antibody-latex]} + k_b \quad (8)$$

The k_f value obtained is 1.3×10^7 $M^{-1}s^{-1}$ at pH 9.2 and 25^oC, and it is much smaller than the value obtained for the dimeric association of oppositely charged latex particles (1.9×10^9 $M^{-1}s^{-1}$ for the association of G-5301 with MATA-2). The rate constant of the forward reaction obtained here is a little larger than the value observed for the association of the anti-HSA-IgG and the HSA detected by the conductance stopped-flow-technique (18)($k_2 = 1.9 \times 10^6$ $M^{-1}s^{-1}$ at pH 8.0 and 25^oC), because the local concentration of the reactants is much higher for the latex-bound case than for the free systems. The experimental value for the free HSA - anti-HSA-IgG system is in good agreement with the values reported for other immunological systems (1.0×10^6 $M^{-1}s^{-1}$ (25^oC) and 1.0×10^6 $M^{-1}s^{-1}$ (20^oC) for ovalbuminin - anti- ovalbumin(19) and cytochrome C -anti-cytochrome C systems(20), respectively).

On the anti-HSA-IgG-spacer-AL-2 latex, 16,000 antibodies were attached. The forward rate constant reduced for one antibody molecule ($k_f' = k_f$/number of antibodies on the latex surface) was estimated to be 810 $M^{-1}s^{-1}$, which is much smaller than the observed reaction rate constant for the antigen-antibody reaction in the homogeneous system. Wolff et al. reported that 1.3 molecules of IgG per vesicle is enough for interaction of the vesicle with antigen-carrying cells (21).

We also immobilized a fragmented antibody ($F(ab')_2$, an antibody without a F_c chain), onto the fluorescent latex particle (22). Using a fluorescence miscroscope, we could confirm that the association took place only between antigen- and fragmented antibody-carrying latex particles, and could rule out any possibility of self association. The association rate constant of the fragmented antibody-latex - antigen latex system was 6.0×10^8 $M^{-1}s^{-1}$, which is

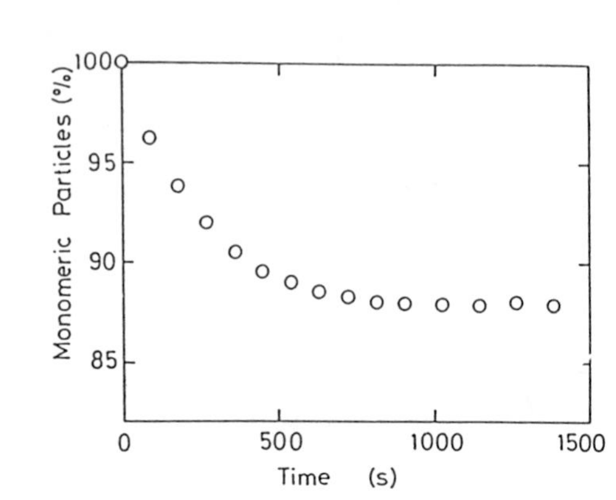

Figure 5. Time dependence of the percent of monomeric latex particles at 25°C. pH 8.7. [anti-HSA-IgG-spacer-AL-2]=360 pM. [HSA-spacer-AL-2]=40 pM.
Reproduced from Ref. 6. Copyright 1987 American Chemical Society

much larger than that for an antibody-latex system. However the reduced k_f value was still much smaller than that for the free antigen-free antibody system.

These results suggest that the disadvantageous orientation of the antigen and the antibody bound onto the latex surface, and the restriction of conformational changes of the immobilized proteins (23,24) for their association largely diminish the binding ability of the antibody (25).

The tendencies of the activation parameters of the inter-latex reaction examined here (Table IV) are slightly different from those of the association of oppositely charged polymer latex particles, and greatly different from those calculated for the diffusion-controlled association of latex particles, which suggests that there must be rate-determining factors such as (1) the conformational changes of the charged side chains near the immobilized proteins on the latex surface, and (2) the conformational changes of the proteins themselves accompanying the association of the latex particles.

Acknowledgments

This work was supported by the Ministry of Education, Science and Culture (Grant-in-Aid for Specially Promoted Research, 59065004 and 63060003).

Literature Cited

1. Rembaum, A.; Yen, S.P.S.; Cheong, E.; Wallace, S.; Molday, R.S.; Gordon, I.L.; Dreyer, W.J. Macromolecules 1976,9,328-336.
2. Ottewill, R.H.; Shaw, J.N. Disc.Faraday Soc. 1966,42,154-163.
3. Watillon, A.; Joseph-Petit, A.M. Disc.Faraday Soc.1966,42, 143-153.
4. Kitano, H.; Iwai, S.; Ise, N. J.Am.Chem.Soc. 1987,109,1867-1868.
5. Kitano, H.; Iwai, S.; Ise, N.; Okubo, T. J.Am.Chem.Soc. 1987,109,6641-6644.
6. Kitano, H.; Iwai, S.; Okubo, T.; Ise, N. J.Am.Chem.Soc. 1987,109,7608-7612.
7. Ito, K.; Nakamura, H.; Ise, N. J.Chem.Phys. 1986,85,6136-6142, ibid. 1986,85, 6143-6146.
8. Bird, R.B.; Stewart, W.B.; Lightfoot, E.N. Transport Phenomena; Wiley: New York, 1960; p 780.
9. Einstein, A. Ann.d.Phys. 1905,17,549.
10. Eigen, M. Quantum Statistical Mechanics in the Natural Science; Plenum: New York, 1974.
11. Debye, P. Trans.Electrochem.Soc. 1942,82, 265-272.
12. Alberty, R.A.; Hammes, G.G. J.Phys.Chem. 1958,62, 154-159.
13. Ito, K.; Ise, N.; Okubo, T. J.Chem.Phys. 1985, 82,5732-5736.
14. Harding, H. J.Colloid Interface Sci. 1972,40,164-173.
15. Johnston, G.A.; Lecchini, A.M.A.; Smith, E.G.; Clifford, J.; Pethica, B.A. Disc.Faraday Soc. 1966,42, 120-133.
16. Lichtenbelt, J.W.T.; Pathmamanoharan, C.; Wieserma, P.H. J.Colloid Interface Sci. 1974,49,281-285.
17. Ono, T.; Ito, K.; Kitano, H.; Ise, N. unpublished results.
18. Okubo, T.; Ishiwatari, T.; Kitano, H.; Ise, N. Proc.Roy.Soc.Lod. 1979,A366,81-90.

19. Levison, S.A.; Jancsi, A.N.; Dandliker, W.B. Biochem.Biophys. Res.Commun.1968,33,942.
20. Noble, R.W.; Reichlin, M.; Gibson, Q.H. J.Biol.Chem. 1969,244, 2403-2407.
21. Wolff, B.; Gregoriadis, G. Biochim.Biophys.Acta 1984,802,259-263.
22. Kitano, H.; Yan, C.-H.; Maeda, Y.; Ise, N. Biopolymers in press
23. Goldstein, L. Methods Enzymol. 1976,44,435.
24. Kitano, H.; Nakamura, K.; Ise, N. J.Appl.Biochem. 1982,4,34-40.
25. Nakamura, H.; Sugiura, T. In Methods in Immunological Biochemistry; Osawa T.; Nagai, K. Eds.; Tokyo Kagaku Dojin: Tokyo, 1986; pp 126-129.

RECEIVED August 10, 1988

Chapter 18

Colloidal Properties of Surface-Active Cellulosic Polymer

K. P. Ananthapadmanabhan, P. S. Leung, and E. D. Goddard

Specialty Chemicals Division, Research and Development, Union Carbide Corporation, Tarrytown, NY 10591

A new class of water soluble cellulosic polymers currently receiving attention is characterized by structures with hydrophobic moieties. Such polymers exhibit definite surface activity at air-liquid and liquid-liquid interfaces. By virtue of their hydrophobic groups, they also exhibit interesting association characteristics in solution. In this paper, results are presented on the solution and interfacial properties of a cationic cellulosic polymer with hydrophobic groups and its interactions with conventional surfactants are discussed.

Aqueous solutions containing both polymers and surfactants are encountered in a number of industrially important areas such as detergents,cosmetics, pharmaceutics, paints, EOR, metal working/ hydraulic fluids, and mineral/ceramic/material processing systems. In these systems, the polymers and surfactants can interact leading to marked changes in such properties as viscosity, solubilization capacity, interfacial tension, wettability, foam stabilization, adsorption,etc. A detailed review of polymer-surfactant interactions can be found in references 1 to 4. Depending upon the actual system, the interactions between the polymers and surfactants may or may not be desirable. For example, while the enhanced thickening and solubilization effects which can be encountered are often desirable, the precipitation of polymer-surfactant complexes leading to depletion of reagents would generally be undesirable. However such complexes can be useful in controlled release systems. In short, polymer-surfactant complexes have several interesting actual (and potential) applications.

Polymer-surfactant aggregates formed "in-situ" in solution can dissociate or change their structure depending upon the solution conditions. Such changes can be avoided by either polymerizing appropriately chosen structures of surfactants or by using preformed entities which combine the two features in the same molecule. We refer here to polymeric surfactants. The concept of polymeric surfactants is not new. To some extent proteins themselves embody this principle. Strauss and co-workers studied "polysoaps" derived from

0097–6156/89/0384–0297$06.00/0

polyvinylpyridine in the early 1950's (5). Since then there has been considerable work with surface active polymers for stabilization of particles in non-aqueous media, but little further work on water soluble hydrophobic polymers, until very recently when there has been a surge of activity (6-14). While on one hand attempts are being made to polymerize surfactant aggregates to obtain interesting controlled release/membrane type structures, on the other hand major efforts are underway to synthesize novel hydrophobe modified polymers with interesting rheological, interfacial and solubilizing properties.

In this paper, the results on solution and interfacial properties of a cationic cellulosics polymer with hydrophobic groups are presented. Interaction of such polymers with added surfactants can be even more complex than that of "unmodified" polymers. In the past we have reported the results of interactions of unmodified cationic polymer with various surfactants investigated using such techniques as surface tension, precipitation-redissolution, viscosity, solubilization, fluorescence, electrokinetic measurements, SANS,etc.(15-17). Briefly, these results showed that as the concentration of the surfactant is increased at constant polymer level significant binding of the surfactant to the polymer occurred leading to marked increases in the surface activity and viscosity. These systems were able to solubilize water insoluble materials at surfactant concentrations well below the CMC of polymer-free surfactant solutions. Excess surfactant beyond that required to form stoichiometric complex was found to solubilize this insoluble complex and information on the structure of these solubilized systems has been presented.

In this paper, the results on the interactions of hydrophobe modified cationic polymer with surfactants is presented and the results are compared with those for the unmodified polymer.

EXPERIMENTAL

QUATRISOFT® LM 200, a cationic cellulosic polymer with hydrophobic groups, is a product of Union Carbide Corporation as is the cationic cellulosic polymer, Polymer JR. Both have a molecular weight in excess of 100,000. Tergitol NP 10, a nonyl phenol ethoxylate used in this study is also a product of Union Carbide Corporation. Sodium dodecyl sulfate is a high purity sample purchased from EM Science Corporation. Pyrene which is used as a fluorescence probe was obtained from Aldrich Chemicals Co.

The surface tension of various solutions was measured by the Wilhelmy plate technique with a sand blasted platinum plate as the sensor. Fluorescence measurements were carried out using a Perkin-Elmer LS 5 spectrophotometer. The solutions for fluorescence measurements were prepared using pyrene saturated (10^{-6}kmol/m^3) water. The measurement itself was done by exciting the samples at 332 nm and monitoring the emission in the range of 360 to 520 nm.

The viscosity of the solutions was measured using the Cannon-Fenske viscometer. The foam tests were done by the conventional cylinder shake test in which a fixed amount of the solution is shaken vigorously for a minute and the foam height is monitored with time. The surface pressure measurements were carried out using a conventional Langmuir trough set-up, again with a platinum plate sensor. See text for further details.

RESULTS AND DISCUSSION

SURFACE ACTIVITY. The surface tension results for aqueous solutions of Polymer JR and Quatrisoft are given in Figure 1. The hydrophobe modified polymers clearly show more surface activity than the unmodified polymer. The surface activity of the modified polymers as measured by the surface tension criterion is only moderate compared to conventional surfactants which exhibit ultimate surface tension values in the range of 20-40 mN/m. The effect of the molecular changes resulting in this moderate surface activity can, however, be considerable on other properties of the polymer, as will be shown in subsequent sections.

SURFACE PRESSURE. An auxiliary method to examine films adsorbed at the air/water interface used the Langmuir trough. Although normally applied to spread insoluble monolayers, we took advantage of the well known characteristic of adsorbing high molecular weight polymers that, even from dilute solutions, e.g. 0.01%, substantial accumulation will occur at the surface if sufficient time is allowed(18). It can be seen in Figure 2 that over a 3-hour period the surface pressure rises from about 2 to over 10 mN/m subsequent compression of the film, corresponding to a three fold reduction of surface area, increases the surface pressure of this dilute polymer solution to as much as 20 mN/m. This behavior has implications regarding foaming where, owing to continual rearrangements such as bubble rupture and consequent restructuring of the foam lamellae, local stresses will result in continuous compressions and extensions of these lamellae. Given sufficient time, the relatively non-surface active Polymer JR also accumulates at the air/water interface, and also develops additional surface pressure on compression - but to a much lower extent than its surface active counterpart.

ASSOCIATIVE INTERACTIONS. Fluorescence characteristics of pyrene in solutions of Quatrisoft and Polymer JR are given in Figure 3. The ratio of the intensity of the first peak(I_1 at 373 nm) to that of the third peak (I_3 at 384 nm) in pyrene fluorescence spectra has been used extensively to detect the formation of micelles and other hydrophobic aggregates in solution (19). This ratio has a value of 1.65 in water, 1.1 in surfactant micelles, and 0.6 in hydrocarbon oil. The results given in Figure 3 show that for Quatrisoft, the ratio decreases from 1.65 at 0.01% level to 1.25 at the 1% level. In the case of Polymer JR, the I_1/I_3 ratio does not show any significant reduction in the above concentration range. In hydrophobe modified polymer solutions, pyrene is solubilized in a relatively hydrophobic environment somewhat comparable in polarity to surfactant micelles. Comparison with results for sodium dodecylsulfate given in the same figure show that, unlike micelle formation, aggregation in polymer solutions takes place over a wide range of concentration. It is therefore not a "critical" phenomenon of the phase transformation type.

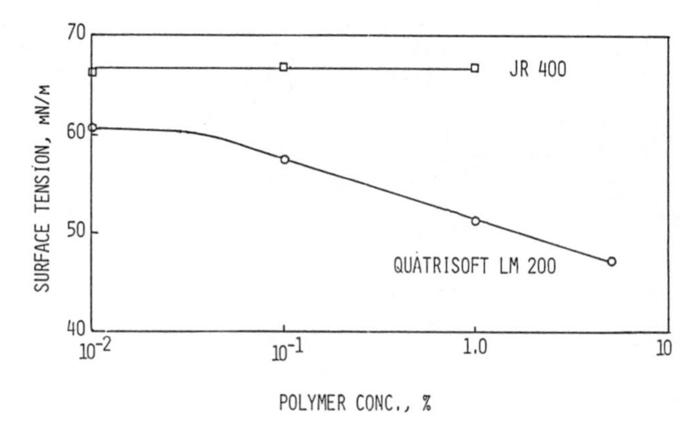

Figure 1. Surface activity of unmodified (Polymer JR 400) and hydrophobe modified (Quatrisoft LM 200) cationic cellulosic polymers as a function of polymer concentration.

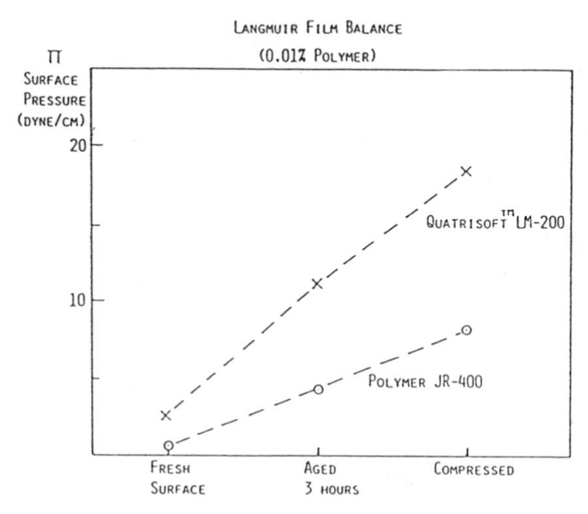

Figure 2. Effect of ageing and compression of adsorbed layers of unmodified and hydrophobe modified cationic cellulosic polymers on their surface pressure. (Reprinted with permission from ref. 21. Copyright 1985 Allured Publishing.)

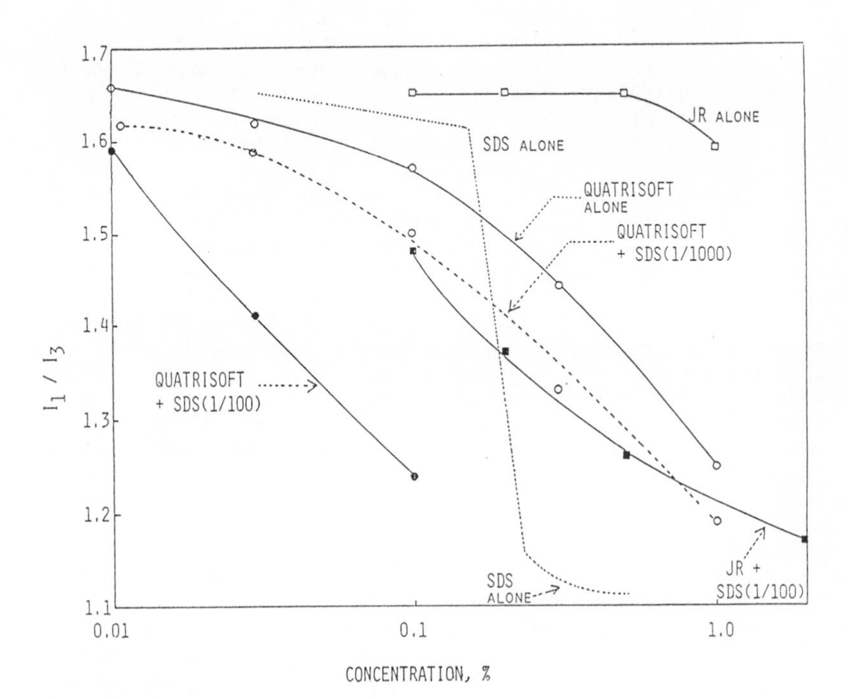

Figure 3. The fluorescence characteristics of pyrene in solutions of hydrophobe modified and unmodified cationic cellulosic polymers both in the presence and absence of SDS.

The concentration of pyrene in the above tests corresponded to its saturation solubility in water which is less than 10^{-6} kmol/m^3. Our past studies have shown that at such low levels the probe does not influence the properties of the system (20). The concentration of the probe can be increased in solution by contacting the micellar or other solutions containing associated hydrophobic structures with excess probe material. Pyrene and other organic materials in such systems are solubilized in the hydrophobic region of the aggregates. The amount of pyrene solubilized is influenced by such factors as the size and number of micelles. When more than one pyrene molecule is solubilized in a micelle and such systems are excited at a particular wavelength, excited pyrene molecules tend to interact with ground state molecules to form excimers. Thus the extent of excimer formation is an indication of the size of the aggregate as well as the fluidity of the region in which the molecules are solubilized. In the present study, excess pyrene was contacted with Quatrisoft solutions.

Interestingly, the solutions did not show the formation of any excimer in the system. The fluorescence spectra of solutions made up of pyrene saturated water, with or without subsequent contact with excess pyrene, exhibited essentially the same characteristics. This indicates that the aggregates present in Quatrisoft solutions are significantly smaller than conventional surfactant micelles. These observations indicate that cellulosic polymers, in view of their relative rigidity, have limited ability to coil up, maximize hydrocarbon-hydrocarbon interactions, and form aggregates capable of solubilizing water insoluble materials. It is not, however, clear from these results as to whether the hydrophobic interactions detected by pyrene are due to inter-molecular or intra-molecular interactions. This aspect is reexamined in a subsequent section.

POLYMER-SURFACTANT INTERACTIONS. Interactions of the above types of polymers with surfactants are of importance in a number of practical applications. The interactions of surfactants such as sodium dodecylsulfate (SDS) with Polymer JR and Quatrisoft LM 200, as measured by changes in the fluorescence characteristics of pyrene, are shown in Figure 3. Note that in these tests the ratio of surfactant to polymer (by weight) was 1 to 100, and was maintained constant along the curve, unless otherwise specified. It is evident that in the case of Quatrisoft strong interactions between the polymer and the surfactant exist even at concentrations as low as 10^{-2}% polymer and 10^{-4}% SDS. The interactions between the polymer and the surfactant follow the order: Quatrisoft LM 200 >>Polymer JR. In fact, Quatrisoft showed definite interaction with SDS even when the surfactant was present at levels three orders of magnitude lower than the polymer (see Figure 3). In contrast to this, Quatrisoft exhibited only marginal interactions with a non-ionic surfactant, Tergitol NP-10. It is clear from these results that while the primary force of interaction between the anionic surfactant and the polymer is electrostatic in nature, the presence of both cationic and hydrophobic groups in the polymer makes the Quatrisoft LM 200/SDS interaction much stronger than that between Polymer JR and SDS. In Figure 4, the interactions between SDS and the polymers are shown at a particular level of polymer as a function of the concentration of the surfactant. Note that, unlike the earlier plot, here the ratio of the polymer to surfactant decreases as the surfactant concentration increases. The sharp reduction in I_1/I_3 at concentrations well below the CMC of SDS clearly show the presence of aggregates at such low levels, which are capable of solubilizing pyrene. Again, it is clear that, while both Quatrisoft and Polymer JR interact strongly with SDS, the former with both cationic and hydrophobic groups exhibits interactions at lower concentrations than Polymer JR. As the concentration of SDS is increased, both systems show a precipitation stage followed by complete redissolution. The mechanism of precipitation-redissolution in Polymer JR-SDS system has been examined in detail (17) and similar mechanisms can be expected to operate in Quatrisoft-SDS systems as well. While precipitation is due to neutralization of the cationic sites on the polymer by the added surfactant, redissolution is considered to be due to the interaction of the polymer with conventional surfactant micelles formed at higher concentrations(4). The cationic polymer can be viewed as being "wrapped around" the anionic micelle or a group of micelles in the latter case. Note that the fluorescence characteristics of pyrene in the redissolution region tends to become similar to that observed in simple SDS micellar solutions. A clearer picture of the configuration of the polymer in solution can be obtained from the discussion given below on the viscosity of polymer-surfactant solutions.

VISCOSITY. Changes in the viscosity of Quatrisoft LM 200 (0.5%) solutions upon adding SDS are shown in Table I. In the pre-precipitation region an increase in the concentration of SDS is found to increase the solution viscosity markedly. In the post precipitation region the viscosity is found to decrease with increase in the surfactant concentration. The changes in the viscosity characteristics of Quatrisoft solutions are similar to those of Polymer JR solutions reported elsewhere (16) but the viscosity reduction is not as great. As mentioned earlier, adding SDS in the pre-precipitation region results in the neutralization of charges on the polymer. Under conditions of "complete" neutralization, precipitation of the polymer-surfactant complex occurs. In general, neutralization of charges of a polyelectrolyte can be expected to lead to a collapse of the extended polymer configuration and result in a reduction in viscosity. The presence of hydrophobic groups in aqueous polymer-surfactant complexes in the pre-precipitation region, on the other hand, promotes associative interactions between polymer molecules which would lead to enhanced thickening. In the present systems, these interactions are predominantly intermolecular in nature, since intra molecular interactions would have resulted in a decrease rather than an increase in solution viscosity. For Polymer JR, the configuration of the polymer-surfactant complex in the post precipitation region is more analogous to "polymer molecules wrapped around micelles"(17). For Quatrisoft more elongated structures with micelles bound through the polymer's hydrophobic groups, appear to be involved in view of the relatively higher viscosities observed.

The discussion so far has been limited to polymer-surfactant interactions in the bulk solution. In the section to follow, interactions at the air-liquid interface are examined.

FOAMING. The foaming of Quatrisoft LM 200 was studied by the conventional cylinder shake test. While the immediate foam height of Quatrisoft LM 200 was much higher than that of Polymer JR (30 ml vs. 10 ml), the foam decay results of Figure 5, show that the foam stability of the former polymer is also much greater.

Quatrisoft LM 200 in the adsorbed state will have its hydrophobic groups in "air" and its hydrophilic polymeric loops and tails submerged in the subsolution. Because of the polymeric nature of the hydrophilic moiety, the viscosity of the solution in the lamellae region can be expected to be considerably higher than that observed in conventional surfactant lamellae. Furthermore, the hydrophobic groups of the adsorbed polymers may form hemimicelle type structures at the liquid-air interface. The resultant viscous surface and subsurface regions are evidently responsible for the slow drainage and the consequently long-term stability of the foams generated by the hydrophobic polymer.

THE DRAINING LAMELLA TEST. A specially designed test cell, described in reference 21, was used to study the drainage characteristics of polymer stabilized films. In this procedure a single, planar lamella of solution is formed in a glass frame by tilting the cell from the vertical to the horizontal position and then re-erecting it (21). The lamella can be regarded as the ultimately simple model of a foam. A study of its stability and drainage characteristics can be viewed as reflecting these properties for a corresponding foam. Several characteristics observed for lamellae formed from 1% solutions of the new polymer are noteworthy: 1. Lamellae are comparatively thick and uneven. 2. They drain very slowly. 3. They are long-lived.

Lamellae from comparable solutions of weakly surface active Polymer JR are very much less stable. Characteristics 1 and 2 above were noted by examining the films for light interference patterns. Most of the area of a particular film was so thick that no interference patterns were discernible. At best, structureless, irregular "swirls" could be seen in the uppermost region of the film. For comparison, a "fast draining" surfactant film shows a series of fairly widely spaced, parallel bands moving downwards quite rapidly: A "slow draining" surfactant film shows a pattern of more closely, but still regularly spaced, and "wavy" interference bands moving more slowly downward as the film drains.

In conclusion, lowering surface tension, which translates into some surface accumulation, while necessary for foaming, is only part of the process. A companion effect is the formation of viscous surface and subsurface layers which can stabilize the lamellae. The new polymer would seem ideal for the latter effect. A depiction of the situation prevailing is attempted in Figure 6 which shows inter- and intramolecular bonds between hydrophobic groups in the polymer chains. Polymers which exhibit these types of interaction are referred to as associating polymers. It should be noted that considerable levels of surface visco-elasticity were also detected in adsorbed films of the polymer.

The single lamellae formed from 1% solutions of the new polymer had remarkable stability. They generally lasted for several hours and in some cases for a day or longer. The lamellae seemed completely static, giving little indication of drainage despite the

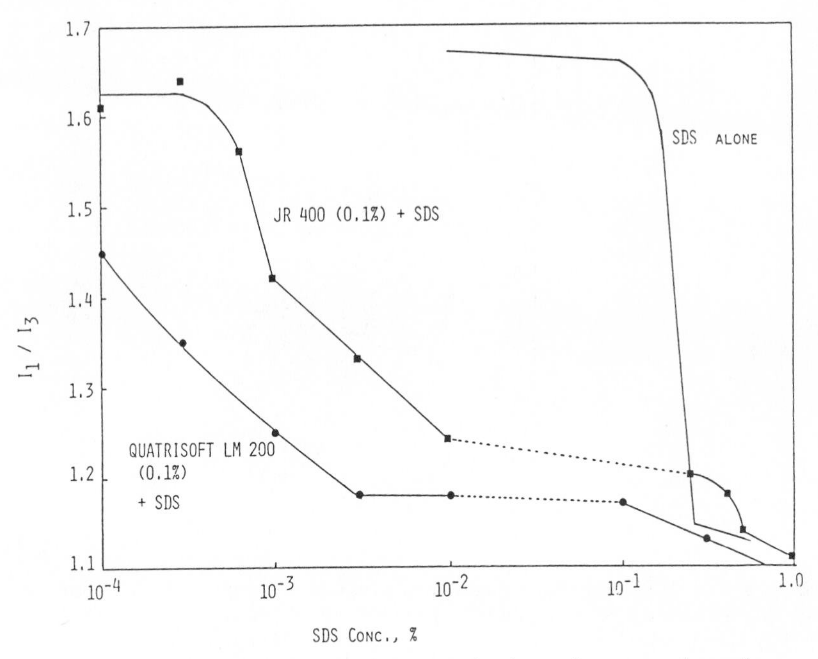

Figure 4. The fluorescence characteristics of pyrene in solutions of SDS in the presence and absence of hydrophobe modified and unmodified cationic cellulosic polymers.

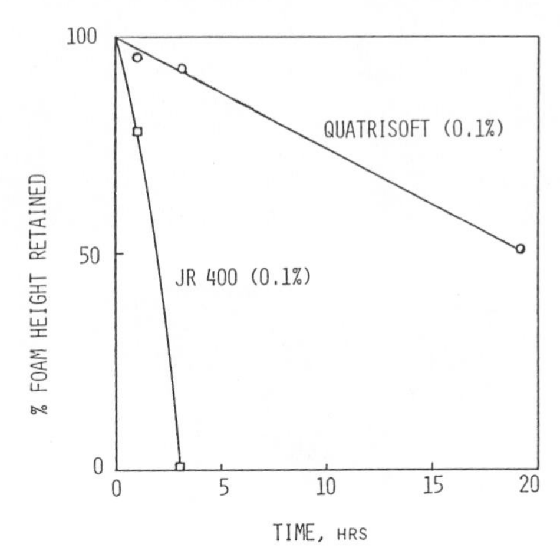

Figure 5. The stability of foams formed in the presence of hydrophobe modified and unmodified cationic cellulosic polymers.

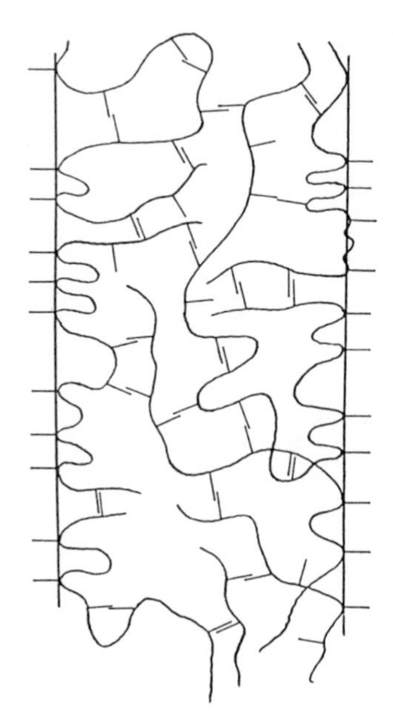

Figure 6. A schematic of the draining of a foam lamellae formed in the presence of a hydrophobe modified polymer. (Reprinted with permission from ref. 21. Copyright 1985 Allured Publishing.)

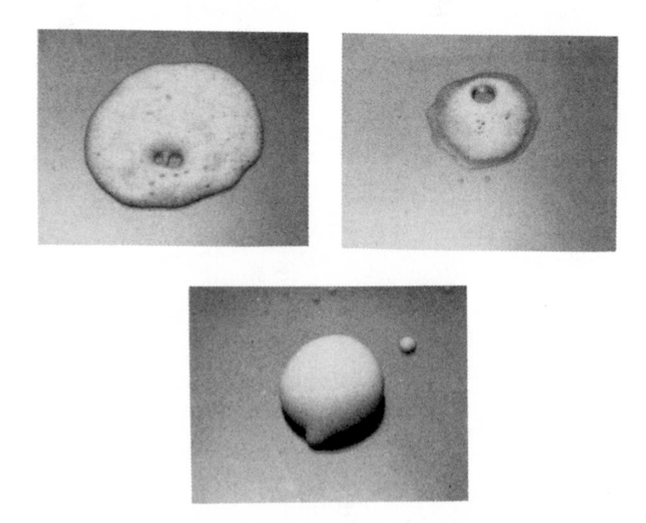

Figure 7. The mousse test one minute after dispensing shows the stability of foam generated by the hydrophobe modified polymer. (Reprinted with permission from ref. 21. Copyright 1985 Allured Publishing.)

TABLE I

EFFECT OF SDS ON THE SOLUTION VISCOSITY OF QUATRISOFT LM 200

QUATRISOFT LM 200 = 0.5%

SDS Conc.,%	Viscosity, cs
None	3.6
$1x10^{-4}$	4.7
$3x10^{-4}$	4.7
$1x10^{-3}$	5.0
$3x10^{-3}$	350.0
$1x10^{-2}$	Precipitation
$1x10^{-1}$	Precipitation
$3x10^{-1}$	7.9
1.0	3.25

TABLE II

SYSTEM: QUATRISOFT LM 200 0.1% (30 ml)

Additive	Foam Height, ml			
	Initial	1 hr.	3hr.	24 hr.
None	30	26	22	17
SDS (0.001%)	32	28	28	22
TEALS (0.001%)	30	30	30	26
NH_4LS (0.001%	30	29	29	25
$L(EO)_2SO_4$ (0.001%)	32	32	32	30
LAS (0.001%)	35	30	30	25

TEALS = Triethanolamine lauryl sulfate, $L(EO)_2SO_4$ = Lauryl ethoxylated sulfate, LAS = Linear alkylbenzene sulfonate.

fact that the films were so thick. By contrast, single lamellae formed from the Polymer JR solutions, although initially fairly thick, showed incessant activity (i.e., until film rupture), with swirls continually rising along the vertical rods of the frame. These films seldom lasted more than 5-20 seconds. The dramatic difference in behavior of the two polymers in the above test undoubtedly helps to explain their considerable difference in ability to stabilize an extruded foam. This is obviously due to hydrophobic groups in the new polymer.

Yet another method of testing foamability employs a pressurized aerosol container and the results have a direct applicability to a new cosmetic form known as "mousse". This concept was employed with the hydrophobic and other cationic polymers. In these experiments, a 1.0 wt.% aqueous solution of the polymer, with no extra additive other than the isobutane/propane propellant A-46, was charged to aluminum containers in the ratio of 80 wt.% aqueous solution and 20 wt.% propellant. Photographs of generated foam after 1 minute standing show the foam produced using the Quatrisoft LM 200 was stable;when using Polymer JR and another cationic polymer, the foam collapsed (see Figure 7). The foam generated using the Quatrisoft was found to be stable up to about 90 minutes when it collapsed. This aspect of the work is described in detail elsewhere (6).

FOAMING OF POLYMER-SURFACTANT SYSTEMS. As discussed earlier, Quatrisoft LM 200 exhibits strong interactions with surfactants such as SDS at extremely low concentrations. It is therefore resonable to expect polymer-surfactant combinations to have a marked influence on the foaming characteristics of the individual components. Foam data for Quatrisoft LM 200 in the presence of a number of surfactants, as measured by the cylinder shake test, are given in Table II. Note that the concentration of the anionic surfactants is two orders of magnitude lower than that of the polymer. At this level, they do not seem to affect the initial foam height significantly. The effect of surfactants is to increase the long term stability of the foam and this effect appears to be significant. In the presence of anionic surfactants, it is reasonable to expect that the hydrophobic groups of the polymer and of the surfactant would combine to form a mixed film at the liquid-air interface. The interactions between the cationic groups of the polymer and the anionic groups of the surfactant would further strengthen the interactions in the monolayer. These effects can be expected to increase the surface and sub-solution viscosity in lamellae and in turn enhance their stability.

LITERATURE CITED

1. Saito, S. Colloid Polym. Sci. 1979, 257, 266.
2. Breuer, M. M.; Robb, I.D. Chem. Ind. 1972, 530
3. Robb, I.D. In Anionic Surfactants in Physical Chemistry of Surfactant Action; E. H. Lucassen-Reynders, Ed.; Dekker: New York, N.Y. 1981, p 109.
4. Goddard, E. D. Colloids and Surfaces 1986, 19,255; 301.
5. Strauss, U.P.; Jackson, J. Polym. Chem. Ed. 1951, 6, 649.

6. Landoll, L. M. Polym. Sci., Polym. Chem. Ed. 1982, 20, 443
7. Camp, R. L., U. S. Patent 4,354,956, Oct. 19. 1982
8. Nagai, K.; Ohishi, Y.,; Inaba, H.; Kudo, S. J. Polym. Sci Polym. Chem. Ed. 1985, 23, 1221.
9. Fendler, J. H. Israel J. of Chemistry 1985, 25, 3
10. Schulz, D. N.; Kaladas, J.J.; Maurer, J. J.; Bock, J.; Pace, S. J.; Schulz, W. W. Polymer 1987, 28, 2110
11. Hamid, S. M.; Sherrington, D. C. Polymer 1987, 28, 325.
12. Jahns, E.; Finkelmann, H. Colloid and Polymer Sci. 1987, 265, 304
13. Valint, P. L.; Bock, J. Macromolecules 1988, 21, 175.
14. Binana-Limbele, W.; Zana, R. Macromolecules 1987, 20, 1331
15. Goddard, E. D.; Hannan, R. B. J. Colloid Interface Sci. 1976, 55, 73
16. Leung, P. S., Goddard, E. D.; Han, C.; Glinka, C. J. Colloids and Surfaces 1985, 13, 47
17. Ananthapadmanabhan, K. P.; Leung, P. S.; Goddard, E. D. Colloids and Surfaces ; 195, 13, 63
18. Lion, S. J.; Fitch, R. M. In Polymer Adsorption and Dispersion Stability; Goddard E. D.; Vincent B., Eds.; ACS Symposium Series No.240; American Chemical Society: Washington, DC, 1983; pp 185-202.
19. Thomas, J. K. The Chemistry of Excitation at Interfaces; ACS Monograph 181; American Chemical Society: Washington DC, 1984.
20. Ananthapadmanabhan, K. P.; Goddard, E. D.; Turro, N.J.; Kuo, P. L. Langmuir 1985, 1,352-355.
21. Goddard, E. D.; Braun, D. B. Cosmetics and Toiletries 1985, 100 (7), 41-47.

RECEIVED September 12, 1988

POLYMER–SURFACTANT ASSOCIATIONS

Chapter 19

Macromolecules for Control of Distances in Colloidal Dispersions

C. Cabane, K. Wong, and R. Duplessix[1]

Centre d'Etudes Nucleaires de Saclay, DPC–SCM–UA331, 91191 Gif sur Yvette, France

Dissolved macromolecules are used as spacers in colloidal dispersions; their function may be to keep surfaces apart from each other (bumpers), or to hold them together (stickers). In this paper we consider the effect of adsorbed macromolecules on the interactions between surfactant micelles, microemulsion droplets, or solid spheres which are dispersed in a liquid. Bridging will lead to the growth of aggregates where the particles are bound in linear arrays (necklaces) or in 3-dimensional ones (gels). We discuss the structures of these aggregates, as observed through neutron small angle scattering experiments, and the conditions for their formation.

A dispersion is a system made of discrete objects separated by a homogeneous medium; in colloidal dispersions the objects are very small in at least one dimension. Colloidal sizes range from 1 to 100 nm; however these limits are somewhat arbitrary, and it is more useful to define colloids as dispersions where surface forces are large compared to bulk forces. Here we are concerned with systems where the dispersion medium is a liquid; examples are droplets in emulsions or microemulsions (oil/water or water/oil), aggregates of amphiphilic molecules (surfactant micelles), foams, and all the dispersions of solid particles which are used as intermediates in the manufacture of ceramics. At this stage we are not too concerned with the nature of the constituents, but rather with the structures which they form; this is a geometrical problem, where the system is characterized by its surface area A, by the shapes of its interfaces (curvatures = b^{-1}), and by the distances between opposing surfaces (d = concentration parameter).

[1]Current address: Institut Charles Sadron, 6 rue Boussingault, 67083 Strasbourg, France

0097–6156/89/0384–0312$06.00/0

Of these three parameters, the first two are notoriously difficult to manipulate in surfactant systems. Indeed they depend upon surface pressures at various depths in the surface film: electrostatic repulsions between head groups, steric repulsions between uncharged groups, and the hydrophobic attraction which gives rise to the water/oil surface tension. The odds for control are much better with the third parameter, the distance between opposing surfaces. Here the aim may be to keep them from approaching at distances d shorter than a minimal distance d_0, in order to prevent the sticking of particles or droplets. Or, to the contrary, it may be to bring their distance d down to a lower value d_0, and expel the excess solvent; in this way a concentrated gel will be separated from a more dilute solution.

Still this type of control is difficult to achieve through the manipulation of existing surface forces, and it makes more sense to just add some large spacers whose dimensions are comparable to the desired distance d. Macromolecules can be appropriate spacers; they are effective in small amounts ($\simeq$1 mg per m^2 of surface), because the forces between distant surfaces are weak, much smaller than those within a surface (Figure 1).

The first step is to ensure that the macromolecules are bound to the surfaces. For this purpose they can be grafted chemically; however this is an expensive process, requiring the use of reactive polymers or tailor made surface active species such as block copolymers. A less expensive method is to use macromolecules which spontaneously adsorb to the surfaces. Then experimental wisdom states that surfaces which are each completely saturated with macromolecules will repel each other (stabilization of the suspension), while surfaces which are unsaturated will attract each other because the macromolecules will bridge them together (flocculation) (1).

This argument is buffered by direct measurements of the force between opposing surfaces (2); it has recently been analyzed theoretically by Rossi and Pincus (3). However it is meant to apply to surfaces which are large compared to the macromolecules. Here we shall argue that the opposite situation, where the macromolecules are larger than the adsorbing particles, is considerably more complex; in particular, the topologies of systems of small spheres and large macromolecules can vary according to the parameters characterizing the configurations of the macromolecules around the spheres (4).

Controlling Distances Between Oxide Particles in Water

For a start let us consider the standard bridging situation, which is best shown with inorganic particles in the range 10-100 nm and weakly adsorbing macromolecules. This is encountered daily when polymers are used as binders in the manufacture of ceramics (5), as flocculants in the separation of mineral particles (6), and as rheology control agents for all sorts of dispersions (7). In such cases the distances between the opposing surfaces can indeed be controlled by adsorbing the appropriate amount of dissolved polymers, and balancing the attractions which they create with direct interparticle repulsions.

To be specific, take a dilute dispersion of silica spheres with radii in the range 10 - 100nm; in water at pH values $\simeq$ 6 to 9 such spheres repel each other because their surfaces are negatively charged. Many types of water soluble polymers will adsorb on these surfaces; for instance ethylene oxide polymers (PEO) will form H bonds between their ether oxygens and the surface SiOH groups (8); or cationic polymers such as the AM-CMA copolymers (9,10) will bind their positively charged CMA monomers to the negatively charged SiO^- groups. Here the basic observations are:

(i) If the polymers are long enough to reach beyond the electrostatic barriers, they will bridge the particles together;

(ii) These bridges collect all the particles into a gel where the volume fraction of particles is in the range 5 - 10%: This is called a floc.

Still our goal was not only to collect the particles into a gel but also to set the distances between them at some specified value. To check on this point, we determine through small angle scattering the distribution of interparticle distances within the floc (11-13). This is obtained as a Fourier transform of a scattering curve (scattered intensity I versus scattering vector Q), and it is usually better to discuss the features of the experimental scattering curve. For systems where the basic unit is repeated at intervals $\simeq d_o$, the distribution of distances has a strong variation near d_o, and the intensity shows a peak near $Q_o = 2\pi/d_o$, at Q values below Q_o the intensities are depressed because the material appear uniform at large scales (translational invariance). For systems which are built by the repetition of a growth process (scale invariance), I(Q) follows a simple power law decay according to the distribution of distances around a reference particle: dense clusters yield a fast decay of I(Q) (Q^{-4} or Q^{-3}), while tenuous objects yields a slow decay of I(Q) (e.g. Q^{-2} for platelets and random walks, Q^{-1} for rods).

Figures 2 and 3 show the scattering curves of flocs, compared with those from free spheres (11-13). In Figure 2 flocculation is obtained by adsorbing a cationic polymer which neutralizes the charges borne by the silica surfaces. The floc scattering curve shows a high intensity at low Q (large aggregates), and a slow power-law decay. This decay is characteristic of tenuous objects which are full of voids at all length scales. At larger Q values (matching the sphere diameter), there is a weak shoulder which corresponds to the first neighbors of a sphere: this coordination shell is also made mostly of voids, and the few first neighbors are stuck directly at contact. The real space picture of such objects is shown next to Figure 2. It belongs to the class of structures built by kinetic aggregation, which are determined not by a balance of forces but by rules for diffusion and sticking of the particles (14). These structures represent a case where the attempt to

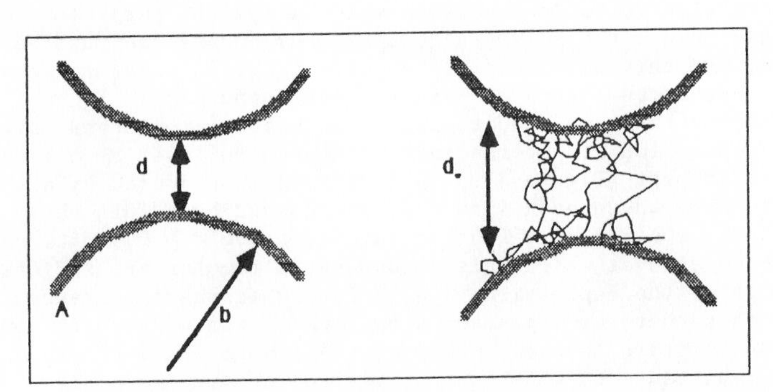

Figure 1. The distances between opposing surfaces in a colloidal dispersion can be controlled by the addition of dissolved macromolecules which adsorb on the surfaces.

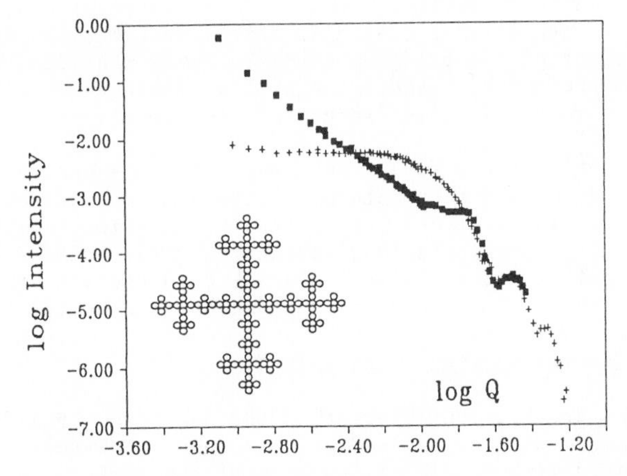

Figure 2. Scattering from silica particles bound with strongly cationic polymers (■), compared with that from free silica particles (+). The particles are spheres of precipitated silica with a radius of 19 nm; in water at pH near 7 they bear 0.3 negative charge per nm^2 of surface, most of which is compensated by adsorbed counterions (15). The polymers are AM-CMA copolymers with a ratio of cationic to total monomers equal to 0.3; the total amount of polymer in the floc approximately compensates the chemical charge borne by the silica particles (9).

control distances has failed, because the repulsive forces which ought to keep the particles apart have been suppressed.

If distance control is to be achieved, it is necessary to retain an electrostatic repulsion which keeps the particles separated; hence the polymers must not compensate the surface charges, and they also need to be quite large in order to form bridges beyond the range of electrostatic repulsions. When flocs are obtained in these conditions, their scattering curves have a strong depression at intermediate distances followed by a short range order peak (Figure 3). This pattern is produced by a liquid like stucture where each sphere is surrounded by a full shell of neighboring spheres at a face to face distance d $\simeq$ 0.5 diameter; the distances over which this structure is regular are indicated by the range of the depression, i.e. a few intersphere distances (11).

Such structures represent a case where the attempt to control distances appears to have succeeded. All opposing surfaces are at the same distance d, which can be varied to some extent by manipulating the electrostatic repulsions (12). However this control is possible only in a narrow range of conditions:

- If the particles are large, the adsorbed polymers tend to collapse on their surfaces, unless these surfaces are nearly saturated with polymer (16). In this situation, calculations of the bridging force (3) show that it is either large but very short range (low coverage) or longer range but very weak (near saturation coverage).

- If the particles are small, each one will only capture a small section of a macromolecule. Then the aggregates tend to be isolated necklaces rather than 3-dimensional gels. This point is particularly relevant to surfactant systems, where micellar sizes are indeed small compared with those of macromolecules; it is discussed in the next section.

Controlling Distances Between Surfactant Micelles

Now consider the problems which occur in the surfactant field, e.g. in detergency, cosmetics or pharmaceuticals, where polymers are usually associated with detergents to control the surface activity of the formulations (17). In these systems the polymers are known to interact with surfactant micelles or microemulsion droplets; the interaction can be beneficial, but most often it is a real nuisance.

In such systems as well, we would like to control through polymer bridges the distances between surfactant micelles or microemulsion droplets. There are cases where such control may indeed be possible; for instance the interaction of rigid polymers with anionic surfactant micelles may lead to the separation of a gel phase (18), which could be similar to the flocs mentioned above. However in most cases, where the micelles remain strongly repulsive, the gel phase is never found. For instance PEO macromolecules will bind to the surfaces of SDS micelles, and if they are long enough (M > 2E4) they can bridge many micelles together (19-21). Yet these bridged micelles never separate out as a macroscopic gel from the solution.

Small angle scattering experiments confirm that the polymer-bound micelles form distinct necklaces (19). Figure 4 shows the scattering curves of micelles bound to a PEO of M = 1.35E5. Here the shoulder corresponds to a separation between bridged micelles, with a low coordination number and a separation d $\simeq$ 3 diameters. This distance can be controlled by changing the number of micelles and their repulsions. These curves resemble those of flocs (Figure 3); however, at lower Q values (larger distances) the structure is that of a necklace rather than a 3-dimensional network. Indeed the low Q slope varies between 1 and 2 depending on the added salt, and its Q-->0 limit yields the number of micelles in one aggregate, which matches the number of micelles bound to one macromolecule. Hence the macromolecules form distinct necklaces, i.e. no micelle is ever shared by 2 macromolecules.

This rule is surprisingly strong: even polymers which can bind up to 300 micelles do not join to form a 3-d network. Moreover, these necklaces will not bind to each other even if salt is added to the solution to screen the repulsion between micelles (precipitation occurs first). Because of this problem we must return to the mechanisms through which a system of particles with adsorbed polymers may build a 3-d network (4), and examine why these mechanisms do not operate when the particles are small.

Gel Formation By Excess Polymers

Necklaces form when the spheres are small, when it is possible for one macromolecule to completely saturate one sphere and then continue to another one, without leaving space for other macromolecules to connect to this necklace. This makes it impossible to build a 3-d network. Still, with large excesses of polymer it should be possible for some free macromolecules to form the additional bridges which are necessary to connect the necklaces. The requirement is that on a few surface sites of a particle some monomers of the original macromolecule will be released and replaced with monomers of another one (Figure 5).

This mechanism is favored by the entropic gain associated with the swapping of monomers on the surface of a sphere: it is an exchange interaction. It is opposed by the osmotic pressure of the adsorbed polymer layer. Indeed, the original (adsorbed) macromolecule has a very high segment density near the surface, and in good solvents these segments will repel those of other macromolecules. Hence it is necessary for the free polymer to have a higher osmotic pressure than that of the adsorbed layers. This problem has been discussed by Marques and Joanny, who find that the number of adsorbed chains per sphere will remain equal to one unless the sphere is larger than the mesh size ξ of the polymer solution, or large enough that it cannot be saturated by one macromolecule (22).

A typical application is the use of small colloidal particles as fillers in elastomers, where cross linking is achieved because many macromolecules can bind to one particle. Another potential application would be for water-in-oil microemulsions where the oil would be polymerized around and between the water droplets. In all those cases, we will have gel formation through percolation rather than phase separation. The spheres do not attract each other, but they remain crosslinked at their original spacing d.

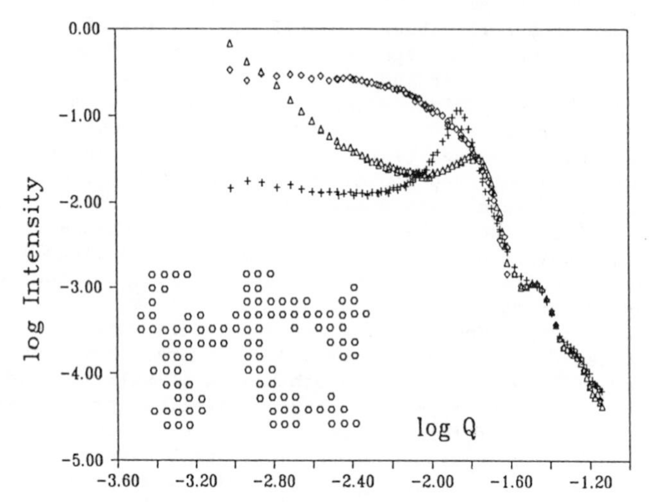

Figure 3. Scattering from silica spheres bound with PEO macromolecules in water (△), compared with free silica spheres (◇) and with a concentrated suspension where the spheres repel each other (+). The solvent is water at pH = 8; in this solvent the spheres bear about 0.3 SiO^- group per nm^2 of surface, and the macromolecules bind to the remaining SiOH groups (15). Very long macromolecules (M ≃ 2E6) are used to promote bridging and flocculation; with shorter ones no flocs are obtained unless the surface charges are neutralized or screened (16).

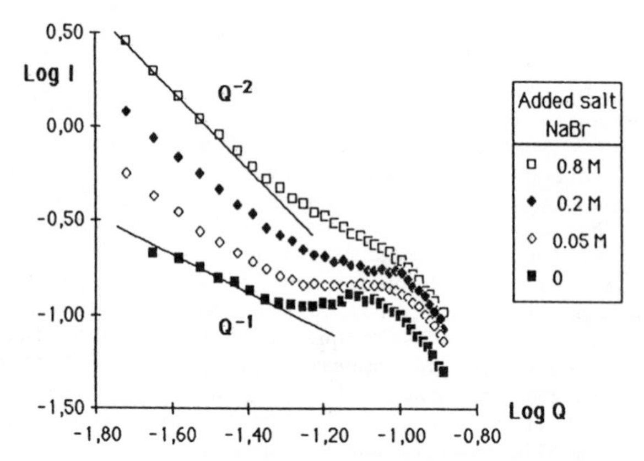

Figure 4. Scattering curves of SDS micelles bound to PEO macromolecules, for various amounts of added salt (19).

Gel Formation By Excess Spheres

Now consider the more general situation where the polymer concentration is rather low. With oxide particles, we find that bridges will be formed if their surfaces are unsaturated, and that the systgem of bridges can grow into a 3-dimensional network. With micelles, we find the growth stopping at the stage of independent necklaces. In order to understand what stops the growth in this last case, we consider a thought experiment where a gel is made in 2 successive steps (4): First, each macromolecule collects all the spheres which it can saturate and makes a necklace with them; then free spheres are added, and if they bind to the necklace polymers they will bridge them together (Figure 6.)

For large spheres, similar procedures are known to lead to flocculation in a very reliable way (23). The reasons for this success are that the first step saturates the surfaces, giving thick adsorbed layers, while the second step brings the unsaturation necessary for bridging. Hence the thought experiment mentioned above is indeed a good way of testing whether or not necklaces can be crosslinked.

Similar experiments have been performed with surfactant micelles, first through surface tension or dialysis experiments (24.25), then through neutron scattering (19.20). The result is interesting and important: excess micelles are rejected. This is the reason why the necklaces do not build a 3-d network.

It is instructive to compare these behaviors in a phase diagram (Figure 7). For all systems of polymers and particles, there is a line of compositions where the polymers exactly saturate the surfaces of all particles. If the spheres are large, it will take many macromolecules to saturate the surface of each one; if they are small, one macromolecule will saturate many spheres and hold them in a necklace. On either side of the stoichiometric line, the behaviors of oxide particles and surfactant micelles diverge:

(i) Oxide particles: Above the stoichiometric line, excess polymer coexists with fully covered spheres. Below the stoichiometric line, excess spheres cause unlimited bridging and the separation of a concentrated gel from the pure solvent. Then at even lower polymer concentrations, below the optimum flocculation concentration (o.f.c.), the gel can no longer accommodate all the spheres, and some are rejected.

(ii) Surfactant micelles: Above the stoichiometric line, loose necklaces are formed where all the excess length of polymer is in the bridges (19-21). Below the stoichiometric line, the necklaces are overloaded with bound spheres, and excess spheres are rejected; they coexist in the solution with the tight necklaces. No phase separation is observed.

Discussion

Aggregates of small particles bridged together by adsorbed polymers may vary considerably in their topology. The examples listed above include liquid-like gels, fractal structures, and

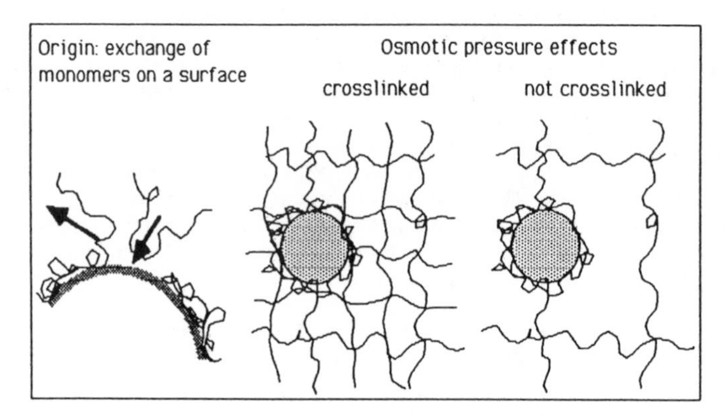

Figure 5. Excess polymer can be used to bridge polymer-covered particles - but its concentration must be high enough to allow it to penetrate the adsorbed layers.

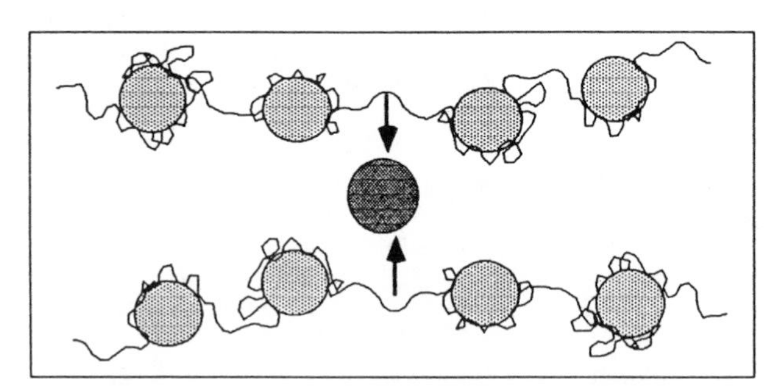

Figure 6. Bridging of necklaces though excess spheres. The thin lines represent the original configurations of the macromolecules, and the arrows point to their new configuration when they bind to an added free sphere.

independent necklaces which resist cross-linking. The criteria which separate these different modes of aggregation have been listed above. They appear to relate to the configurations of macromolecules adsorbed on small spheres (fraction of a sphere's surface which is covered by the macromolecules, thickness of the adsorbed layer), and to the interactions of polymer-covered spheres (range of electrostatic repulsions compared with the thickness of adsorbed layers). Thus there is reason to hope that a suitable description of macromolecules adsorbed on small spheres may have the power of predicting which topologies will be found when the covered spheres stick to each other.

One Sphere. We start from the simplified picture proposed by de Gennes (26) for polymer layers adsorbed on macroscopic, flat surfaces, and curl it around a sphere of radius b (Figure 8). In this model, the adsorbed layer is divided into cells which grow larger as the distance from the surface is increased; this represent the self-similar decay of the polymer concentration. As a result, the total number of strands which stick out at a given distance from the surface also decreases in a selfsimilar way.

Here the unusual point is the behavior of one macromolecule at the outer edge of the adsorbed layer. If the chain is much longer than the adsorbance capacity of the sphere, we must follow the argument of Pincus, Sandroff and Witten (27), and assume that the excess length of the chain forms two free tails extending away from the sphere. If the chain is shorter than this adsorbance capacity, then the adsorbed layer will be thinner, and there will be no tails sticking away from its outer edge. A similar description should hold for many macromolecules adsorbed on one sphere, except that the adsorbance capacity of the sphere cannot be exceeded in this case (the excess macromolecules will not be bound), so that there will be no long tails extending away from the outer layer -for a better description of this limit, see the paper by Marques and Joanny (22).

Two Spheres. The interaction of spheres which are covered with adsorbed polymer layers is expected to follow that of macroscopic surfaces. If the surfaces of the spheres are saturated, the force between them will be repulsive, and if they are unsaturated there will be an attraction caused by bridging (2,3). However special effects are expected to arise due to the finite sizes of spheres and macromolecules. Firstly, when the spheres are small, their radii may be comparable with the range of the bridging attraction. Secondly, when the macromolecules are large, one of them may saturate two spheres; then it may keep them bound to each other even though they are both saturated.

In addition, if the spheres are in water, there will be a repulsive electrostatic force between them. Under usual conditions of stability, this force is quite strong and its range is on the order of the sphere's radii. It is easy to see that the location of this repulsive wall with respect to the thickness δ of the adsorbed layers will determine the type of binding which will be possible. If the wall is within the adsorbed layer ($\sigma \lesssim \delta$), unsaturated spheres will experience a short range attraction and

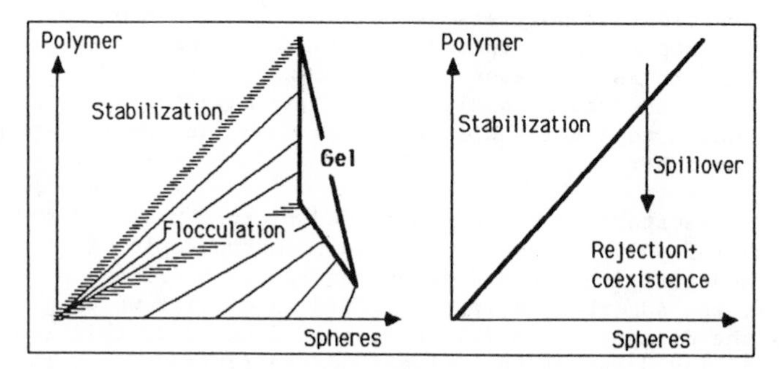

Figure 7. Solution diagrams for spheres with adsorbed polymer. Left: for oxide particles in water; the triangular region marked "gel" is a single phase which separates from the solvent (upper part of the flocculation region, above the o.f.c.) or from a dispersion of excess spheres (below the o.f.c.). The boundary between stabilization and flocculation corresponds to the saturation of the surfaces with polymer. Right: for surfactant micelles, all these boundaries are replaced by a stoichiometry which divides a region containing necklaces only from a region containing necklaces plus excess micelles (no phase separation).

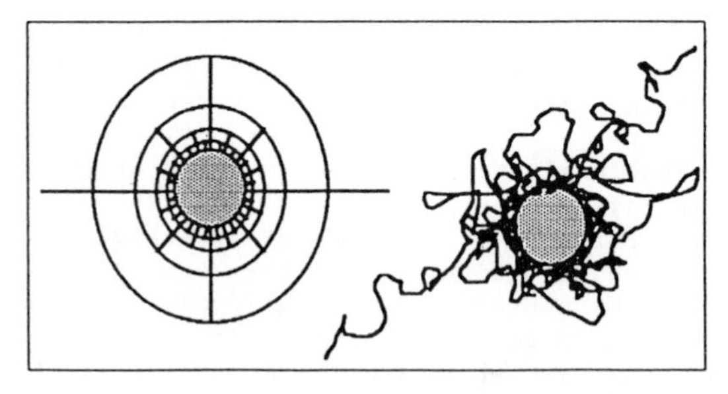

Figure 8. A selfsimilar picture for an adsorbed polymer layer around a sphere, derived from the model of de Gennes for flat surfaces (left), and the corresponding real life picture (right). We define the thickness δ of the adsorbed layer by the location of the last layer which contains loops.

remain bound just beyond the electrostatic wall. If the wall is beyond the adsorbed layer ($\sigma > \delta$), binding will be possible by macromolecules which are too large for the adsorbance capacity of one sphere and have tails extending out of the adsorbed layer.

For this long range binding, the simplest situation is that where a macromolecule is long enough to saturate both spheres and make long bridges between them. In this case the distance between the spheres will be determined by the length of polymer available for the bridge. However there will also be a more interesting situation where the macromolecule is long enough to saturate one sphere and bridge to another one, but not to saturate the second one as well (Figure 9). Then the length of the bridge is determined by a balance of forces. Bringing the spheres closer together increases their electrostatic repulsion, while pulling them apart stretches the chains, and forces them to desorb some monomers.

In a first step, we calculate the equilibrium span d for a bridge made of n_f monomers, each of length a. The balance of forces (repulsions and stretching) is written:

$$\frac{Z^2}{\epsilon\, d^2} = kT \frac{d}{n_f\, a^2}$$

where Z is the net charge borne by a sphere (we ignore counterions in the diffuse layer), ϵ the dielectric constant of the solvant, and the elongation d is supposed to be large with respect to the sphere radius b, large compared with the random walk length $a(n_f)^{1/2}$, but small compared to the fully stretched length an_f. Hence:

$$d^3 = n_f\, a^2 \frac{Z^2}{e^2} \frac{e^2}{\epsilon kT}$$

where e is the charge of an electron and $e^2/\epsilon kT$ is the Bjerrum length l_B which is microscopic (0.7nm).

In a second step we minimize the free energy F with respect to n_f, taking into account the adsorption energy E per adsorbed monomer, and the above value of d:

$$F = \frac{Z^2}{\epsilon\, d} + \frac{kT}{n_f} \frac{d^2}{a^2} - E\, n_b$$

With the constraint that $n_b + n_f = N$, the equilibrium length of the bridge is found to be:

$$n_f = \frac{Z}{e} \left(\frac{kT}{E} \right)^{3/4} \left(\frac{l_B}{a}\right)^{1/2}$$

Typically the last term is of order unity, Z/e is 10 to 100, and kT/E is between 1 and 10, hence n_f is between 10 and 1000 monomers. The equilibrium distance is then:

$$d = \left(\frac{Z}{e} \right) \left(\frac{kT}{E} \right)^{1/4} a^{1/2}\, l_B^{1/2}$$

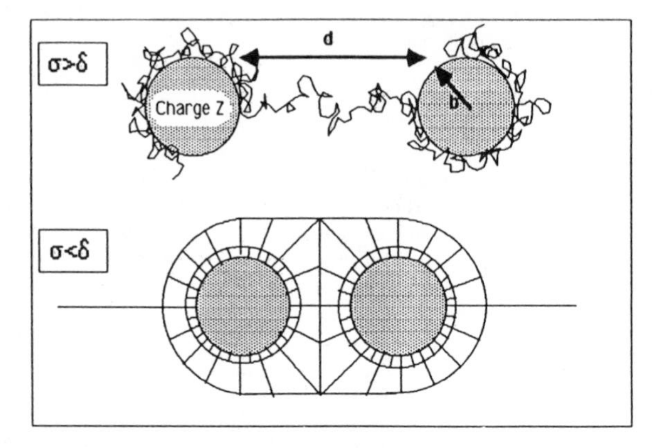

Figure 9. Bridging of two spheres by a macromolecule which does not saturate both of them.

Hence d is essentially Z/e times a microscopic length, fairly insensitive to the adsorption energy E, and independent from the chain length N, provided that N is large enough for bridging. For SDS micelles bound with PEO, Z = 15, hence d $\simeq$ 7.5 nm, for silica spheres also bound with PEO, Z/e = 200, hence d $\simeq$ 100 nm. Such distances have indeed been observed, but in systems where there are many spheres per macromolecule (19,21); such systems are discussed next.

Many Spheres. Setting the distance between bridged spheres does not determine the structure of the resulting aggregates. Once 2 spheres are bound, the course of growth for the aggregates depends on where the third, fourth, and later spheres will bind. This will be determined by the locations of polymer strands available for bridging. Here again, the determining feature will be the location of the electrostatic wall with respect to the adsorbed polymer layers. Let us start with spheres which must remain far apart, and bring the wall progressively closer to the surfaces.

- $\delta < \sigma$: As above, bridging is possible only by macromolecules which are too large for the adsorbance capacity of one sphere and have long tails extending away from the adsorbed polymer layer. In this limit, there are only 2 tails extending away from the outer shell, hence the aggregates which are formed with many spheres must be necklaces. Depending on the number of available spheres, these necklaces can have long loose strands between consecutive spheres, or they can be saturated with spheres and coexist with excess spheres in the solution.
- $b < \sigma < \delta$: The electrostatic wall has been pushed within the adsorbed layer, at a depth where this layer has many strands and loops all around the sphere. Then many spheres can bind at this distance, provided that the polymer does not saturate all the surfaces. The actual coordination number will depend on the thickness of the adsorbed layer with respect to the sphere radius b: thick adsorbed layers ($b < \delta$) allow many free spheres to bind to a covered sphere without getting in each other's way, as in the heterocoagulation of small spheres with large ones. This range of parameters is relevant to the "liquid-like" flocs observed with repelling silica spheres and very large macromolecules (11-13).
- σ and $\delta < b$: In this case the first few spheres which bind to the adsorbed layer will prevent others from reaching it. This will favor the growth of branches in the aggregate structure, as in most kinetic aggregation mechanisms (14). Because polymer adsorption is irreversible on the time scales of aggregate growth, voids in the aggregate structure will not be filled, and the structure can grow selfsimilar without collapsing. This range of parameters is relevant to the fractal structures of non repelling silica spheres bound at or near contact by shorter macromolecules (11,12).

Conclusion

Polymer covered particles are unusual objects, and they can interact to form a variety of original structures. The examples which have been presented in this paper can be regarded as a cautionary tale, or as a set of tools.

Caution is required if we intend to design a structure as a set of spheres and bridges: even if all spheres end up bridged together, the topology of the units which they form may vary. Depending on the subtleties of the adsorbed polymer layers, we have found necklaces, liquid like gels, or solid fractal aggregates.

The tools are the clever games which can be played once the subtleties of polymers adsorbed on small spheres are understood. For instance, consider the case where the dispersion forms necklaces instead of a gel because the spheres are too small to allow crosslinking. Such dispersions could be seeded with a few larger spheres which would connect the necklaces together. Alternatively, consider the case where the spheres are large, and their adsorbed polymer layers are too thin because the surfaces are not saturated. It might then be better to use a non equilibrium situation where all the macromolecules are on a smaller number of spheres, making thick layers because they saturate the surfaces. These saturated spheres could then be mixed with the remaining free spheres, giving a better chance of bridging at large distances.

In summary, the configurations of polymers around small objects are unusual, and it is better to know them rather than ignore them.

Acknowledgments

Some of the ideas presented here originate from discussions with S. Alexander, J. F. Joanny, P. A. Pincus, G. Rossi, and T. A. Witten. It is a pleasure for us to acknowledge their share in this field.

Literature Cited

1. La Mer, V. K.; Healy, T. W. Rev. Pure. Appl. Chem. 1963, 13, 112
2. Klein, J. J. Colloid Interface Sci. 1986, 111, 305-313
3. Rossi, G.; Pincus, P. Europhys. Lett. 1988, 5, 641-646
4. Alexander, S. J. Physique (France) 1977, 38, 977-981
5. Woodhead, J. L. J. Physique Colloque C1 1986, 47, C1-3 to C1-11
6. Atesok, G.; Somasundaran, P.; Morgan, L. J. Colloids Surfaces 1988, 32, 127-138
7. Otsubo, Y.; Watanabe, K.; J. Colloid Interface Sci. 1988, 122, 346-353
8. Killmann, E.; Maier, H.; Kaniut, P.; Gutling, N. Colloids Surfaces 1985, 15, 261-276
9. Wang, T. K.; Audebert, R. J. Colloid Interface Sci 1987, 119, 459-465
10. Wang, T. K.; Audebert, R. J. Colloid Interface Sci. 1988, 121, 32-41
11. Cabane, B.; Wong, K.; Wang, T. K.; Lafuma, F. Colloid Polymer Sci. 1988, 266, 101-104
12. Wong, K.; Cabane, B.; Duplessix, R. J. Colloid Interface Sci. 1988, 123, 466-481
13. Wong, K.; Cabane, B.; Somasundaran, P. Colloids Surfaces 1988, 30, 355

14. Martin, J. E.; Hurd, A. J. J. Appl. Cryst. 1987, 20, 61-78
15. Iler, R. K. The Chemistry of Silica; Wiley: New York, 1979
16. Lafuma, F.; Wong, K.; Cabane, B., to be published
17. Jost, F.; Andree, H.; Schwuger, M. J. Colloid Polymer Sci 1986, 264, 56-64
18. Thalberg, K.; Lindman, B. J. Phys. Chem., in press
19. Cabane, B., Duplessix, R., J. Physique (France) 1982, 43, 1529-1542
20. Cabane, B., Duplessix, R. Colloids Surfaces 1985, 13, 19-33
21. Cabane, B.; Duplessix, R. J. Physique (France) 1987, 48, 651-662
22. Marques, C. M.; Joanny, J. F. J. Physique (France) 1988, 49, 1103-1109
23. Fleer, G. J.; Lyklema, J. J. Colloid Interface Sci. 1974, 46, 1-12
24. Jones, M. N. J. Colloid Interface Sci. 1967, 23, 36-42
25. Shirahama, K.; Ide, N. J. Colloid Interface Sci. 1976, 54,
26. de Gennes, P. G. Adv Colloid Interface Sci. 1987, 27, 189-209
27. Pincus, P. A.; Sandroff, C. J.; Witten, T. A. J. Physique Letters 1974, 45, 725-729

RECEIVED September 12, 1988

Chapter 20

Interactions of Water-Soluble Polymers with Microemulsions and Surfactants

D. B. Siano[1], J. Bock, and P. Myer[1]

Corporate Research Science Laboratories, Exxon Research and Engineering Company, Clinton Township, Route 22 East, Annandale, NJ 08801

The interaction of water soluble polymers with microemulsions and with surfactants will, when the components are sufficiently concentrated, often result in a phase separation or change in the phase boundaries of the mixture as a function of external variables, such as temperature or salinity. In order to arrive at a better understanding of this technologically important phenomenon, a series of experimental studies was carried out using a variety of water soluble polymers in conjunction with model microemulsion systems. The polymers used included polyethylene oxide, polyvinylpyrrolidone, dextran, xanthan, polyacrylamide, and hydrolyzed polyacrylamide. The surfactants used were either nonionic or anionic and included a variety of alkyl ethoxylates such as members of the commercially available Brij and Tween surfactants, and monoethanolamine salts of alkyl arene sulfonates. Measurements of the phase behavior, as a function of temperature and polymer molecular weight, revealed a number of features that were consistent with a coacervation model of compatibility that is qualitatively similar to that known to occur for two incompatible polymers in a common solvent. Measurements of dilute solution viscosities of polymer-microemulsion mixtures also could be adequately described by this approach. These experiments showed, for example, that the weight average molecular weight (as opposed to the number or z-average molecular weight) of the polymer is the dominant variable in the dependence of the cloud point temperature of the microemulsion-polymer

[1]Current address: Exxon Chemical Company, 1900 East Linden Avenue, Linden, NJ 07036

0097–6156/89/0384–0328$06.00/0

mixture in cases where no specific interactions between the polymer and surfactants are believed to occur. In all of the cases studied, the region of incompatibility was larger for increasing polymer molecular weight, and in several of the systems studied, other qualitative features of the phase diagram could be rationalized on the basis of the coacervation model.

Polymers interact with surfactants and microemulsions in diverse, interesting and technologically important ways(1,2). The mechanisms that are responsible for the interactions include the usual panoply of forces involved in the interaction of any two different molecules: ion-ion, ion-dipole, dipole-dipole, and van der Waals forces all modulated by the presence of solvent and/ or other species such as dissolved salts. All may play a role. The special factors involved in surfactant/polymer and polymer-/microemulsion interactions that form the basis for their particular interest lies in their tendencies to form a variety of supermolecular clusters and conformations, which in turn may lead to the existence of separate phases of coexisting species. Micelles may form in association with the polymer, polymer may precipitate or be solubilized, microemulsion phase boundaries may change, and so on.

Previous studies in this area have mostly focussed on two of these phenomena: shifts in critical micelle concentration of the surfactant due to the presence of polymer, and polymer precipitation/redissolution phenomena. The former case is epitomized by the well-studied sodium dodecyl sulfate (SDS)-polyethylene oxide system(3-5), where two breaks are observed in a property sensitive to the state of aggregation or micellization, such as the equivalent conductivity as a function of SDS concentration at constant polymer concentration. The second case is commonly observed in systems where the polymer and surfactant carry opposite charges(6,7). When sufficient anionic surfactant is added to a cationic polymer, the system may become insoluble because of charge neutralization. Excess surfactant may adsorb further, imparting the opposite charge and therefore causing the polymer to redissolve. A variety of techniques may be employed to quantify these interactions, and have provided insights to greater detail of the relative balance of the forces involved.

Systems in which polymers interact with microemulsions can be even more complex, not the least because of the increased number of components present. The presumed coexistence of surfactant micelles with the larger aggregates of surfactant, cosurfactant and oil, the microemulsion particles, together with their attendant polydispersity, might be expected to provide for a wide variety of system-dependent behaviors. One sort of behavior, believed to be fairly common, is exemplified by the introduction of a water-soluble nonionic polymer into a continuous microemulsion. This often results in a well-known incompatibility, manifested by a deleterious phase change and therefore accompanied by large changes in viscosities of the phases. The major observation is that on the addition of sufficient polymer to the microemulsion, two coexisting liquid phases form: one is more concentrated in polymer, the other in surfactant. This phenomenon presents an obstacle to the viscosification of microemulsions by high molecular weight polymers and to their use together in such applications as enhanced oil recovery.

This has been explained (for microemulsions having a lower cloud point) on the basis of a theory that is very similar to that involved in $polymer_1$-$polymer_2$-solvent incompatibility (8). Ternary phase diagrams generated at a constant temperature for polymer-microemulsion mixtures show a similar shape and behavior with changing polymer molecular weight when the system can be described in terms of pseudo-components. Here the oil, surfactant and cosurfactant are considered to comprise a single pseudo-component, the polymer is a second component, and the water or brine is a third component. The phenomenon is controlled by the weight average molecular weight of the polymer, rather than the z- or number average molecular weight.

The incompatibility is very easily followed by simply measuring the decrease in the cloud point of the microemulsion by the dissolution of varying amounts of the polymer. The cloud points of the polymer-microemulsion mixtures were measured in a flat bottom 15 ml culture tube containing about 5 ml of sample. The tube and its contents were briefly immersed in a beaker of hot water just until it clouded, then the tube was removed and stirred with a thermometer until the turbid solution become clear. The temperature at which the opaque mixture abruptly cleared was noted and taken as the cloud point. This was usually repeated three times and the average of the readings was taken. For the system used, this was reproducible to about 0.2°C. In the measurement of the cloud point slope, the amount of added polymer was varied to yield a 3-5°C temperature change. The rate of change in the cloud point with added polymer concentration was found to be correlated with the molecular weight of the polymer. A plot of this sort for polyethylene oxide (PEO), polyvinylpyrrolidone (PVP), dextran and the sodium salt of polystyrene sulfonate is shown in Figure 1. All four lines are reasonably straight, even the case of polyethylene oxide, which covers almost four orders of magnitude. Although the slopes vary a little from one polymer to the next, it is apparent that the molecular weight of the polymer is a controlling variable for the incompatibility. This sort of plot is reminiscent of the widely used Mark-Houwink intrinsic viscosity-molecular weight correlation. It has been proposed (9,10) that the cloud point plot of the sort shown in Figure 1 can serve a similar role for determining polymer molecular weight. The measurement requires very simple equipment (only a thermometer) and is also fast and easy to carry out.

In an effort to explore more broadly the applicability of cloud point phenomena and the coacervation model to describe polymer-microemulsion interaction, additional polymers and microemulsion systems were selected for study. Because polyacrylamide polymerizes to very high molecular weight and therefore is one of the best polymers for efficiently thickening or viscosifying brine, it is of some interest to carry out the same sort of experiments with it. This has not been previously done because of the lack of a sufficient number of samples with differing molecular weights. In terms of the microemulsion, an anionic surfactant system was developed to provide a means of observing the effect of a systematic variation of the hydrophile-lipophile characteristics on the interaction with water soluble polymers.

Experimental

The polyacrylamides were synthesized at 55°C in water, using isopropyl alcohol as a chain transfer agent to give samples with varying molecular weight. Potassium persulfate was the initiator, and the polymers were purified by precipitation into acetone. A partially hydrolyzed polyacrylamide, HPAM, was prepared by treating polyacrylamide with base, NaOH, at 50°C to introduce charge in the form of sodium carboxylate groups into the polymer. Titration of these polymers indicated that the charge level was approximately 20 mole %.

The nonionic microemulsion used is a well characterized (11-14) nonionic microemulsion with Brij-96 (oleyl-10 ethoxylate) as the primary surfactant, n-butanol as the cosurfactant and hexadecane in the weight ratio of 5.32/2.78/0.90. When diluted with water to 9%, oil + surfactant + cosurfactant, this forms a water clear microemulsion having a sharp cloud point of about 58°C. The cloud point of this microemulsion is depressed to about 51°C when the external phase is 2 wt % NaCl in water.

Although the oil uptake of the surfactant and cosurfactant in this system is rather low, we have preferred to use the term "microemulsion" to describe it. The properties of the system (as assessed by light scattering and viscometry), when nearly saturated with oil, resemble those expected for a microemulsion. When smaller amounts of oil are taken up, the system behaves like a swollen (anisotropic) micelle(11).

An anionic microemulsion system was based on blends of monoethanolamine salts of 'bilinear dodecyl' benzene sulfonic acid and branched pentadecyl o-xylene sulfonic acid. The 'bilinear' structure results from the alkylation of benzene with a linear α-olefin. The former acts as a surfactant hydrophile (H) while the latter acts as a surfactant lipophile (L) at room temperature for the oil and water phases used in this study. The hydrophile tends to form water-continuous emulsions while the lipophile forms oil-continuous emulsions. The hydrophile-lipophile characteristics were varied by changing the weight ratio of H/L from 0.5 to 0.8. Decane was used as the oil phase and 2.0 wt. % NaCl in water as the aqueous phase. The water-oil ratio was fixed at 95/5 and the total surfactant content was fixed at 2 g/dl.

The viscosities were measured on a Contraves LS 30 low shear couette viscometer at a shear rate of 1.3 s^{-1} thermostatted at 25°C.

Results and Discussion

Cloud Points

The cloud points of the Brij-10 microemulsion with varying amounts of added polyacrylamide for different intrinsic viscosities of the polymer are shown in Figure 2. Intrinsic viscosity is a measure of polymer molecular weight. At these low concentrations of added polymer the fitted lines are linear, as shown previously for other nonionic polymers. Also, the lines have a steeper slope for higher intrinsic viscosity or molecular weight of the polymer. This demonstrates that the incompatibility is more severe as the molecular weight of the polymer increases.

The negative of the slope of the lines are shown plotted against the measured intrinsic viscosity of the polymers in Figure 3. The previously described coacervation model (8) predicts that the slope of this line should be unity. A line with this slope accurately represents the data, as expected. These results indicate that polyacrylamide has no attractive interaction with the microemulsion particles (or with its components) and the interaction is a repulsive, excluded volume one. This leads to the conclusion that polyacrylamide is similar to the other nonionic water soluble polymers, PEO, PVP and dextran in its behavior toward water external microemulsions, possibly by a "volume restriction" mechanism(15).

Another way to view the effect of the polymer on the microemulsion is to translate the cloud point change in terms of the response of the surfactant system. The cloud points of nonionic surfactants increase as the surfactant becomes more hydrophilic. This can be achieved by increasing the degree of ethoxylation for a given lipophile structure or by decreasing the lipophile chain length for a given amount of ethoxylation. Thus a decrease in the surfactant cloud point or microemulsion cloud point indicates a lipophilic shift in the behavior of the system. For all of the nonionic polymers investigated, the direction of the cloud point shift indicates that these polymers are creating a lipophilic shift in the Brij-10 microemulsion.

In an effort to investigate the universality of this type of polymer-microemulsion interaction, an anionic surfactant system was studied. Two sulfonate surfactants were chosen to enable variation of the hydrophile-lipophile characteristics of the surfactant couple and in turn of the microemulsion. The hydrophilic surfactant was the monoethanolamine salt of bilinear dodecylbenzene sulfonic acid while the lipophile was the monoethanolamine salt of branched pentadecyl o-xylene sulfonic acid. Thus by varying the relative amounts of these surfactants, the H/L characteristics of the microemulsion could be systematically probed. The other compositional components of the microemulsion were fixed as discussed in the Experimental section.

At room temperature, the microemulsion is single phase and balanced when the wt. % of the hydrophile is 67 % of the total surfactant mixture. At higher concentrations of the hydrophile, the system splits into two phases: a lower microemulsion phase in equilibrium with excess oil. The first manifestation of the phase separation results in a cloudy solution which can be generated by either a change in the H/L properties or by increasing the temperature. It should be realized that the temperature response of anionic surfactants and microemulsions is opposite to that of

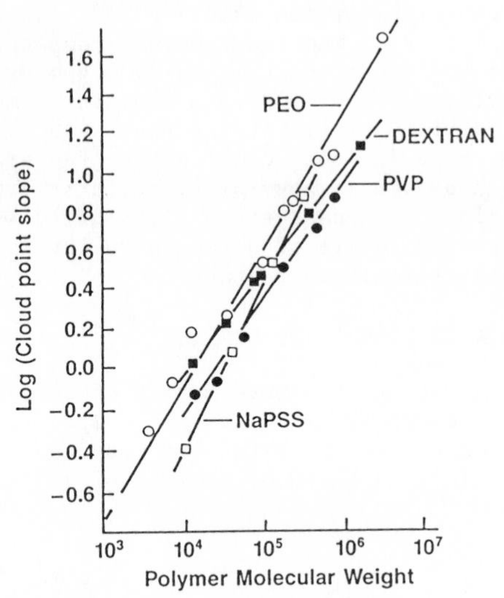

Figure 1. Slope of the Brij-10 microemulsion cloud point versus polymer concentration for different water soluble polymers as a function of polymer molecular weight.

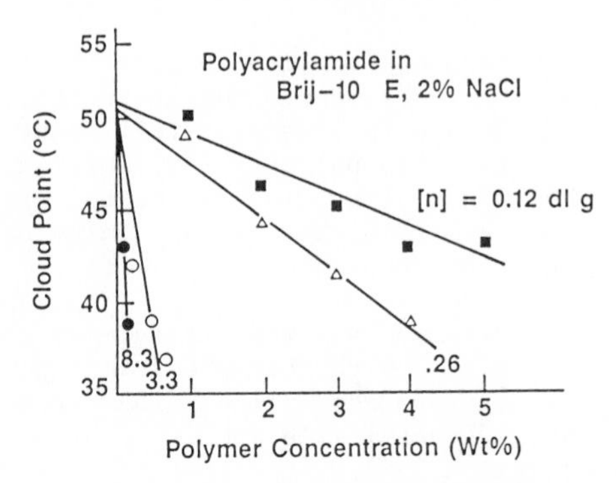

Figure 2. Cloud point of Brij-10 microemulsion (with 2 % NaCl in water phase) as a function of polymer concentration for different molecular weight polyacrylamides.

nonionic surfactants and microemulsions. Anionic systems become more hydrophilic with an increase in temperature. In other words, the cloud point temperature of anionic systems decreases as the surfactant becomes more hydrophilic. Thus an increase in the cloud point is an indication of a lipophilic shift.

The effect of water soluble polymers on the phase behavior of the anionic microemulsion system was studied as a function of surfactant H/L properties. The cloud point temperatures for the neat microemulsions and those containing 1500 ppm HPAM, partially hydrolyzed polyacrylamide, and 1000 ppm Xanthan gum are given in figure 4. The addition of either xanthan biopolymer or HPAM results in an increase in the cloud point temperature of the micreomulsion. Both polymers have similar interactions with the microemulsion. Again one observes a lipophilic shift of the microemulsion system indicative of a repulsive interaction between the polymer and these anionic surfactants.

Viscosity of Polymer-Microemulsion Mixture

Another method of assessing the interaction of polyacrylamide and the model Brij-10 microemulsion is to compare the specific viscosity of mixtures of the two with that expected on the basis of ideal mixing (16). This is given by

$$(\eta_m - \eta_0)/\eta = [\eta]_p C_p + k_p [\eta]_p^2 C_p^2 + [\eta]_\mu C_\mu + k_\mu [\eta]_\mu^2 C_\mu^2 + 2k_{\mu p}[\eta]_p C_p [\eta]_\mu C_\mu \quad (1)$$

where η is viscosity, C is concentration in g/dl, k is the Huggins' coefficient, the square brackets denote intrinsic viscosity and the subscripts m, 0, μ and p stand for, respectively, mixture, solvent, microemulsion and polymer. The polymer-microemulsion term is the last term on the right hand side. The deviation of the Huggins' interaction coefficient, $k_{\mu p}$ from the "ideal" value of $(k_\mu \ k_p)^{1/2}$ is a measure of the interaction of the two species.

A plot of the reduced viscosity, $(\eta_m - \eta_0)/\eta_0 C$, for the microemulsion, a polyacrylamide and the mixture of the two (solid dots) is compared in Figure 5 with that predicted upon the basis of eq. 1 using the "ideal" mixture assumption. The agreement is quite close, indicating that polyacrylamide and this microemulsion have only a repulsive interaction.

Conclusions

The cloud point slope correlation with the intrinsic viscosity of the nonionic Brij-10 microemulsion with added polyacrylamide shows that polyacrylamide does not have any specific interaction with the microemulsion, but rather a nonspecific repulsive interaction similar to that shown by PVP, PEO and dextran. The nature of the interaction between HPAM and anionic water-continueous

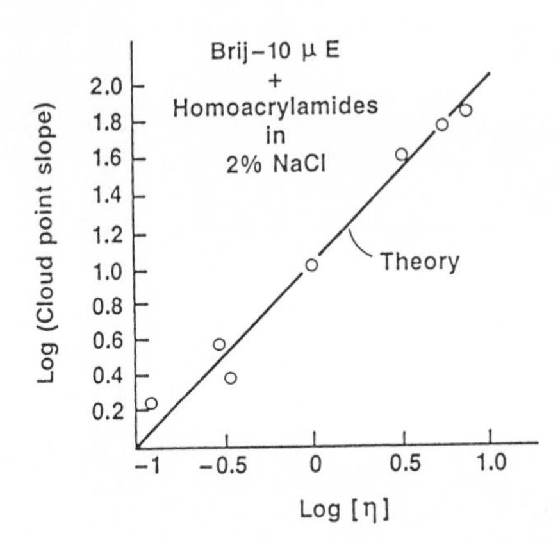

Figure 3. Cloud point slope (of Brij-10 microemulsion with 2 wt.% NaCl) as a function of intrinsic viscosity for different polyacrylamides.

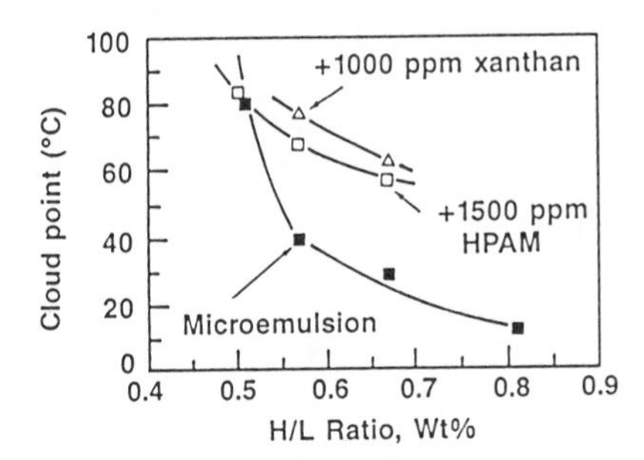

Figure 4. Cloud point temperature as a function of surfactant H/L ratio for microemulsions with and without water soluble polymer.

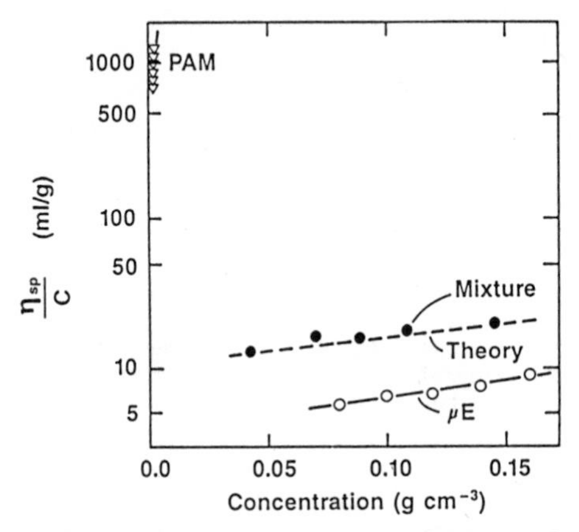

Figure 5. Reduced viscosity as a function of concentration for polymer, microemulsion and their mixture.

microemulsions in terms of cloud point behavior is similar in that the polymer creates a lipohilic shift of the microemulsion due to the repulsive interaction. Dilute solution rheology of the nonionic microemulsion supports this picture of the interaction based on a comparison of the Huggins' interaction coefficient with the "ideal" value. These experiments also imply that there is probably a repulsive interaction between polyacrylamide and nonionic ethoxylated surfactants and between partially hydrolyzed polyacrylamide and anionic surfactants.

Acknowledgments

The authors wish to thank Dr. M. Robbins for many useful discussions of microemulsion phase behavior.

References

1. Trushenski, S. P.; Dauben, D. L.; Parrish, D. R. Soc. Pet. Eng. J. 1974, 633.
2. Pope, G. A.; Tsaur, K.; Schecter, R. S.; Wang, B. The Effect of Several Polymer Fluids on the Phase Behavior of Micellar Fluids. 1980, SPE 8826.
3. Nagarajan, R. Colloids and Surfaces 1985, 13, 1.
4. Turro, N. J.; Baretz, B. H.; Kuo, P-L. Macromolecules 1984, 17, 1321.
5. Cabane, B.; Duplessix, R. Colloids and Surfaces 1985, 13, 19.
6. Ananthapadmanabhan, K. P.; Leung, P. S.; Goddard, E. G. Colloids and Surfaces 1985, 13, 63.
7. Robb, I. D. in Surfactant Science Series; Lucassen-Reynders, E. H., Ed.; Dekker: New York, N.Y., 1981; Vol. 11, p 109.
8. Siano, D. B.; Bock, J. J. Coll. Interface. Sci. 1982 90, 359.
9. Siano, D. B.; Bock, J. J. Polym. Sci., Polym. Lett. Ed. 1982, 20, 151.
10. Mathias, L. J.; Moore, D. R. J. Chem. Educ. 1985 62, 545.
11. Siano, D. B. J. Coll. Interface Sci. 1982 93, 1.
12. Siano, D. B.; Bock, J.; Myer, P.; Russel, W. B. Colloids and Surfaces 1987, 26, 171.
13. Siano, D. B.; Bock, J.; Myer, P. J. Coll. Interface. Sci. 1988, 90, 274.
14. Hermansky, C.; MacKay, R. A. J. Colloid Interface Sci. 1980, 73, 324.
15. Sperry. P. R.; Hopfenberg, H. B.; Thomas, N. L. J. Coll. Interface. Sci. 1981, 82, 62.
16. Siano, D. B.; Bock, J. Colloid and Polymer Sci. 1986, 264, 197.

RECEIVED September 12, 1988

Chapter 21

Chain Interactions in Thin Oil Films Stabilized by Mixed Surfactants

Mark S. Aston[1], Carlier J. Bowden[1], Thelma M. Herrington[1], and Tharwat F. Tadros[2]

[1]Department of Chemistry, University of Reading, Reading RG6 2AD, United Kingdom

[2]ICI Plant Protection Division, Jealotts Hill Research Center Station, Bracknell, Berks RG12 6EY, United Kingdom

The thickness of horizontal films of n-decane sandwiched between two water (or aqueous electrolyte) droplets has been determined by a light reflectance technique. The films were stabilised by three surfactants: an XYX block copolymer of poly(ethyleneoxide) and poly(12-hydroxystearic) acid; soya bean lecithin; 'Arlacel 83' (sorbitan sesquioleate). Results obtained for two and three component mixtures of the surfactants were compared with those for the single surfactants. The results showed that, provided sufficient polymer is present in the film, the thickness is determined by the longest oleophilic chain, namely the poly(12-hydroxystearic) acid.

Polymers have long been exploited to stabilise emulsions; for example, naturally occurring macromolecules such as gums and proteins are used in the food and pharmaceutical industries (1,2). In milk the fat globules are stabilised against coalescence by adsorbed proteins. The remarkable stability of these emulsions towards coalescence can be attributed to their mechanical properties, namely the viscosity and elasticity of the interfacial film (3). Biswas and Haydon (4) systematically investigated the rheological characteristics of various proteins at the oil/water interface. No significant stabilisation occurred in the case of non-viscoelastic films. However, the presence of viscoelasticity alone is not sufficient to confer stability. For example, it was found that the highly viscoelastic film of bovine serum albumin could not stabilise a water-in-oil emulsion. They concluded that the requirements for stability to coalescence are the presence of a film of high viscosity and of appreciable thickness. It is also necessary that the principal contribution to the film thickness should be located on the continuous - phase side of the interface. 'Thick' adsorbed layers of nonionic polymers impart emulsion stability through steric interactions. A nonionic polymer chain in an aqueous environment is

0097–6156/89/0384–0338$06.00/0

insensitive to the presence of electrolytes. It has been shown that the best steric stabilisers are amphipathic block or graft copolymers (5).

In this work block copolymers of the type poly(12-hydroxystearic acid)-poly(ethyleneoxide)-poly(12-hydroxystearic acid), i.e. XYX, are used to stabilise a water-in-oil emulsion where the aqueous phase is concentrated ammonium nitrate. Sorbitan sesquioleate (Arlacel 83) and soya bean lecithin are also added to aid emulsion stability. To understand the behaviour of the interfacial region it is necessary to examine the physicochemical properties of the interface. A typical study would involve interfacial tension, interfacial rheology and film thickness and stability measurements. Also surfactant packing at the interface will play an important role in determining film stability and this can be assessed using spread monolayer (i.e. Langmuir trough) techniques. All of these measurements have been carried out as well as some neutron scattering studies. In this paper, film thickness and stability measurements are presented for a water/n-decane/water film stabilised with one of the XYX copolymers, in combination with sorbitan sesquioleate and soya bean lecithin. Results of studies on the morphologies of the neat XYX polymers are also briefly described. The surfactants are dissolved in the oil phase and it is their presence in the thin oil film between water droplets which prevents coalescence and hence breaking of the emulsion in practice. The surfactants orient with their hydrophilic groups in the aqueous phase and hydrocarbon tails in the oil phase, giving rise to a sterically stabilised system. In the case of the XYX block copolymer, the Y group is hydrophilic and must orient itself at the interface, whilst the two oleophilic X groups extend into the oil. It should be noted that electrostatic interactions are negligible in these systems, in contrast to aqueous systems where they play a very important role.

Materials and Methods

As noted above block copolymers had a poly(ethyleneoxide) head group Y and tails X of poly(12-hydroxystearic) acid. The polymers are formed in a one-step polymerisation from 12-hydroxystearic acid and poly(ethyleneoxide) polymers of various molar masses. The values of the number and mass average molar masses shown in Table 1 were obtained by GPC. Vapour pressure osmometer measurements gave the

Table I. Molar Mass of the Block Copolymers

Polymer	M_n/g mol^{-1}	M_m/g mol^{-1}	PEO/g mol^{-1}
B1	3500	7000	1500
B2	7000		4000
B3	5000		1500 + 4000

number average molar mass of B1 as 3543±30 g mol^{-1} and this was the value used in all calculations; the number average molar mass was considered to be the most relevant to the study of monolayers where the number of molecules in the interface is the important factor.

All three block copolymers are waxy yellow solids at 20°C, soluble in aliphatic hydrocarbons. B1 is a wide spectrum water-in-oil emulsifier with an HLB of 6. They are all anisotropic, transmitting light between crossed polars and all give positive uniaxial conoscopic figures typical of the lamellar phase. The phase diagram of B1 with n-decane is shown in figure 1. Pure B1 changes to an isotropic liquid at 33.2°C. The enthalpy of transition is 147 kJ mol^{-1} consistent with a gel to isotropic liquid phase change. We are interested in the narrow isotropic region up to 2 wt% of B1 at 25°C. Low angle X-ray diffraction results on B1 gave a lamellar interplanar spacing of 17.3 nm; this is equivalent to two chains of poly(12-hydroxystearic) acid, each containing five monomer units, placed end to end. An electron micrograph of pure B1 is shown in figure 2; this was obtained by freeze-fracturing the polymer and shadowing by platinum/carbon at a known angle. From the electron micrograph a stratified morphology is clearly visible, with a repeat unit of the same order of magnitude as the X-ray diffraction data. B1 and B2 were heated on the hot stage of a polarising microscope to the isotropic phase and allowed to cool; they both showed the mosaic textures shown in figures 3 and 4. B2 forms a more definite mosaic texture and is more highly birefringent. B3 which has the Y block of poly(ethyleneoxide) intermediate in size to that of B1 and B2 shows behaviour in between that of B1 and B2 under the microscope. The magnification in both these photomicrographs is only times ten, so that there are much larger areas between disclinations than for a typical monomeric lamellar liquid crystal.

For the film stability studies three surfactants were used singly and as mixtures: B1, sorbitan sesquioleate and soya bean lecithin. The surfactant was dissolved in n-decane and the thickness of the oil film between two water droplets was determined by a light reflectance technique. The basic cell which was used to form the films is shown in figure 5. Film thickness measurements were carried out at 25±0.5°C; temperature control was achieved using a container through which water circulated from a thermostat bath. Approximately equal-sized droplets of the aqueous phase were produced at the upper and lower orifices by adjusting the Rotaflow taps. Using adjusting screws, it was possible to accurately align the upper orifice above the lower one at a separation distance that was approximately equal to the orifice diameter, so that on contact the droplets would be approximately hemispherical. In order to prevent the droplets from wetting the glass and spreading around the orifices, the immediate surrounding areas were made hydrophobic by treatment with a 2% solution of dimethyldichlorosilane in carbon tetrachloride.

The optical set-up for determining film thickness is shown in figure 6. Light from a 1 mW helium-neon laser passed through the lower orifice of the cell (not shown) and was reflected back at both the lower and upper film interfaces where refractive index boundaries occurred. The net reflected light passed back through the lower orifice again onto the telescope system. The angle Θ was kept small (<5°) to satisfy the requirements of optical theory. The split-image

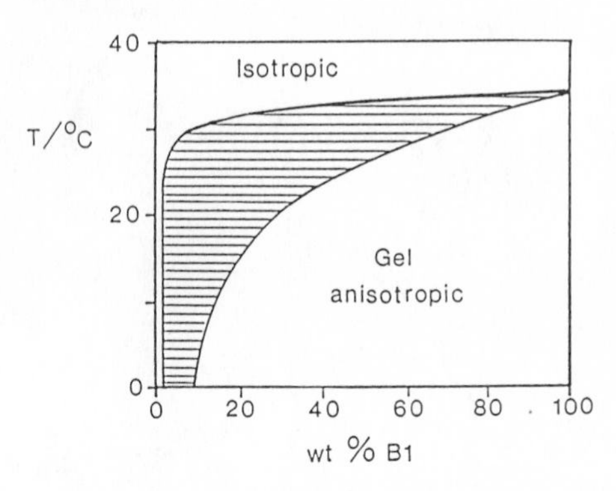

Figure 1. Phase diagram of B1+n-decane

Figure 2. Electron micrograph of pure B1; 1cm = 200 nm.

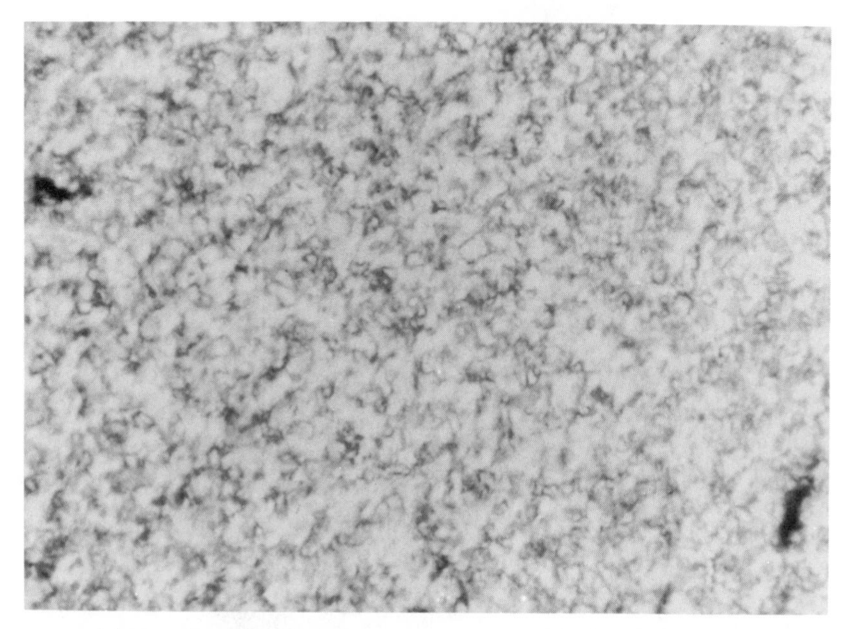

Figure 3. Photomicrograph of B1; magnification x 10

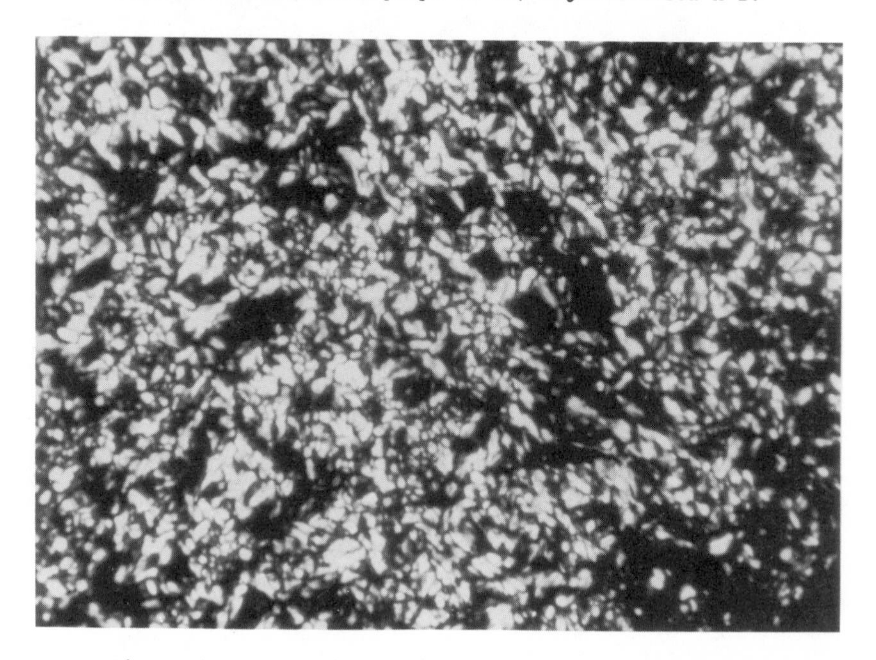

Figure 4. Photomicrograph of B2; magnification x 10

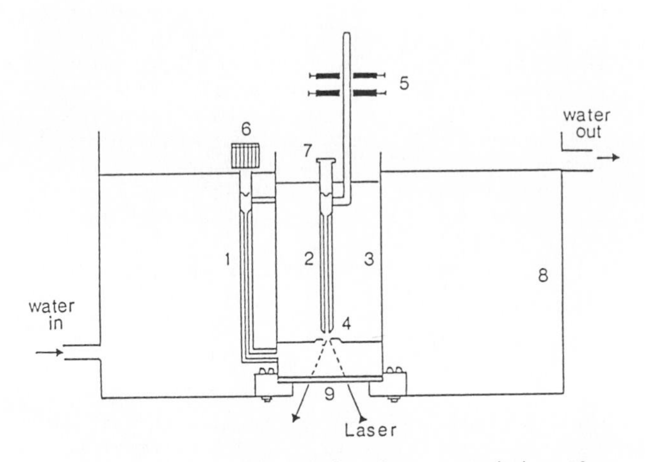

Figure 5. The cell assembly: 1,2 tubes containing the aqueous phase; 3 cylindrical cell containing oil plus surfactant; 4 upper and lower orifices where droplets are formed; 5 adjusting screws; 6 'Rotaflow' tap; 7 'Rotaflow' tap with outer piece removed; 8 perspex water tank; 9 optical flat.

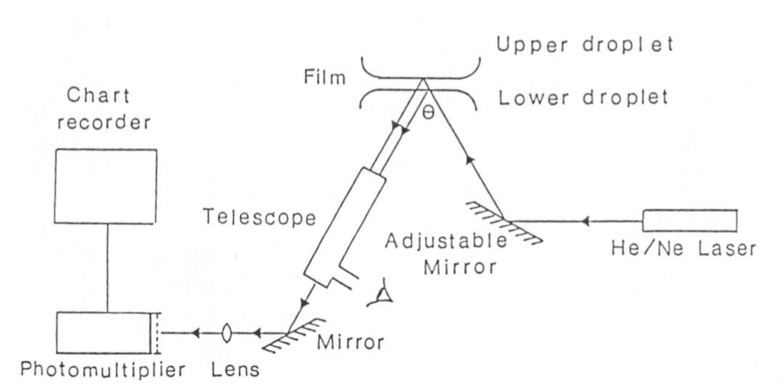

Figure 6. Optical components of the apparatus

telescope provided a visual image of the film, allowing a portion of the reflected light to pass to the photomultiplier detector, which was linked to a chart recorder. A graticule in the telescope eye piece enabled the film diameter to be measured. The circular film image was focussed at the detector aperture; only the central area of the film was recorded. On contact of the two droplets the film thins. For a water film between two oil droplets, the reflected intensity passes through maxima and minima attaining a steady value at the equilibrium thickness of the film. The film thickness can then be calculated using the intensity of the last maximum as a reference (6). However, the 'lipid' films of this work did not thin uniformly so that it was necessary to use an alternative reference intensity. The reference intensity used was that obtained from a water-air boundary with the interface at the same level as that at which the thin film is formed. The advantage of this over, for example, a quartz plate is that the liquid level is automatically horizontal.

The film thickness was calculated from the measured intensities using the equation

$$h = \frac{\lambda}{2\Pi n_1} \sin^{-1} \left[\frac{R_o}{2} \frac{J^{\frac{1}{2}}}{R_{o1}} \right] \quad (1)$$

The film refractive index (n_1) is that for the surfactant concentration in the film, and this will not be equal to the refractive index of the neat oil phase (n-decane). n_1 was calculated using an iterative procedure, from refractive index versus concentration data for the bulk surfactant solutions (in the oil phase) together with close-packed area per molecule values for the surfactants obtained from Π-A measurements. The area per molecule enables the surfactant concentration in the film to be calculated for a given film thickness. The method assumes that a) the packing of the surfactant molecules at the film interface is identical to that found in a close-packed monolayer in a Langmuir trough Π-A experiment and b) the bulk solution will have the same refractive index as a thin film. The latter assumption is only valid if the films can be treated as isotropic. Calculations based on refractive indices parallel and perpendicular to the optic axis for the hydrocarbon chains showed that the error in film thickness would be less than 5% (7). The results obtained for the final film thickness were in agreement with literature values, where available, and in agreement with theoretical values obtained from a consideration of the steric interactions.

Results and Discussion

Single Surfactant Systems. Relative intensity results for an equilibrium film of the block copolymer B1 in n-decane sandwiched between two water droplets at 25°C, are shown in Table II. The intensity was independent of the bulk polymer concentration within the accuracy of measurement. Assuming a constant film refractive index this implies that the film thickness is independent of surfactant concentration, and an average value of J was used for the calculation of film thickness. Coalescence occurs below a concentration of 0.1 g dm^{-3}, presumably because there is insufficient polymer present to form a coherent monolayer at the interface. Above 20 g dm^{-3} the system is in the two-phase region (isotropic solution + mesomorphic phase) and the solution is too cloudy to align the droplets. The iterative calculation of film thickness is shown in

Table II. Equilibrium Film Intensities for Polymer B1 in n-decane at 25°C

Surfactant g dm^{-3}	Concentration mol dm^{-3}	Relative Intensity $J=I/I_0$
0.05	1.41×10^{-5}	Unstable film
0.10	2.82×10^{-5}	0.019_2
1.00	2.82×10^{-4}	0.017_8
5.00	1.41×10^{-3}	0.018_5
10.00	2.82×10^{-3}	0.018_0
20.00	5.65×10^{-3}	0.018_9

Table III. The area per molecule of B1 is 110 ± 10 $Å^2$ (8). The equilibrium film thickness calculated by the iterative

Table III. Iterative Calculation of Film Thickness for Polymer B1 in n-decane at 25°C

Cycle	h/nm	c/g dm^{-3}	n_1	h/nm
1	15.34	697.3	1.4612	14.62
2	14.62	731.6	1.4635	14.35
3	14.35	745.4	1.4645	14.23
4	14.23	751.7	1.4649	14.19
5	14.19	753.8	1.4650	14.18

method is 14.2 ± 1.1 nm.

The refractive index versus concentration plot for B1 in n-decane is shown in figure 7. From this it can be inferred that there is only about 20% 'free oil' in the film. Thus the refractive index of the film is considerably greater than that for n-decane alone, confirming that it is invalid to use the n-decane refractive index in the film thickness calculation. The low oil content of the film suggests that the film is predominately sterically stabilised.

The film thickness obtained by the light reflectance technique is compared with that from low angle X-ray diffraction and a theoretical value in Table IV. The larger value from X-ray

Table IV. Experimental and Theoretical Values of the Film Thickness for Polymer B1

Theory (M_n : 2 x 3½ 12-hydroxy stearic acid chains).	13 nm
Low angle X-ray data	17.3 nm
Light reflectance	14.2 nm

diffraction of the gel phase suggests that the presence of the oil phase leads to some overlapping of the opposing hydrocarbon chains in the bilayer. The number average molar mass corresponds to three and a half 12-hydroxystearic acid residues at each end of the poly(ethyleneoxide) chain, giving an estimated bilayer of 13 nm. However this does not allow for any contribution of the poly(ethyleneoxide) chain. Also with a polydispersity of around 2 it may well be that the longer hydrocarbon chains determine the interlamellar distance in the neat gel phase. Certainly the implications are that the film is sterically stabilised and the polymer chains are fully extended rather than in random coil form.

The structures of the two monomeric surfactants used in this work, 1,4 sorbitan sesquioleate (Arlacel 83) and L,α-dipalmitoyl phosphatidyl choline, used in the impure form of soya bean lecithin, are very different from the block copolymer B1; both can be considered to be V-shaped and about 3 nm in overall length. A comparison of the film thicknesses obtained for the three surfactants is given in Table V. The film thickness for sorbitan sesquioleate is

Table V. Equilibrium Values of the Film Thickness for the Single Surfactant in n-decane at 25°C

Surfactant	h/nm	c/g dm^{-3}	n_1	% oil in film
B1	14.2	754	1.4650	20
'Arlacel 83'	5.9	773	1.4628	23
Soya bean lecithin	90.0	55	1.4167	94

of the order of magnitude expected from molecular geometry, and the film again contains only a relatively small proportion of n-decane (about 23%). This is reflected in the high refractive index of the film which is also similar to that for the polymer. For lecithin the film thickness is extremely large at 90 nm; the concentration of surfactant in the film and the film refractive index are both relatively low and the oil content high at 94%. The thick film is almost certainly due to impurities as a film thickness of 5-6 nm would have been anticipated. Measurements on pure lecithin were unsuccessful as the droplets distorted.

Table VI gives a comparison of the sorbitan sesquioleate film

Table VI. Values for the Oleate Chain Bilayer Thickness

Theoretical thickness (2 x chain length)		3.7 nm
Light reflectance in n-decane (this work)		5.9 nm
Light reflectance in n-decane (reference 9):		
Glycerol monooleate	a) aqueous phase NaCl	5.8 nm
	b) aqueous phase sucrose	6.7 nm
Capacitance technique in n-decane (reference 9):		
Glyerolmonooleate	a) aqueous phase NaCl	4.8 nm
	b) aqueous phase sucrose	4.8 nm

thickness with theoretical and experimental values for the oleate chain. The thickness of 5.9 nm obtained by light reflectance is greater than twice the theoretical length of the oleate hydrocarbon chain (3.7 nm). This can be attributed to a contribution to the film thickness from the headgroup due to the method of measurement used. Dilger (9) found 5.8 nm by the light reflectance method for the oleate chains for an n-decane film containing glycerol monooleate in aqueous NaCl solution. However, he found a reduced film thickness by the capacitance method confirming that the contribution of the head group depends on the technique used.

Binary Surfactant Systems. Film thickness measurements were performed using binary mixtures of the three surfactants in n-decane. A total surfactant concentration of 5 g dm^{-3} was used throughout. The plot of film thickness against mol fraction of B1 is shown in figure 8 for B1 plus sorbitan sesquioleate stabilised films. The film thickness increases with increasing concentration of polymer reaching a thickness of 14 nm, characteristic of the single component polymer film, at a polymer mol fraction of 0.2 ± 0.05. Addition of polymer to the 91 nm soya bean lecithin stabilised film caused a dramatic reduction in film thickness to that of the single-component polymer film (figure 9) at a mol fraction of 0.17 ± 0.03. Thus, in both cases there is approximately the same threshold mol fraction (x_{B1} = 0.2) above which the polymer determines the film thickness.

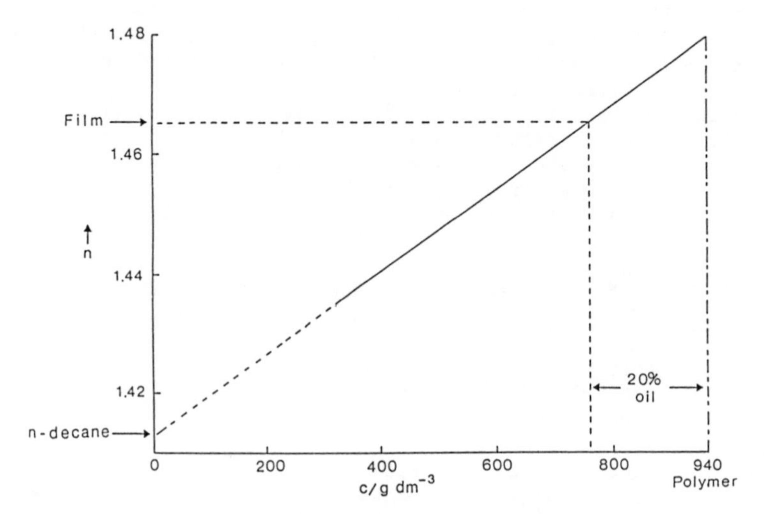

Figure 7. Refractive index (bulk) versus concentration for the block copolymer in n-decane at 25°C. Ordinate: n; Abscissa: c/g dm^{-3}.

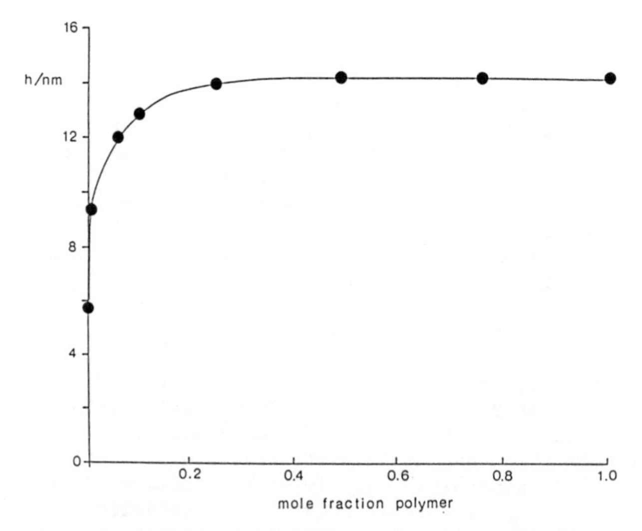

Figure 8. Equilibrium thickness of B1+'Arlacel 83' mixed films in n-decane at 25°C. Total surfactant concentration = 5 g dm^{-3}; aqueous phase:water. Ordinate: h/nm; Abscissa: mol fraction polymer.

In both cases there was a notable decrease in film stability as the polymer mol fraction was decreased below this value. This could be important with regard to potential emulsion stability. From the molecular areas, the fraction of total area occupied by polymer is 0.41 for the B1 plus sorbitan sesquioleate film and 0.35 for the B1 plus soya bean lecithin film. Thus in terms of surface coverage a large proportion of polymer is required in both cases to achieve the characteristic polymer film thickness, although the number of molecules of polymer is relatively small in view of its larger molar mass. Mixed films containing sorbitan sesquioleate and soya bean lecithin were unstable, except at high or low mol fraction of sorbitan sesquioleate (figure 10). At a sorbitan sesquioleate mol fraction of < 0.15, the film thickness was characteristic of soya bean lecithin, and at a mol fraction > 0.75, it was approximately equal to that of sorbitan sesquioleate (i.e. 6.0 ± 1.0 nm). The instability of the mixed sorbitan sesquioleate and soya bean lecithin films correlates with the inefficient packing found for the Π-A isotherms (8). At high surface pressures the unsaturated oleate chains do not pack well with the saturated lecithin chains.

Ternary Surfactant System. The results for the three-component surfactant mixtures are summarised in figure 11. Clearly there are four main regions:

Region A Stable films approximately equal in thickness to that of the B1 single-component film (about 14 nm);

Region B Stable, thinner films approximately equal in thickness to that of the sorbitan sesquioleate single-component film (about 6 nm);

Region C Stable, very thick films, approximately 100 nm in thickness, characteristic of the single-component soya bean lecithin film;

Region D Highly unstable films leading to immediate droplet coalescence.

Electrolyte in the Aqueous Phase. Measurements in the presence of ammonium nitrate showed the film thickness to be independent of electrolyte concentration up to 0.0625 mol dm^{-3}. Unfortunately measurements could not be made at high concentration due to distortion of the upper droplet. The presence of electrolyte in the aqueous phase would not in fact be expected to affect significantly the thickness of a decane film of low relative permittivity.

Conclusions

If the relationship between the film thicknesses of the three surfactant system (fig.11) and potential emulsion stability is considered, perhaps the most significant feature is the widening of region D (the unstable film region) with increasing lecithin mol fraction. This means that lecithin has a destabilising effect and it would be predicted (bearing in mind the instability of the

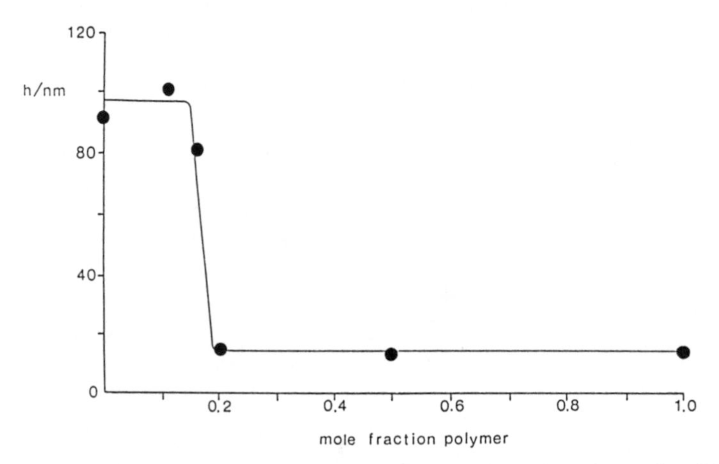

Figure 9. Equilibrium thickness of B1 + soya bean lecithin mixed films in n-decane at 25°C. Total surfactant concentration = 5 g dm^{-3}; aqueous phase : water. Ordinate: h/nm; Abscissa: mol fraction polymer.

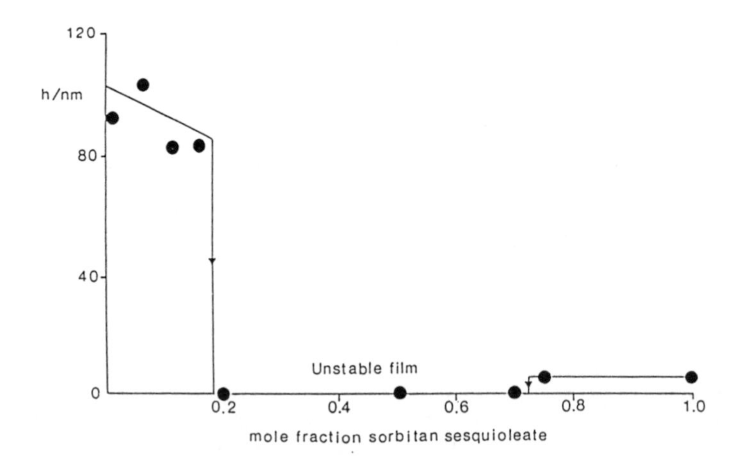

Figure 10. Equilibrium thickness of 'Arlacel 83' + soya bean lecithin mixed films in n-decane at 25°C. Total surfactant concentration = 5 g dm^{-3}; aqueous phase: water. Ordinate: h/nm; Abscissa: mol fraction sorbitan sesquioleate.

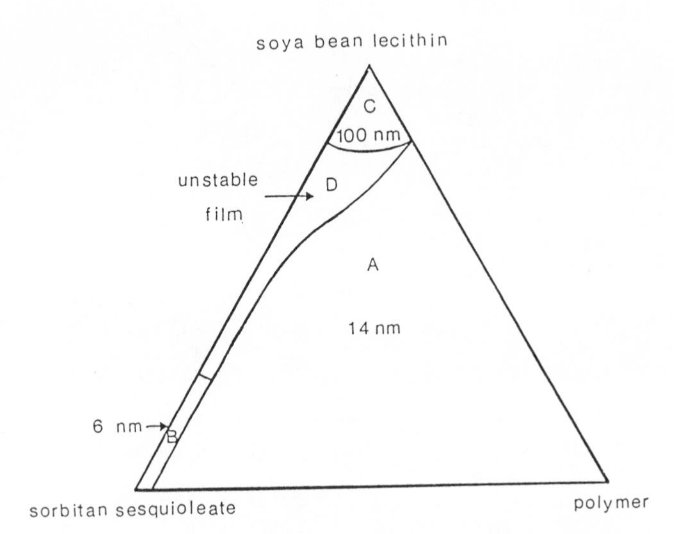

Figure 11. Film thickness for the three component mixed surfactant films in n-decane at 25°C. Total surfactant concentration = 5 g dm^{-3}; aqueous phase:water.

thick lecithin films) that emulsions containing high proportions of lecithin will be particularly unstable to coalescence. This was indeed found in practice (8). In all cases the thickness of stable films was independent of the surfactant concentration. The film thicknesses calculated for B1 and sorbitan sesquioleate confirmed that these films were predominantly sterically stabilised. The results emphasise the dominance of polymer chain interactions in mixed surfactant systems; the film thickness is determined by the block copolymer even at relatively low mol fractions.

Legend of Symbols

h		thickness of film
I		intensity of reflected beam
I_o		reference intensity
J		relative intensity
n_o		refractive index of aqueous phase
n_1		refractive index of film
R_o	=	$(1-n_o)/(1+n_o)$
R_{o1}	=	$(n_1-n_o)/(n_1+n_o)$
		wavelength of laser

Literature Cited

1. Food Emulsions; Friberg, S., Ed.; Marcel Dekker: New York, 1976.
2. Schwartz, T.W. Amer.Perfum.Cosmet. 1962, 77, 85.
3. Clayton, W. In The Theory of Emulsions; Churchill: London, 1954; p92.
4. Biswas, B.; Haydon, D.A. Kolloid Z. 1962, 185, 31.
5. Barrett, K.E.J. Dispersion Polymerisation in Organic Media; Wiley: London, 1975.
6. Herrington, T.M.; Midmore, B.R.; and Sahi, S.S. J.C.S. Far.Trans.1 1982, 78, 2711.
7. Cherry, R.J.; Chapman, D. J.Mol.Biol. 1969, 40, 19.
8. Aston, M.S. Ph.D. Thesis, Reading University, Reading, U.K., 1987.
9. Dilger, J.P. Biochemica and Biophysica Acta 1981, 645, 357.

RECEIVED August 10, 1988

INDEXES

Author Index

Affiliation Index

Subject Index

Production by Barbara J. Libengood
Indexing by Deborah H. Steiner and A. Maureen Rouhi

Elements typeset by Hot Type Ltd., Washington, DC
Printed and bound by Maple Press, York, PA